PICACHO
and the
CARGO MUCHACHOS

Gold, Guns, and Geology of Eastern Imperial County, California

2011 SAN DIEGO ASSOCIATION OF GEOLOGISTS FIELD TRIP PARTICIPANTS

Left to right: *George Johnson, John Teasley, Brian White, Rosalind Wirths, Greg Cranham, Izzy Wirths, Todd Wirths, Annabelle Wirths, Steve Jacobs, Harry Snowden, Maria Zigan, Steve Zigan, Joe Corones, Mary Walke, Jim Peeler, Carole Ziegler, Chuck Houser, Bert Vogler, Sandy Jack, Marina West, Monte Murbach and Ruby, (unknown, in back), Ed Kiessling, Monte Marshall, Diane Murbach, Vic Camp, Dave Bloom, Beckie Biagini, Gary Girty (head down), Perry Crampton, Ian Swarr, Chris Livesey, Chuck Hargrave (kneeling), Brian Olson, Bill Elliott (in front), Donna Gooley, Sue Tanges, Chris Perkovac, Joe Walsh, Nicole Velasco, Marc Ledergerber*

Not pictured: *Nick Klados, Randy West, Jim Manion, Lowell and Diana Lindsay*

Photo by *Carol Teasley.*

PICACHO
and the
CARGO MUCHACHOS

GOLD, GUNS, AND GEOLOGY OF EASTERN IMPERIAL COUNTY, CALIFORNIA

A Guidebook for a Field Trip by the San Diego Association of Geologists • November 4–6, 2011

2ND EDITION ❋ TODD A. WIRTHS, EDITOR

SAN DIEGO ASSOCIATION OF GEOLOGISTS

Distributed by Sunbelt Publications
www.sunbeltpublications.com

Picacho and the Cargo Muchachos: Gold, Guns, and Geology of Eastern Imperial County, California

Second edition 2015. Fourth printing 2024.

Edited by Todd A. Wirths
Cover Concept by Todd A. Wirths and Philip T. Farquharson, CG-Squared Productions
Production management by Deborah Young
Design by Michael Schrauzer
Printed in United States of America

San Diego Geological Society
3130 North Evergreen Street
San Diego, California 92110
www.sandiegogeologicalsociety.org

Available through:

Sunbelt Publications
664 Marsat Ct., Suite A
Chula Vista, CA 91911
(619) 258-4911, fax: (619) 258-4916
www.sunbeltpublications.com info@sunbeltpub.com

27 26 25 24 5 4 3 2

Library of Congress Cataloging-in-Publication Data

Picacho and the Cargo Muchachos : gold, guns, and geology of eastern Imperial County, California / edited by Todd A. Wirths. -- Second edition.

pages cm

Includes bibliographical references.

ISBN 978-1-941384-07-7 (softcover : alkaline paper) 1. Geology--California--Imperial County. 2. Gold mines and mining--California--Imperial County--History. 3. Imperial County (Calif.)--Description and travel. 4. Chocolate Mountains Region (Calif.)--Description and travel. 5. Imperial County (Calif.)--History, Local. 6. Imperial County (Calif.)--Social life and customs. I. Wirths, Todd A. II. San Diego Association of Geologists.

QE90.I46P53 2014

557.94'99--dc23

2014035324

San Diego Geological Society, Inc., gratefully acknowledges the generous contributions of our sponsors to the publication of this book.
Front cover photograph courtesy of: Ernie Mendivil
Back cover illustration courtesy of: The Desert Magazine, Volume 3, Number 5, March 1939, pg. 10.
Front inside photographs courtesy of: (top) Ernie Mendivil; (bottom) Arizona Historical Society, Yuma
Frontispiece photograph by Carol Teasley
Back inside photographs courtesy of: Ernie Mendivil

Table of Contents

The editor with a copy of the 1st edition at the Picacho mill

Preface to the Second Edition

PICACHO AND THE CARGO MUCHACHOS **WAS** reckoned by the San Diego Association of Geologists as worthy of a facelift, and therefore several contributions to the volume were upgraded, revised, and redesigned. With time now abundant, articles by Bill Elliott, Stephen Van Wormer, Monte Marshall, and myself benefited from various improvements in research, content, illustrations, and hindsight. John Young's gem, "A place of enchantment on the lower Colorado," was upgraded from a barely legible newspaper clipping to a full article. Three other key reprints, "Geologic and geochemical study of the Picacho gold mine," "Saga of old Picacho," and "Mysterious Cargo Muchachos" have been reformatted to fit the style of the book. Finally, thanks go to Michael Schrauzer for his professional touch designing this volume.

Since the publication of the first edition of this volume in March of 2012, the California State Parks budget crisis — and I use the term *crisis* genuinely — was resolved, albeit in an unconventional fashion. Consequently, the Picacho State Recreation Area, facing closure, was removed from the bureaucratic chopping block, and is now back to its fully operational status. The volume's Introduction has been revised to include the startling details.

Todd Wirths, editor
May 20, 2014

Introduction

WELCOME TO THE SAN DIEGO ASSOCIAtion of Geologists (SDAG) Field Trip for 2011. This year, we will tour and investigate the extreme southeastern corner of California: Picacho and the Cargo Muchacho Mountains. This geologic field trip and volume marks the farthest eastern exploration conducted by the organization, as well as the 40th anniversary of our founding.

In recognition of our 40th year of existence, San Diego Association of Geologists' co-founders and past presidents Charles "Mike" Bemis and Mike Hart have contributed a humorous account of our humble beginnings as part of the front matter for the volume. But make no mistake-these two gents and several others made tremendous contributions to the understanding of SDAG's core focus, the geology of the San Diego Region.

Background

I like to say that I threw a dart at a map as the selection process for the destination of this field trip, but in reality I discovered Picacho in the mid-2000s while on assignment installing groundwater monitoring wells in nearby Bard as an environmental geologist. After the job, I followed the signs on Picacho Road, the brief paved portion, indicating the direction to the Picacho State Recreation Area. I was directed to a broad dirt road ... that led to seemingly nothing. Nothing, except for piles of Pleistocene gravels and scrubby creosote, that is. I got out of the company pickup and looked around. In the words of astronaut Buzz Aldrin, a pleasant sense of "magnificent desolation" was upon me. Then I left, thinking someday I might be back, but in reality, probably not.

Fast forward: Well, I came back to Picacho! For many people, the extreme southeast corner of California is *terra incognita* and is probably better known to Arizonans than Californians. Indeed, Picacho State Recreation Area is a popular vacation spot for many residents of Yuma. This recreation area, and the surrounding southeastern Chocolate and Cargo Muchacho Mountains, is remote with few amenities, but for some people, this is a paradise of sun, solitude, and monster catfish. The focus of this volume is the geology of the region, and the deposits of gold found in the sun-baked rocks that begat its colorful history.

Road Trips

You have to drive to participate in this adventure; therefore, the first section in this book describes the geology and history surrounding the main roads that access the desert interior, Picacho Road and Ogilby Road. They are both north-south roads, but the similarities end there. The majority of Picacho Road is not paved, and it should remain that way, to maintain a sense of adventure and exclusivity. The road ends at the sleepy Picacho SRA. However, the "real" entrance to the Picacho SRA is not at the park's entry fee kiosk, but at the transition of the road from asphalt to dirt. This is the "Picacho Experience," and it begins with the drive. The geology that you will see is as complex as it is wondrous, and the geologic road log for Picacho Road is an attempt to make sense of the sights.

While Picacho Road passes by the famous old Picacho Gold Mine, the *raison d'être* for Ogilby Road is gold, pure and simple. Several famous mines lie to the east of Ogilby Road within the Cargo Muchacho Mountains, and countless smaller forgotten claims and diggings lie throughout the range. Mining gold in these hills has been an almost constant activity since the early 19th Century. The road log for Ogilby Road is extended from its northern end, heading west down Highway 78 to encompass the Mesquite Gold Mine, a thoroughly modern, and huge, open-pit gold mine.

Early travel in the desert southwest was particularly difficult. Geologist Bill Elliott (another eminent co-founder of SDAG) offers a unique look at ourselves as travelers during the past 100-plus years, and provides accounts of how people coped and adapted while traveling in extreme heat. In the first of four contributions to this volume, Bill adds a personal touch reflecting on his own experiences since the 1940s, vacationing with his family in the deserts of California.

Perhaps you've heard of the Mint 400, the Baja 1000, and the "Barstow to Vegas" off-road races. Believe it or not, cross-country desert racing is nothing new. One hundred years ago, America had the Cactus Derby—a race of survival and luck more than anything else, from Los Angeles to Phoenix. This was at a time when automobiles were in their infancy, when Arizona was still a United States territory, when some racers also owned the car company. One particular race car driver was so popular they named a road after him. He is practically forgotten now, but a happy coincidence has brought together a dirt road, a historic race, a charismatic race car driver, and a little geology.

Geology

An esteemed geology professor from San Diego State University has been mapping the vicinity of Picacho continuously for over ten years, along with a succession of geology undergraduate and graduate students. Dr. Gary Girty is probably more familiar with this region than any prospector, miner, or park ranger ever was. When I approached Gary and asked if he could write a contribution to the volume, he inquired what I had in mind. I responded by asking for everything including the proverbial kitchen sink: if he could please write an up-to-date summary of all the geology of the whole area, that would be just fine, thanks. And he obliged, and the results are presented in this volume. The article by Gary and several of his students encompasses a summary of the region's geological formations and structures (including the newly described Copper Basin Fault), the latest analytical data and age dates, and a synopsis of how Picacho fits into the big picture: the tectonic evolution of the southwestern Cordillera of North America. Should you go exploring Picacho on a fine winter weekend, don't be surprised if you bump into Gary, he'll probably be there, with the latest flock of geology majors. Otherwise, he usually camps at site #35 in the Picacho SRA!

Within the Picacho area, outcrops of the enigmatic Orocopia Schist can be seen after a short hike from northern Picacho Road. This rock unit has undergone a complex tectonic history and has been the subject of numerous studies. For this volume, I have selected a recent paper that examines the origin of the Orocopia Schist, and its intriguing San Diego connection. Likewise, the origin of gold at the Picacho Gold Mine has been a focus of scientific scrutiny for nearly three decades—how did the gold get here? The authors of "Geologic and Geochemical Study of the Picacho Gold Mine, California" investigated the mine over a dozen years ago during active mining, and had the privilege to observe the chemistry and structure of the gold-bearing rock as mining activities progressed deeper into the earth. Their description of Picacho's complex "plumbing system" is impressive and thorough.

Fundamentally, the mindset of a geologist is to investigate, to explore, to keep learning. In a sense, we are forever students of our craft. In this volume, San Diego State University emeritus professor Monte Marshall contributes a historical background analysis of how the use of paleomagnetic studies shaped our understanding of the structural landscape of southeastern California and southwestern Arizona. Utilizing the paleomagnetic signature locked within the minerals of rocks, a history of microtectonics, crustal rotation, and oblique subduction unfolds to piece together a story of tectonic connections, starting from the west coast and leading into the desert interior. In a similar approach, Monte's compatriot Bill Elliott investigates a little-known suite of Quaternary geologic formations around the old Araz Stage Depot, just north of Interstate 8 and the Quechan Casino Resort. Simply titled "Araz Outcrop," Bill provides a brief historical background of the spot, and summarizes the geologic investigative history of these outcrops of Quaternary deposits. From there, Bill brings the reader along as an observer of a geologist at work, posing questions and hypotheses based on observations and what is known. While Monte and Bill's articles cannot be further apart in subject matter, they are similar in that they conclude, using a geologic mantra, "more data is needed!"

Did I mention Bill and Monte are compatriots? With "In search of Colorado River 'B' Gravel," they team up to pursue the origin of mysterious fluvial gravel deposits near Highway 78. In the same vein of the Araz conundrum, an investigation into whether an ancient course of the Colorado River left deposits of gravels of the type that would be expected, they came away with something unexpected, and in the process, enriched us all.

And now for something completely different.... The Cargo Muchacho Gold Mine is one of the oldest, if not the oldest, underground gold mine in the Cargo Muchacho Mountains. Bill Elliott took a curious journey to the mine, then to the photographic archives of the Arizona Historical Society in Yuma, and then back to the mine. His serendipitous account is presented in a wonderful photographic essay.

Mining History

There was a mining camp called "Potholes" that has the distinction of being, quite likely, the first site of gold mining attempted in North America north of present-day Mexico. Potholes lies at the very southeastern edge of Imperial County. Indeed, the location of Potholes is farther east than Yuma, Arizona. I am very pleased that California State Historian, archeologist, and author Stephen Van Wormer has contributed *the* authoritative historical narrative of the mining district, entitled "Dry Diggings: A History of the Potholes Mining District." Potholes is mentioned in many references and accounts, but almost always as an aside or anecdote. Here, Stephen provides a thoroughly full and enjoyable chronicle of the district's development, from its Spanish inception to an American enterprise.

From Potholes, head up the Colorado River on your steamboat to Picacho. The Picacho Mine has been dubbed the most famous gold mine in Imperial County. I selected a few articles originally published over a half-century ago to provide, not only nostalgic histories of Picacho the mine and Picacho the community, but for first-hand accounts of what life, and death, for a miner was like a hundred years ago. "Saga of Old Picacho," originally published in *The Desert Magazine* in 1939, presents the tale of Picacho and introduces us to José Maria Mendivil, the man who discovered gold at what is now the Picacho Mine. José's grandson, Michael "Mike" Mendivil, grew up in Picacho and recorded his experiences in "Gold, Guns and Fiestas in Old Picacho." By the same token, Ed Rochester, the honorary "Mayor" of Picacho, paints a vivid picture of Picacho's business, politics, and way of life in "Blame the Gold or the Climate — We Can't Get Away from Picacho." These two articles appeared in the February 1952 issue of the *Calico Print*, recording "Tales and Trails of the Desert West." The editors of the *Calico Print* were the husband and wife team of Harold and Lucile Weight, who dedicated themselves to preserve a bygone era by traveling and interviewing any and all that had a story to tell. It should be noted that the Weights did not ask Mike and Ed to write down their thoughts about Picacho; rather, they had Mike and Ed speak into a tape recorder — in 1952! The Weights then laboriously transcribed it all.

Ernie Mendivil *at home (Photo by Todd A. Wirths)*

For Mike and Ed's articles, I retained some photos from the *Calico Print* issue, while other historic photos are published here for the first time. Mr. Ernest "Ernie" Mendivil — son of Michael Aguilar, son of Ysidro, son of José Maria — is the keeper of the spirit of Picacho. Ernie graciously allowed the San Diego Association of Geologists to reprint photos I selected from his personal archives. The cover photo is among these photos, and is probably the best. The image is described as "taken on Main Street, town of Picacho about 1903." The dual-pistoled hombre on the far right has the signature "J. Maria Mendivil" over him — possibly identifying the only known photo to exist of the man, José Maria Mendivil.

If you enjoy Stephen Van Wormer's "Dry Diggings," then the next two articles by Stephen and co-author James Newland will, to borrow a mining term, put a "fire in your hole." "The History of Hedges and the Cargo Muchacho Mining District" is so comprehensive it was originally written as two parts. Hedges was a mining town in the Cargo Muchacho Mountains, and is better known today as the ghost town of Tumco. These articles describe a dominant issue present during the heyday of gold mining at Hedges: the interaction between Mexicans and Anglo-Americans, with the underlying themes of race, culture, and the division of labor. Without a doubt, these topics could easily apply to any other southwestern mining community of the era. These articles are truly fascinating and present a definitive account of mining life in the region.

Today's Picacho

When the first edition of this volume was published in March of 2012, the status of the Picacho State Recreation Area was in doubt. Picacho was listed among 69 other California state parks slated for closure as a plan to solve the state's budget woes. Up and down the state, grassroots non-profit organizations sprouted to raise funds in support of their cherished parks, and Picacho was no exception. Friends4Picacho was created, not only to raise cash to keep the park open and running, but as a means to get the word out that Picacho is a great place to visit. By June of 2012, thousands of dollars had been pledged to Friends4Picacho by the Picacho faithful. In the meantime, a quick succession of events uncovered the existence of a "special fund" within California's Department of Parks and Recreation, where over $50 million was hidden. Within a few of months after the discovery of the special fund, much of these moneys were appropriated for use by the state parks, thus sparing the 70 parks from shuttering.

You have probably heard the saying, "You don't know what you have until you lose it," or some similar version. In this case, "it" wasn't lost, but the fear was real. While State Parks and Picacho SRA officials, and the public, were spared of the ignominy of losing a precious resource, a sense of relief, as well as disappointment, was experienced by those involved in the rescue effort. The relief is obvious, but disappointment in that so many fought so hard to preserve Picacho, all because park operating funds were not located where they should have been. But a lesson was learned, and fortunately, not the hard way. Today, the organization Friends4Picacho lives on as a legacy of those tumultuous times, and functions as a close-knit, Picacho-loving group of people dedicated to improving, preserving, and promoting the park. Interested enthusiasts can go to http://www.friends4picacho.org to get informed and involved.

Todd A. Wirths, M.S., P.G.
Vice President, 2011
San Diego Association of Geologists

Acknowledgements

THIS FIELD TRIP AND GUIDEBOOK WAS AN ambitious project on my part, the success of which depended on the enthusiasm and support of many people. To begin, I want to thank the sponsors, especially Hargis + Associates and Cascade Drilling, for essential financial assistance. Much of these donations allowed for the acquisition of the rights to reprint several necessary archival photos and a professional paper that aided this guidebook immeasurably.

For the field trip lecture stops at the Picacho State Recreation Area, the Mesquite Mine, and the Mesquite Regional Landfill, thanks go to trip speakers Dr. Gary Girty, Don Wagstaff, and Michele Ochs, respectively, for a job well done and time away from their busy schedules. Thanks go to SDAG past president Greg Cranham for procuring local craft brewing refreshment and bringing Brian Papurello & Sons Expedition Catering all the way from Salinas to provide mouth-watering geology food.

Speaking of past presidents, thanks go to Monte and Diane Murbach, who were instrumental with their advice, support, and guidance for coordinating the field trip and guidebook. Dave Bloom was influential in making sure I was on the right track, and offered valuable advice as well. I want to recognize Chuck Houser for coordinating Sunday's field trip stop localities and speakers — in a word, Mesquite! And thanks go to the Picacho State Recreation Area park rangers, Robin Greene and Sue Barney, for outstanding camp facilities and service.

This book is the result of the hard work by many people, and appreciation is due. I want to thank all the authors for excellent essays and professional papers. In particular, I want to recognize Dr. Gary Girty for giving up his precious time to write the important Picacho geology paper, and to Bill Elliott for his four contributions. Thanks go to Carole Ziegler, SDAG's guidebook managing editor, who strove mightily to complete the seemingly Sisyphean task of editing all papers, and succeeded. I want to thank Joe Corones and Bill Elliott for providing valuable, out-of-print guidebooks from past field trips and other print matter that greatly assisted my research. Thanks are in store for Phil Farquharson, our talented and fearless book designer, and Lowell Lindsay, publisher extraordinaire. I want to thank Lisa Pierce of Cogstone Resource Management for creating the map figure for the Barney Oldfield Road article, and Carol Teasley for her professional photographic contributions.

I had the great pleasure of meeting Mr. Ernie Mendivil, the great-grandson of José Maria Mendivil, the discoverer of gold at Picacho. I want to thank Ernie for his welcoming hospitality and for allowing me the use of his uniquely wonderful photographs from his family collection that now adorn this book.

I want to thank my mom, Rosalind Wirths, for many things, but one thing in particular is relevant here. In the mid-1970s, she gave my brother Craig and me the greatest gift a parent could give to a child: an encyclopedia set, specifically the World Book Encyclopedia. I actually read it. Thanks Mom, it looks like it paid off. Last but least, I want to thank my wife, Izzy Tihanyi, for all her support, encouragement, understanding, and (usually) extreme patience!

Todd Wirths

Dedication to Dr. Gary H. Girty

THIS VOLUME, THE 30TH PUBLISHED BY THE San Diego Association of Geologists, recognizes the commitment and contributions to geology by Dr. Gary H. Girty, Professor of Geology at San Diego State University. For over thirty years, Gary has focused his professional life on two fronts, researching the geology of the southwestern Cordillera of North America and teaching.

A native of California's Central Valley, Gary received his BA and MA degrees in geology from Fresno State University, both with honors. For his Master's thesis, Gary studied the regional deformation of the Boyden Cave roof pendant in the central Sierra Nevada. During this time he was also engaged as a marine science technician at University of California, San Diego's Scripps Institution of Oceanography aboard the *Glomar Challenger*. By the end of the 1970s, Gary had enrolled at Columbia University in New York City. At Columbia, he finalized his Master of Philosophy degree in 1981, followed shortly thereafter with a Ph.D. His dissertation, "The Culbertson Lake Allochthon, a newly identified structural unit within the Shoo Fly Complex: Sedimentologic, stratigraphic, and structural evidence for extension of the Antler Orogenic Belt to the northern Sierra Nevada, California," was completed in 1983.

After two years at the University of Kansas as an Assistant Professor, Gary landed at San Diego State University in 1984, and has flourished there ever since. While his research focuses on the geologic history and tectonics of the southwestern Cordillera, the subject matter within that sphere has evolved. Concentrating mostly on the pre-Mesozoic geology of the northern Sierras and the tectonic history of the northern Peninsular Ranges during the 1980s and 1990s, Gary shifted some of his attention toward creating and developing online teaching courses and software programs for geologic applications by the beginning of the millennia. It was during this time he also began exploring the Picacho State Recreation Area. Gary has diversified his expertise recently, researching the development of saprolite and saprock in the San Diego region, and contributing to the historical implications of the composition of ancient Nubian and Egyptian ceramics. His current research projects are focused on the evolution and characteristics of the fault zone architecture of the San Jacinto fault, and he continues to investigate the Tertiary structural and tectonic evolution of the lower Colorado River region between Indian Pass and Picacho State Recreation Area.

Gary at the newly described Copper Basin fault, east of Arrastra Wash in the Picacho State Recreation Area. The dark-colored wedge is fault gouge along the trace, dipping steeply to the south. This outcrop is described at Stop 6 in the road log for Picacho Road. Photo by Carol Teasley

Gary served as the Chair of the San Diego State University Department of Geological Sciences from 1998 to 2001 and again from 2003 to 2008. He also has served as the president of the Pacific Section SEPM (1996), as the president of the Cordilleran Section of the Geological Society of America (2006), and was selected by the Academic Senate at San Diego State University for the *Excellence in Teaching Award — Recognizing Excellence as a Teacher-Scholar* for 2006–2007. In 2009 he was selected for the *President's Top 25 Award* at San Diego State University. Colleagues and students alike recognize Gary's "remarkable" excellence and dedication to scholarship and teaching. He has supervised the successful completion of 71 Master's theses (including the editor's in 1995) and 84 senior theses, and counting. From all of us at the San Diego Association of Geologists, thank you Gary for your tireless enthusiasm and influence!

In Memoriam

Dr. Michael J. Walawender

1939–2011

I WOULD LIKE TO GRATEFULLY ACKNOWLEDGE Dr. Mike Walawender's contributions to our understanding of southern California's complex geologic history. Retired from a rewarding teaching and research career that spanned from 1972 to 2001 at San Diego State University, Mike passed away at age 71 in the Finger Lakes region of upstate New York on September 13, 2011. We are grateful to his wife, Sonja, and family for being a special part of our extended geo-family.

One of my fond memories of Mike's enthusiasm for hands-on education goes back to our annual San Diego State University Alumni Field Trip in late February 1994. A couple of days before we departed for the Cargo Muchacho Mountains, our much anticipated tour of the American Girl Gold Mine was abruptly canceled. So, we "made lemonade," and spent the weekend prowling the nearby ghost town of Tumco and abandoned gold mines.

Recognizing our collective disappointment, Mike arranged for a follow-up desert trip in mid-April that same year to tour the Mesquite Gold Mine located a few miles north of the American Girl Mine. Being able to go down into the open pit and put our fingers on visible flecks of gold in several very rich quartz veins was an experience not soon to be forgotten, either by his Ore Deposits Class or the alumni.

What an exciting geo-journey we have all had the privilege to share! Thanks Mike, you've so enriched our lives, we will miss you!

Bill Elliott

Lyne Perry

1962–2011

LYNE PERRY WAS A DEAR FRIEND OF MANY geologists in southern California. She was an honorary member of the South Coast Geological Society and past president of the San Diego Association of Geologists. After graduating with her bachelor's degree in Geological Sciences from San Diego State University in 1985, Lyne worked for Catlin Engineering, Leighton and Associates, and Kleinfelder, before becoming an independent consulting geologist. Besides enjoying paragliding, marathon running, skiing, scuba diving, hiking mountains, and geology field trips, she loved to travel the world and enjoyed adventures in places like Belize, Costa Rica, the Galapagos, Turkey, Nepal, Australia, and Tanzania. Lyne *loved* anything purple! Please join us as we raise a margarita or a glass of fine wine for a toast, to celebrate a great friend and life well spent!

Sue Tanges
Diane and Monte Murbach

Photo above left: Dr. Michael J. Walawender at Desert View Tower (Interstate 8), Geological Society of America GeoHostel. Photo by W. J. Elliott, March 13, 2006.

San Diego Association of Geologists Beginnings

MIKE HART, C.E.G. AND MIKE BEMIS, C.E.G.*

HOW DID THE SAN DIEGO ASSOCIATION OF Geologists come into being? In the early 1970s there was a dedicated group of young geologists that used to meet occasionally for what was called "lunch hour field trips." These included Mike Bemis, Ray Scaramella, Les Reed, Mike Hart, and a couple of others whose names have receded into some foggy corners of our minds. During those meetings there was some conversation regarding the formation of a local geological society. Mike Hart remembers that the new society was modeled after the South Coast Geological Society that had just formed in Orange County, as organized by John Minch and Don Fife. Mike Bemis begs to differ and thinks the South Coast Geological Society came a bit later.

A couple of organizational meetings were had at a local pancake house, now defunct, by the name of Aunt Emma's (affectionately referred to by us for some reason as Aunt Enema's). It was located near San Diego State University. The first meeting was called by Mike Bemis of Woodward-Clyde & Associates (now URS). In attendance were Mike Bemis, Mike Flynn of Southern California Testing Lab (now Southern California Soil and Testing) and Mike Hart with Wm. S. Krooskos and Associates. The mighty triumvirate of Mikes agreed on the formation of a group to be named the San Diego Association of Geologists. The first few breakfast meetings were held at Ricky's Restaurant in Mission Valley. Our first President was Mike Bemis. The group grew quickly, and the monthly meetings became dinner meetings with guest speakers. Our very first speaker was Dr. William Soldman on the *Treatment of Rattlesnake Bites*. The primary accomplishment of his talk was that for several months many of us were a bit nervous about travelling cross country through the chaparral. Some of us bought snake leggings. Mike Hart preferred the homemade variety, and made his by cutting up two concrete cylinder test cans and inserting eyelets for laces. The looks from his comrades were too much for him, and they were soon abandoned.

For the first few years, the very existence of the group was somewhat in question because of low attendance at some of the meetings. At some dinner meetings the officers were a little embarrassed for the speaker because we had less than ten in attendance. Greg Farrand, one of our first members, recalls that some of our early meetings were at the Charcoal House in La Mesa. He also recalls that one of the speakers, Bill Elliott, was giving his talk when the attendees became distracted by a cockroach that was doing sprints back and forth on a railing in front of the projector. His image, much enlarged, was projected onto the screen. Bill was oblivious to the action, but it provided amusement for the audience.

The first field trip, in 1972, was led by Ernie Artim, who took us on a tour of the La Nacion and Sweetwater faults recently discovered by him and other Woody-Clyde geologists while working on residential subdivisions in the South Bay area. Hard to believe now, but the only fault that was publically known in the San Diego metro area at that time was the Rose Canyon fault and it was firmly believed to be inactive. Ernie faced a lot of skepticism from both his fellow geologists and even our old professor Dick Threet, when he attempted to convince the group at several outcrops in the Paradise Hills area that the previously unrecognized fault he had identified extended for nearly 15 miles from northern San Diego all the way to the Mexican border.

The next field trip really put the San Diego Association of Geologists on the map, so to speak, and assured its success. In 1973, Bob Dowlen, who was then working in the Paleontology Department at the San Diego Natural History Museum, and the Museum paleontologist, Arnold Ross, put together the very first professional-looking San Diego Association of Geologist's guidebook, *Studies on the Geology and Geologic Hazards of the Greater San Diego Area, California*. The field trip was the first by bus for the San Diego Association of Geologists. The two busses hired for the occasion were old City of San Diego busses, so underpowered that they could not make the grades on some of the streets in the Point Loma area. There was a great time had by the passengers as they had to exit the bus and then get behind it to push it up the steepest grades. You could say that the bus made the grade by a beer powered assist. The rest, as they say, "is history."

---

*Mike Bemis passed away since the first edition of Picacho and the Cargo Muchachos was published.

Sponsors

DIAMOND SPONSORS ($1,000)

CASCADE DRILLING, INC.

Cascade Drilling, L.P.
www.cascadedrilling.com

HARGIS+ASSOCIATES, INC.
ENGINEERING • HYDROGEOLOGY

Dr. David R. Hargis
Hargis + Associates, Inc., www.hargis.com

RUBY SPONSORS ($500)

Jim Ashby, Mission Geoscience
(949) 955-9086
www.missiongeoscience.com

Suzie Nawikas, H&P Mobile Geochemistry, Inc.
(800) 834-9888
www.handpmg.com

David Bloom, Tetra Tech
(619) 702-6059
www.tetratech.com

Steve Zigan,
Student Sponsor

EMERALD SPONSORS ($100)

Dr. Pat Abbott
Sally & Dennis Avery
Richard W. Berry
Curtis Burdett
Joe Corones
Greg Cranham
Bill Elliott
Eggers Environmental, Inc.
Greg Farrand
Phil Farquharson, CG-Squared Productions
Carolyn Glockhoff, Caro-Lion Enterprises
Cari Gomes
Sarah Gray
Rob Hawk
Jonas & Associates
Diana & Lowell Lindsay, Sunbelt Publications
Ninyo & Moore
Monte Marshall
Monte & Diane Murbach, Murbach Geotech
Doug Peeler
Les Reed, Geotechnical Exploration Inc. (GEI)
Anne Sturz
Sue Tanges, Southland Geotechnical Consultants
Terra Costa Consulting Group, Inc.
Bob Smillie, Terra Costa Consulting Group, Inc.
David & Jan Steller
Mary Walke
Carole Ziegler

The generosity of these sponsors allows our organization to keep membership dues low, reduce rates for students, produce monthly newsletters, purchase supplies, host the annual field trip, produce our geologic guidebooks, and provide scholarships to local geology students. To contact any of the above sponsors, please visit www.sandiegogeologists.org.

Officers

2011 San Diego Association of Geologists

PRESIDENT: Chuck Houser
VICE PRESIDENT: Todd Wirths
SECRETARY: Brian J. Olson
TREASURER: Cari Gomes
PUBLICATIONS: Lowell Lindsay
WEBMASTER: Carolyn Glockoff

2014 San Diego Geological Society, Inc.

PRESIDENT: Dave Bloom
VICE PRESIDENT: Mike Hart
SECRETARY: Diane Murbach
TREASURER: Greg Cranham
DIRECTOR: Lowell Lindsay
DIRECTOR: Monte Murbach
DIRECTOR (EX OFFICIO): Brian J. Olson

San Diego Association of Geologists is a program of San Diego Geological Society, Inc., a California 501(c)(3) nonprofit corporation.

San Diego Association of Geologists Publications

1973 Studies on the Geology and Geologic Hazards of the Greater San Diego Area, California, Arnold Ross and Robert J. Dowlen, eds.

1974 Recent Geological and Hydrologic Studies, Eastern San Diego County and Adjacent Areas: a Guidebook, Michael W. Hart and Robert J. Dowlen, eds.

1975 Studies on the Geology of Camp Pendleton, and Western San Diego County, California, Arnold Ross and Robert J. Dowlen, eds.

1977 Geology of Southwestern San Diego County and Northwestern Baja California, Gregory T. Farrand, ed.

1978 Natural History of the Coronado Islands, Baja California, Mexico, Herman T. Kuper, ed.

1979 Earthquakes and other Perils, San Diego Region, Patrick L. Abbott and William J. Elliott, eds. *(GSA annual meeting field trip guidebook).*

1981 Geologic Investigations of the Coastal Plain, San Diego County, California, Patrick L. Abbott and Shannon O'Dunn, eds.

1982 Geologic Studies in San Diego, Patrick L. Abbott, ed.

1985 On the manner of Deposition of Eocene Strata in Northern San Diego County, Patrick L. Abbott, ed.

1988 Landslides in Crystalline Basement Terrain, James R. Evans, ed.

1989 Field Trip Road Log: Fault Features between La Jolla, U.S.A. and Ensenada, Mexico, by Diane Murbach in "The Seismic Risk in the San Diego Region: Special Focus on the Rose Canyon Fault System, Proceedings; Glenn Roquemore, Susan Tanges, and Marion Wright, eds."

1990 Geotechnical Engineering Case Histories in San Diego County, John H. Hoobs, ed.

1991 Environmental Perils, San Diego Region, Patrick L. Abbott and William J. Elliott, eds. *(GSA annual meeting field trip guidebook).*

1992 Natural History of the Coronado Islands, Baja California, Mexico (revisited 1992), Lyne Perry, ed.

1993 Colorado Desert and Salton Trough Geology, Joe Corones, ed.

1994 Geology and Natural History of Camp Pendleton, U.S.M.C. Base, San Diego County, California, Phillip S. Rosenberg, ed. *(2nd Edition, 2010).*

1995 Paleontology and Geology of the Western Salton Trough Detachment, Anza-Borrego Desert State Park, California, Paul Remeika and Anne Sturz, eds.

1996 Geology and Natural Resources of Coastal San Diego County, California, Tissa Munasinghe and Phillip Rosenberg, eds.

1997 Santa Cruz Island: Geology Field Trip Guide, James R. Boles and Werner Landry, eds.

1998 Geology and Geothermal Resources of the Imperial and Mexicali Valleys, Lowell E. Lindsay and William Hample, eds.

1999 Water for Southern California: Water Resources Development at the Close of the Century, Gregory T. Cranham, ed.

2000 Geology and Enology of the Temecula Valley, Riverside County, California, Barbara B. Birnbaum and Kerry D. Cato, eds.

2001 Coastal Processes and Engineering Geology of San Diego, California, Robert C. Stroh, ed.

2003 Geology of the Elsinore Fault Zone, San Diego Region, Monte L. Murbach and Michael W. Hart, eds. *(joint publication with SCGS)*

2004 Mining History and Geology of Joshua Tree National Park, Margaret R. Eggers, ed.

2005/06 Geology and History of Southeastern San Diego County, California, David M. Bloom, Philip T. Farquharson, and Carole L. Ziegler, eds.

2006 Mines and Geology of the Randsburg Area, An Historical Gem of the Mojave Desert, by D.D. Trent.

2008/09 Palms to Pines: Geological and Historical Excursions through the Palm Springs Region, Scott Snyder, Bryan Miller-Hicks, and Jeff Miller, eds. *(joint publication with SCGS)*

2010 Geology and Lore of the Northern Anza-Borrego Desert Region, Monte L. Murbach and Charles C. Houser, eds. *(joint publication with SCGS).*

2011 Picacho and the Cargo Muchachos: Gold, Guns, and Geology of Eastern Imperial County, California, Todd A. Wirths, ed. (*2nd Edition, 2015*).

2012 Roadside Geology along Sunrise Highway, by Michael J. Walawender

2012 Waiting for Tsunami: Coastal Hazards of Northern San Diego County, California, Cari Gomes, ed.

2013 San Luis Rey on Display, Geoscience in Northern San Diego County, California, Brian J. Olson, ed.

2014 Coast to Cactus: Geology and Tectonics, San Diego to Salton Trough, California, Jennifer Bauer Morton, ed.

SECTION 1

ROAD LOGS and ROAD LORE

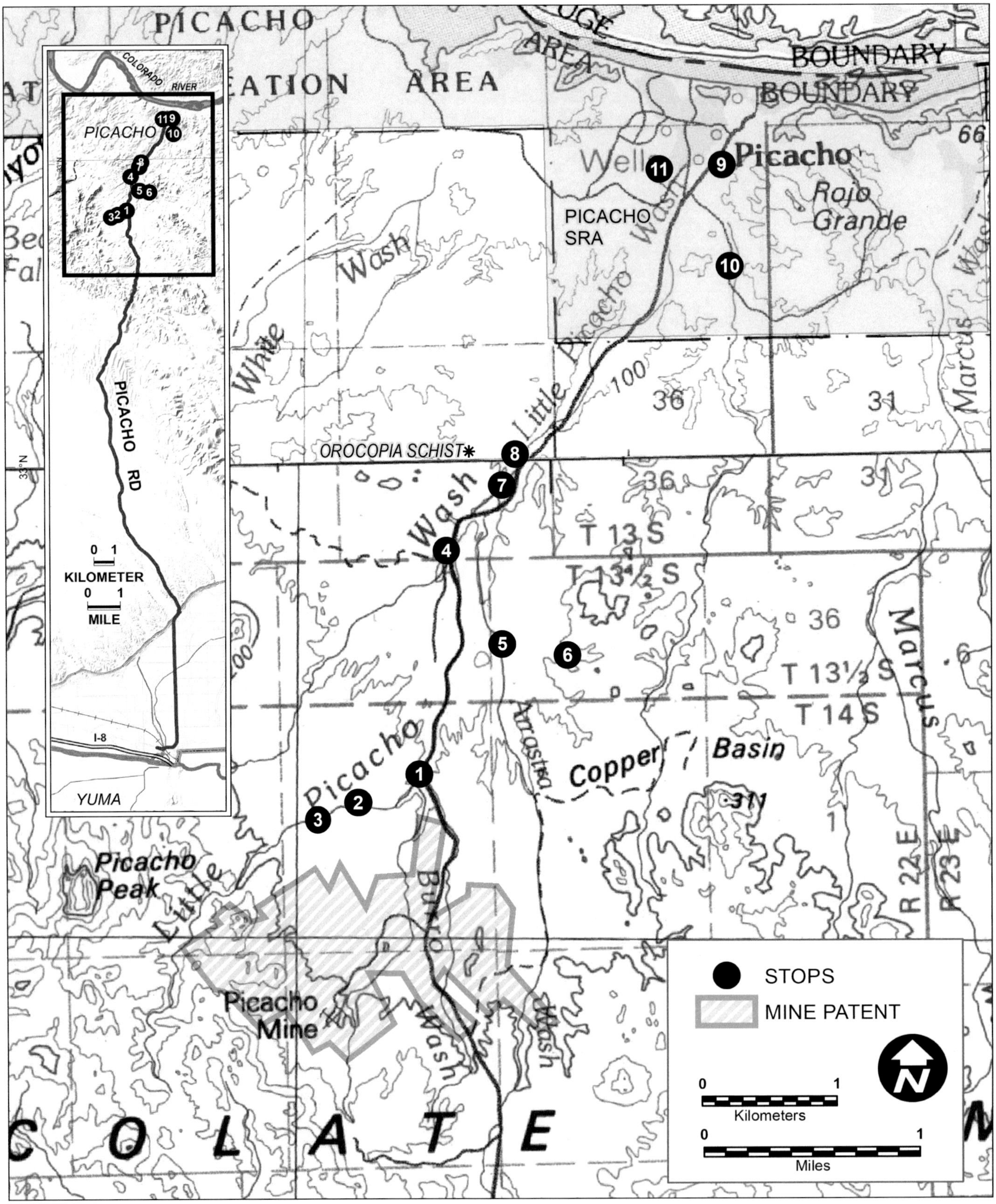

Map of the Northern Portion of Picacho Road and Field Trip Stop Locations. Map by David Bloom.
* Denotes hike destination for Stop 8.

2011 San Diego Association of Geologists Field Trip Road Log
Picacho Road

TODD A. WIRTHS, M.S., P.G.
Wirths Consulting
2160 Avenida de la Playa, La Jolla, CA 92037; todd@wirths.com
GARY H. GIRTY, PH.D.
Department of Geological Sciences
San Diego State University, San Diego, CA 92182

PICACHO ROAD WAS ONCE A DUSTY PATH to gold riches for over 150 years. Today, it is the main entrance to the 6,769-acre Picacho State Recreation Area, and provides access to the Bureau of Land Management's Picacho Peak and Little Picacho Wildernesses, as well as to the Imperial National Wildlife Refuge along the Colorado River. Starting in Winterhaven and heading north, Picacho Road is 24 miles long, but only the first six miles are paved. The remaining 18 miles are developed mostly as a wide, graded dirt road, maintained by the County of Imperial. At times the road is rough, and while 18 miles of dirt may try the patience of any driver, a light right foot is recommended to keep your tires intact. Patience is rewarded on arrival at Picacho State Recreation Area, as the serenity and raw beauty of the countryside will shed any leftover anxiety.

The mining of gold at Picacho began unpretentiously. Beginning in the 1850s, placer and lode deposits were worked by Mexican miners and their families at various locales between the Colorado River and southeastern Chocolate Mountains, including what would later be called Picacho (Odens, 1973; Morton, 1977; Van Wormer, this volume). Legend has it that around 1861 or 1862, a young fellow by the name of José Maria Mendivil, seeking the next opportunity in life, came across family groups placering for gold near the future townsite of Picacho along the Colorado River. He was invited to stay and work for gold, but knowing that potentially richer sources of placer gold lay in the higher ground; he declined and explored the hills to the south. José staked claims but without the means to develop the newly recognized rich deposits, he departed from Picacho, not returning until the 1870s with his small family. By then Anglo-Americans began arriving, investing capital and labor into the extraction and milling of the deposits (Odens, 1973; McKenney, 1939). A few stories have been included herein to provide a glimpse of the triumphs and tragedies of Picacho the town and Picacho the mine since Jose's return — see McKenney's "Saga of Old Picacho" and two *Calico Print* reprints from 1952.

Industrial gold mining at Picacho generally occurred in two phases. The first phase, which could be termed the "historical," occurred from about 1879, when either Dr. DeWitt Jayne of New York or David Neahr of Yuma, depending on the source, built a 15-stamp ore crushing mill beside the Colorado River, to September of 1910, when the Picacho Basin Mining Company halted operations due to a combination of accidents, low quality ore, and interruption of cheap riverboat transport of the ore as a result of the erection of the Laguna Dam. Nevertheless, sporadic attempts at mining continued until World War II. The "modern" phase was initiated by Glamis Gold, Inc., in 1979. Open pit mining commenced in 1981, and was phased out by 1998. Production of gold continued until 2000, with a total of 380,000 ounces of gold extracted. Notably, Glamis Gold was the operator during the mine's entire existence, considered a rare feat from a mining corporation perspective (Reference for Business, 2012). Glamis Gold deemed the mine "exhausted," but recent interest in renewing mining operations at Picacho has begun. At this writing (2012), effort is underway to possibly open the mine once again.

The following road log is designed to be toured in about one day. All the stops are located north of the Picacho Mine, and cover many of the major rock units and structures found in the area. During the 11 stops outlined below you will see rocks representative of the following 5 major tectonic events that have affected the southwestern Cordilleran margin.

1. Jurassic volcanism and sedimentation within an arc-graben depression that developed within the Cordilleran volcanic arc (Stops 4, 7, and 8).
2. Amphibolite-grade rocks of the Orocopia Schist that represent remnants of subduction complex material or forearc basin sediments that were scraped off the Farallon Plate and underplated beneath the western North American margin during the low-angle subduction event associated with the Laramide Orogeny (Stop 8).
3. Lower Miocene volcanic flows and pyroclastic material that formed during the westward sweep of arc volcanism following the low-angle subduction event (Stops 1 through 6, 9, 10, and 11).
4. The development of the Oligocene-Miocene Extensional Belt of Dickinson (2002). The Sortan Fault, a major early Miocene detachment that formed after or during the westward sweep of arc volcanism represents an important component of this belt (Stops 2, 8).
5. A conjugate set of strike-slip faults and a newly discovered reverse fault that record north-south shortening after ~9.45 million years ago. These structures likely represent the initial stages in the development of the Eastern California Shear Zone at the latitude of Picacho State Recreation Area (Stops 1, 5, 6).

Stops 1 through 11 in the following road log were constructed using Topo USA (version 8.0), a software package produced by Delorme. To insure accuracy multiple tracings of each route were made on pertinent 7.5′ quadrangle sheets within Topo USA. However, due to limited time we were unable to drive the routes between some of the stops, and as a result, your mileage between stops may be slightly different.

MILE LOG

0.0 From east-bound Interstate Highway 8, take the Winterhaven Drive/4th Avenue exit. Begin mileage at off-ramp intersection with Winterhaven Drive. Turn left (north) toward Winterhaven, CA.

0.5 Turn right on Imperial County Road S24 toward Picacho State Recreation Area.

0.7 Cross the Picacho Road Bridge over the Yuma Main Canal. This timber stringer-type bridge was constructed in 1925, then rehabilitated in 1947 (Bridgehunter.com, 2011). Construction of the canal was completed over a century ago, in 1912.

0.75 Turn left at the fork, and drive through railroad underpass.

1.0 Road veers left. Drive north through the agricultural fields of the Fort Yuma Indian Reservation of the Quechan People, established in 1884.

4.8 Picacho Road pavement yields to graded dirt. Road veers right along the All American Canal to the left. After construction throughout the 1930s, water was first delivered by the All American Canal in 1940.

5.1 Sign "Picacho State Recreation Area 18 miles." Cross Picacho Bridge, turn left and cross the flume of the All American Canal. Drive on wide, well-graded dirt road. Some washboard.

(Photo courtesy of Isabelle "Izzy" Tihanyi-Wirths)

6.35 On the right, Imperial County's Picacho Landfill, closed since May 2011.

8.35 Road forks, stay right. Left fork is the start of Barney

Oldfield Road (Imperial County Road # 986). For a brief biography of Barney Oldfield and geology along Barney Oldfield Road, see the article by T. A. Wirths in this volume.

8.45 Another fork, stay right.

11.1 On right, a low hill displaying an example of local-origin Plio-Pleistocene alluvial sands, gravels, and cobbles (Olmsted et al., 1973).

14.5 Slow down, approach right curve. View of "Unnamed Wash," the southeastern portion of the Chocolate Mountains, Picacho Peak, and Pebble Mountain at 2 o'clock. Begin descent into wash.

14.9 Hairpin left at "Unnamed Wash." Stay left (north) on Picacho Road. A right turn down the wash (Imperial County Open Route # 984) will lead to The Narrows, a narrow passage of the wash accessible only by motorcycles and small 4x4 vehicles (the passage is said to be too narrow for full-size vehicles). After about a ½ mile from Picacho Road, ride through steep, narrow walls of coarse Miocene sediments 40 to 60 feet deep. A number of chasms and slot canyons can be accessed by foot (In the Desert, 2011a).

15.0 Roadcut of Miocene or older sediments on the right (Olmsted et al., 1973). Note the southerly tilting beds.

15.3 Approaching Pebble Mountain on the right. Pebble Mountain is one of several desert varnish-topped volcanic mesas in the region. Crowe (1978) has mapped Pebble Mountain as a "capping pyroxene andesite" approximately 25.1 million years old. In a field guide by Frost, Drobeck, and Hillemeyer (1986), these pyroxene andesites are said to be overlain by basalts that "probably originated as flows from related feeder dikes, spread over the landscape, and are now preserved as erosional remnants of once widespread flows. These [vesicular] basalts have inclusions of olivine and diopsidic augite, low strontium ratios, and appear to represent mantle-type derivation. A widespread age for the oldest of these flows is about 12–13 million years ago."

15.7 Road cut at base of Pebble Mountain, exposing the underlying white, tuffaceous pyroclastic units (Frost, Drobeck, and Hillemeyer, 1986).

15.95 On the right is Imperial County Open Route # 697. After about 0.7 mile, this 4wheel drive jeep trail will lead you to "Crystal Hill," an outcrop of the ignimbrite of Ferguson Wash with fracture-filling secondary crystals of quartz (Crowe, 1978; In the Desert, 2011b). A quarry developed around the base of Crystal Hill by rock hounds has left fresh exposures of translucent ¼-inch quartz crystals in the pink, lower Miocene ignimbrite.

16.1 Hill on the left composed of the 23.4 million-year-old Quechan volcanic rocks.

16.5 Crossing small wash: on the east, two successive pyroclastic units. Take a short walk down the wash to the east and observe inclusions of mafic volcanic clasts captured in the upper, paler unit during flow stage. To the west of the road, only the darker lower unit crops out.

Close Up of Pyroclastic Deposits with inclusions in the Upper Unit, East of Mile 16.5

16.7 Roadcut of the so-called Picacho Dacite on left (Richard and Spencer, 1996). This massive very fine-grained light gray andesite or dacite contains sparse crystals of dark green pyroxene to white plagioclase, and strongly developed platy jointing is present locally.

16.75 Summit of Picacho Road, crest of the southeastern Chocolate Mountains. Your elevation is approximately 873 feet above sea level.

16.95 Roadcut of blocky Quechan volcanic rocks.

17.0 Two more Quechan roadcuts on the right. The gray-green color of this outcrop is likely attributable to chloritic alteration due to the influence of hydrothermal fluids. Two normal faults approximately 22 feet apart can be seen here. They have a total displacement of about 6.5 feet. Drag structures related to the faulting are visible. Walk to a second outcrop just north of this one to see a debris flow composed of cobbles and sand overlying more light green Quechan volcanic rocks.

Roadcut of Quechan Volcanics at Mile 17.0

17.2 Roadcut on the left of Quaternary surficial deposits (Richard and Spencer, 1996).

17.3 Low hill to the east, with brown Quechan basalts on the left of contact and gray Tertiary "sedimentary(?) breccia" of shattered, fine-grained leucogranite and gneiss on the right (Richard and Spencer, 1996).

17.6 Road crosses the upper reaches of Burro Wash. To the east is a white outcrop of the Jurassic-aged leucocratic Marcus Wash Granite as described by Frost et al. (1986), or the leucogranite member of the biotite quartz-feldspar gneiss and leucogranite basement as mapped by Richard and Spencer (1996). The overlying Quechan volcanic rocks have moderate to steep dips here (presumably as a result of normal faulting), but the contact with the underlying granite is nearly flat (Frost et al., 1986).

18.0 View ahead at 11:30 of Picacho mine tailings surrounded by fencing.

18.3 Good roadcuts of the probable Jurassic mylonitic orthogneiss.

18.55 Side road to the west. This is the entrance to the Picacho Mine. Access to the mine is closed to the public. The following history is from Shumway, et al. (1980). Primary sources for the history are from Odens (1973), Love (1974), and Morton (1977).

"Perhaps the most famous mine in this area, and in all of Imperial County, is the Picacho Mine (also known as the Dewitt C. Jayne Mine). Dr. Jayne was a New York drug manufacturer and one of the first to invest in this mine. His investment may have been profitable, but the Picacho Mine was beset with problems and bad luck every decade of its active existence. David Neahr began construction for the mine of a fifteen stamp mill overlooking the Colorado River in 1879. In 1882, 8,000 tons of ore were mined, yielding an average of $21 per ton. Although the mill was profitable, Neahr was forced into bankruptcy when a dishonest employee stole $7,000. The California Gold King Mining Company, with former Colorado Senator Stephen A. Dorsey as president, consolidated the Picacho mines and operated them until 1906, when the Picacho Basin Mining Company took over. In 1902, a huge 450 ton mill was in operation, and by 1904 it had 700 employees with a monthly payroll of $40,000. A narrow gauge train brought ore from the Picacho Mine to this mill, which was boasted as the largest cyanide plant in America. At this time the town of Picacho, which grew up and around the mill, consisted of some 2,500 souls. In July, 1904, a belt in the mill broke loose due to overloading, and the flywheel disintegrated, showering pieces through the roof and up to one-fourth of a mile down slope. Although workers repaired this damage quickly, construction of the Laguna Dam on the Colorado curtailed the hauling of ore concentrates by steamer to Yuma, and this, plus diminishing ore values, contributed to the final shutdown of the mill in September, 1910. In 1939, the Nipissing Mining Company of Canada hauled in a 200-ton mill from Tonopah, Nevada, in efforts to re-establish mining operations at Picacho, but World War II prevented the company from staging a comeback. Ruins of the mill, the machine shop of the 450-ton mill, and the boiler and tank are among the objects and buildings still standing. The total production estimate for the Picacho Mine [during the period of 1904–1910] is approximately $2,000,000."

The mine cemetery is in view to the west just beyond the arroyo. According to *The San Diego Union*, "The cemetery is probably the richest in the world." This small peninsula of land with the graves of 22 mostly anonymous individuals stands isolated as the surrounding gravels were dug out for precious yellow metal a century ago. Indeed, gravels dug out for a grave were run through the mill just as any other ore would be. Prophetically, the mine's sole occupant at the time, a watchman for the mine's owner, said "the gold is there" and that Picacho would rise again (*The San Diego Union*, 1959).

18.95 Vista point and Picacho monument on right, with Quechan volcanic rocks in view to the east. The "Picacho Mines" monument was dedicated on April 20, 1985, and reads: "Opened by placer miners in 1852, the gold mines expanded into hard rock quarrying by 1872. Picacho employed 700 miners at its peak from 1885 to 1900. Mill accidents, low ore quality, and the loss of cheap river transportation with the building of Laguna Dam led to numerous periods of inactivity. With ores far from worked out, the Picacho Mines, using modern techniques, again resumed operations in 1984. California Historical Landmark No. 193"

19.1 Sign "Picacho State Recreation Area 4 miles." Intersection with Burro Wash again. To the east in the wash, an outcrop of the tilted and faulted pre-Miocene to Miocene Little Picacho Wash Fanglomerate, containing clasts derived from the Winterhaven Formation.

Little Picacho Wash Fanglomerate at Junction with Burro Wash. Unconformably overlying the Little Picacho Wash Fanglomerate are Quaternary alluvial deposits. Height of the cliff face is between 15 to 20 feet.

19.5 **STOP 1.** Stop at the junction with Little Picacho Wash, at N32°58.7138′, W114°38.1938′. If there are a large number of vehicles, then parking will be difficult here. After parking, move to a location near the intersection of Little Picacho Wash and just off the west side of the main road leading to Picacho State Recreation Area. From this location, as you look toward the northwest (about 11 o'clock), note the sub-vertical contact separating the lower Miocene Quechan volcanic rocks on the right from the gray rocks of the Jurassic Winterhaven Formation on the left. This is the typical expression of high-angle faults throughout Picacho State Recreation Area and adjacent areas. In many cases, but unfortunately not at this easily accessible site, exposed fault surfaces like the one visible at Stop 1 display horizontal to sub-horizontal slickenlines and grooves, and thus were produced by strike-slip movements. As shown by map relationships, such structures cut foliations in the metamorphic basement complex, the Sortan Fault, and bedding and flows in the lower Miocene volcanic and epiclastic unit. Hence, they must have formed after approximately 23 million years ago (see Girty et al., this volume).

Outcrop at Stop 1 Showing Winterhaven Formation on the Left of the Fault, and Quechan Volcanics on the Right.

STOP 2. Walk westward up Little Picacho Wash for about 1,370 feet and stop around N32°58.6145′, W114°38.4999′. At this location in the northwest wall of Little Picacho Wash are several normal faults that cut the lower Miocene Quechan volcanic rocks and the Jurassic Winterhaven Formation. Note that they appear to sole downward into the Sortan Detachment, the major fault separating the Jurassic Winterhaven Formation and the Jurassic orthogneiss. Such observations suggest that extension continued or was renewed after or during eruption and emplacement of the lower Miocene volcanic and epiclastic unit (Girty et al., this volume). If you continue to walk up Little Picacho Wash, then you will first pass through an interval of highly altered orthogneiss, and then into the Quechan volcanic rocks at Stop 3.

STOP 3. The fault contact between the Quechan volcanic rocks and the Jurassic orthogneiss is located at N32°58.5784′, W114°38.6158′. As you walk from this point up Little Picacho Wash you will encounter several thick flows of variably vesiculated flow breccia and laharic breccia. Such flows are characteristic of the upper Quechan (Girty et al., this volume). Return to your vehicle and continue toward the entrance of Picacho State Recreation Area.

20.4 Layered Quechan volcanic rocks in view.

20.6 Jeep trail turnoff going left (west). This is Imperial County Open Route #875, a rough jeep trail heading west through White Wash, intersecting with Hyduke Mine Road and ultimately with Ogilby Road north of the Cargo Muchacho Mountains. See "Ogilby Road," Mile 11.8, by T. A. Wirths in this volume. Dark greenish gray Winterhaven Formation volcanic rocks outcrop to the northwest.

The Georgia Placer Gold Mine was located in this area on the west side of Little Picacho Wash, in 1891. The source of the gold is presumed to be from the Picacho Mine area (Morton, 1977; Shumway et al., 1980).

20.75 **STOP 4:** Pull off the main road and park at N32°59.6193′, W114°38.0380′. As you look westward down the road connecting to the road leading from Picacho State Recreation Area, you are more or less looking down the contact between the lower Miocene Quechan volcanic rocks on the left and the Jurassic Winterhaven Formation on the right. The dark gray to almost black rocks immediately to your right are part of the metavolcanic member of the Winterhaven Formation. As discussed in Girty et al. (this volume), the metavolcanic member is largely derived from basalt and andesite. Chondrite-normalized REE (rare earth element) distribution patterns are generally like those common to andesites erupted in volcanic arc settings, but form a wide band with REE concentration levels ranging from 50 to 200 times chondrite values.

20.9 **STOP 5.** From Stop 4, drive north to the intersection with Arrastra Wash. At this intersection turn right (southward) and drive up the wash for approximately 0.75 mile to Stop 5 at N32°59.2086′, W114°37.7843′. After exiting your vehicle, note that in the ridge to your west, a thin (~5–8 meter thick) gray to black section of the Jurassic Winterhaven Formation separates an otherwise thick and continuous appearing sequence of the lower Miocene Quechan volcanic rocks. Also note that the lower contact of the thin section of the Winterhaven Formation appears to dip to the south. Though this contact is not cleanly exposed at this location, it is at Stop 6. Hence, we will walk about 1,860 feet to the SE to view an excellent exposure of the contact at this location.

STOP 6. From Stop 5 walk up the ravine on the east side of Arrastra Wash to the top of the ridge. There is a donkey trail following this route. At the top of the ridge, looking at about 3 o'clock you should see a canyon with high peaks to both the north and south. You are going to walk down and into this canyon. At about N32°59.1847′, W114°37.419′ the canyon will turn relatively sharply to the left, and in the eastern wall of the canyon there is an excellent exposure of the contact that we are looking for (see map in Girty et al., this volume).

At Stop 6, an approximately cross-sectional view of the contact first seen at Stop 5 is cleanly exposed. At this location, lower amphibolite to greenschist facies Jurassic orthogneiss sits atop unmetamorphosed lower Miocene Quechan volcanic rocks along a fault contact dipping approximately 56° SW. At the contact, the orthogneiss in the hanging wall block is brecciated, and the Quechan volcanic rocks in the footwall block have been transformed into a red foliated fault gouge. In pieces of the gouge plucked from this outcrop, we have observed faintly preserved slickenlines plunging down the dip of the foliation which lies parallel to the contact. Walking from the fault contact northward, a wide zone of highly bleached and fractured Quechan volcanic rocks is evident. We refer to this fault as the Copper Basin Fault, and based on the above observations we interpret it to be a reverse fault.

If you were to continue walking to the east for approximately another 1.36 miles, then you would encounter strands of the Taylor Lake Fault, a major dextral-sense strike-slip fault (Girty et al., this volume). These strands cut the Copper Basin Fault, and are interpreted to be components of a regionally developed conjugate set of faults that formed after approximately 9.45 million years ago. The conjugate set of faults is the record of north-south shortening, and the geometry of the Copper Basin Fault is consistent with such an event. If this interpretation is correct, then the Copper Basin Fault also likely formed after approximately 9.45 million years ago. As discussed in Ricketts et al. (2011) and Girty et al. (this volume), the Copper Basin Fault and the conjugate set of strike-slip faults may have formed as part of the Eastern California Shear Zone. If this interpretation is correct, then you are looking at one of the earliest structures that formed as the plate boundary changed from a converging to transform geometry.

Stop 7 Outcrop Showing Fault Striations on Orthogneiss. Arrow highlights direction of slip.

21.05 STOP 7: If possible, please park off the main road. The outcrop that we will be visiting is located in the southern wall of the prominent ravine to your left, at N32°59.8945′, W114°37.7674′. At this location, the contact between the metavolcanic member of the Jurassic Winterhaven Formation and the Jurassic orthogneiss is well exposed. The contact is polished and deeply grooved and striated. Note that the striations rake shallowly within the plane of the contact (<45°) toward the southwest. If the fault surface has not been rotated, then this geometry would suggest that the magnitude of the strike-slip component was larger than the dip-slip component of the displacement vector during the event that produced the grooves and striations. However, map relationships suggest that the fault at Stop 7 is the Sortan Fault, and that it has been folded about the Chocolate Mountains anticlinorium. Hence, prior to folding the striations and grooves may have been produced during northeast-southwest displacements along this ancient detachment.

Photo of Winterhaven Formation near Stop 8, with Picacho Peak in the Background. (Photo by Carol Teasley)

21.3 STOP 8: Drive to N33°0.0023′, W114°37.6912′ and park for a short hike. As you look to the north, we will walk about 2,470 feet up the canyon on your left. During this walk we pass through a relatively thick section of the Metavolcanic Member of the Winterhaven Formation, and then into the metasedimentary member at N33°0.0554′, W114° 37.7672′. At about N33°0.0573′, W114°37.8984′ we will encounter our first outcrops of the Orocopia Schist. We will stop at N33°0.0177′, W114° 38.0458′. At this location the high ridges to your southwest are underlain by Jurassic orthogneiss, and the Chocolate Mountains Fault separates them from the outcrops of the Orocopia Schist that you are standing on. As discussed in Girty et al. (this volume), the Orocopia Schist likely represents the remnants of subduction zone material or forearc basin sedimentary rocks that were scraped off the subducting Farallon Plate as it descended along a shallow inclination beneath the southwestern Cordilleran margin. It was subsequently brought closer to the surface through a combination of extensional faulting along the Chocolate Mountains, Gatuna, and Sortan faults. At Picacho, the earliest clear indication that the schist was exposed at the Earth's surface occurs in sequence II of the Bear Canyon conglomerate. Unfortunately, we will not be able to see the Bear Canyon conglomerate on this trip. The interested reader should refer to Girty et al. (2006) and Ricketts et al. (2011) for additional discussion of this relationship.

Return to your vehicle and drive north for about 1.4 miles. Near the end of this drive you will pass through the entrance to Picacho State Recreation Area. The road that you will turn right on to will be the first right after you pass through the park entrance. Drive a few hundred feet, and park your vehicle there. We will walk approximately 600 to 700 feet to Stop 9, the white to tan outcrops in the distance to the northeast.

21.6 In this area on the west side of the wash, the Crescent Placer Gold Mine was located, in 1891 (Morton, 1977; Shumway et al., 1980).

21.95 Sign for Picacho.

22.3 Red Rojo Grande volcanic rocks on right (east), slopes of sequence III of Bear Canyon conglomerate on the left (west). If you investigate the rocks to the east, you will also locate the trace of the path of the *Picacho and Colorado River*, the century-old railroad line built for the transport of gold ore from the mine to the mill (Jennings, 1977). The railroad started at the mine and was directed north down Little Picacho Wash (Picacho Road). At or about here, the railroad's path turned east, headed to deliver ore to the mill for crushing and processing near the river. See the "Side Trip" at Stop 10 for a short history of the railroad.

22.7 Park entry. $5 for day use; $20 overnight.

22.7 *(plus about 260 feet more)*
STOP 9. Make right turn at this road and park. Do not drive off the graded road, as the park forbids off-road vehicle travel. This east-bound road is Railroad Canyon Road.

At N33°1.1226′, W114°36.7535′ is an outcrop of the white to tan fine-grained ignimbrite that forms the lower part of the White Wash unit. As discussed in Girty et al. (this volume) the white to tan ignimbrite at Stop 9 is characterized by a very negative Eu (europium) anomaly. Many questions remain unanswered about this rock unit. For example, we do not know its source, and we do not know its petrologic relationship to the ignimbrite of Ferguson Wash.

In the ridge to the northeast of Stop 9 are exposures of volcanic rocks of Rojo Grande. If the lighting is correct, you may be able to see the distinctive purple banding characteristic of this unit. We have not yet studied the petrology of the banding or its potential structural significance. Though we refer to this planar feature as flow banding, it may represent a series of dikes. If this speculation turns out to be correct, then the volcanic rocks of Rojo Grande may be the source of the rocks of Marcus Wash, which areexposed well to the south of Stop 9.

Return to your vehicle and drive 0.35 mile south on the dirt road that you parked adjacent to for Stop 10.

22.9 Several yards off to the right (south) of the road lay the foundation and chimney of a post-mining era squatter's cabin.

Cabin remains near start of Railroad Canyon Road

23.1 **STOP 10.** The contact between the Quechan volcanic rocks and the volcanic rocks of Rojo Grande can be observed at N33°0.8002′, W114°36.7145′. At this location, the contact dips northeast. Looking to the north toward Rojo Grande, note that the volcanic rocks of Rojo Grande are characterized by many irregularly shaped areas that appear to bleached white relative to the generally purple color of this unit. These areas are likely the result of hydrothermal alteration during or soon after emplacement.

The Quechan volcanic rocks at Stop 10 are dark gray, and display a well preserved porphyritic texture. Unlike localities to the south, the outcrops of the Quechan south of Rojo Grande lack the well-defined flow layering observed at Stop 3.

SIDE TRIP A drive on Railroad Canyon Road. The final eighth of a mile of this road was originally constructed as the final leg of the narrow-gauge railroad, the *Picacho and Colorado River*, used by a small steam locomotive to haul gold ore from the mines at Picacho Basin down to the 450-ton stamp mill just above the Colorado River (Jennings, 1977). During construction of the railroad, work halted when new placer deposits were discovered and workers ran off to pan gold. Work resumed when the placer was exhausted (Odens, 1973). At about 1.1 miles from Picacho Road, Railroad Canyon Road actually traverses the old route of the *Picacho and Colorado River*. The so-called "upper mill," constructed in 1897, was specially designed to handle large quantities of low-grade ore. The mill ran almost non-stop from 1904 to 1908. In 1908 the mill was moved to the mine site, in part because flash-flooding sporadically washed out sections of the railroad. The town was dismantled and rebuilt near the mine, thus ending rail transport of the ore. When mining, as well as the pumped water supply to the Picacho Basin area, ceased in 1910, those who remained rebuilt Picacho where it previously was, by the river (McKenney, 1939; *Calico Print*, 1952). The 1.9 mile-long road ends with an overlook of the Picacho Stamp Mill ruins. At the lip of the overlook are the remnants of what was the foot of the wooden railroad trestle that spanned the arroyo and ended at the stamp mill.

Return to your vehicle and drive back to the main road leading into Picacho State Recreation Area. Turn left and then drive a short distance to the road leading into the group campground. Drive into the group campground which is Stop 11.

STOP 11. At the group campground are clean well exposed outcrops of the ignimbrite of Ferguson Wash. This rock unit has been dated by Needy et al. (2007) using zircon laser-ablation methods at 23.2 ± 0.2 million years old. Well-developed fiamme (lens-shaped volcanic fragments, such as pumice, flattened by compression or shear, usually millimeter to centimeter in size, seen in some pyroclastic rocks) are present in most outcrops of the ignimbrite, and a moderately to well-developed foliation defined by aligned and flattened pumice fragments is another common feature. Though it is commonly viewed as the youngest unit within the lower Miocene volcanic and epiclastic unit, our field work and mapping suggests that the Walker volcanic rocks may actually postdate its eruption. The Walker volcanic rocks are located approximately 0.87 mile northwest of Stop 10, and also have been dated by the zircon laser-ablation method at 23.5 ± 1.0 million years old.

Both the Walker volcanic rocks and the ignimbrite of Ferguson Wash are cut by the Taylor Lake Fault, which also transects and displaces the Bear Canyon conglomerate, a relationship that is consistent with the idea that they were deformed and tilted after about 9.45 million years.

REFERENCES

Bridgehunter.com, 2011, Picacho Road Bridge facts. Accessed February 11, 2011. http://bridgehunter.com/ca/imperial/58C0028/.

Calico Print, 1952, several articles by former citizens of Picacho, edited by Harold and Lucile Weight, Vol. VII, No. 2, February, on file at Arizona Historical Society Archives, Yuma Arizona. Two articles, by M. Mendivil and E. Rochester, are reprinted herein.

Crowe, Bruce M., 1978, Cenozoic volcanic geology and probable age of inception of basin-range faulting in the southeasternmost Chocolate Mountains, California: *Geological Society of America Bulletin*, v. 89, pp. 251–264.

Dickinson, W.R., 2002, The Basin and Range province as a composite extensional regime: *International Geology Review*, v. 44, pp. 1–38.

Frost, Eric, Peter Drobeck, and Bud Hillemeyer, 1986, Geologic Setting of Gold and Silver Mineralization in Southeastern California and Southwestern Arizona, *in Cenozoic Stratigraphy, Structure and Mineralization in the Mojave Desert, Field Guide for the 1986 Cordilleran Section Meeting of the Geological Society of America*, March 25–28, Los Angeles, California.

Frost, Eric, Peter Drobeck, Bud Hillemeyer, and Gail S. Liebler, 1986, Field Guide to the Picacho Mine and Vicinity, California, *in Field Trip to Mesquite Mine, American Girl — Padre-Madre*

Properties, and Picacho Mine, Imperial County, California, Arizona Geological Society Fall Field Trip, March.

Girty, G.H., Stephenson, D., Canfield, A., Marsh, J., Middleton, T., Rayburn, J., Sisk, M., Verdugo, D., Moniz, R., Gunter, J., Nielsen, J., Gasca, C., Rowland-Smith, A., Smith, M., Lovering, K., Campbell, K., Castenada, F., Gray, R., and Campbell, C., 2006, Geology of the Picacho State Recreation Area, SE California: Implications for the timing of formation of the Chocolate Mountains anticlinorium, *in* Girty, G.H., and Cooper, J., eds., *Using Stratigraphy, Sedimentology, and Geochemistry to Unravel the Geologic History of the Southwestern Cordillera: A volume in honor of Patrick L. Abbott*, Pacific Section SEPM Book 101, pp. 255–274.

Girty, Gary H., Krista K. Muela, Holly Olson, Lauren Sutton, Melanie Biggs, Jon Sainsbury, Jason W. Ricketts, Rachel Errthum, Thomas A. Colby, Taylor Carrasco, Erik Gordon, and Tim Pelbath, 2014, Toward a unified model for the geologic and tectonic history of the region between Indian Pass and Picacho State Recreation Area, southeastern California; this volume.

In the Desert, 2011a, Description and interactive map of The Narrows. Accessed February 2011. http://www.in-the-desert.com/narrows.html.

In the Desert, 2011b, Description and interactive map of Chrystal Hill. Accessed February 2011. http://www.in-the-desert.com/crystalhill.html.

Jennings, Bill, 1977, "Picturesque Picacho: Historic Mining Camp on Lower Colorado now a Recreational Paradise," *The Desert Magazine*, v. 40, No. 2, February.

Love, Frank, 1974, Mining Camps and Ghost Towns: A History of Mining in Arizona and California along the Lower Colorado. Westernlore Press, Los Angeles.

McKenney, J. Wilson, 1939, "Saga of Old Picacho," *The Desert Magazine*, v. 2, No. 5, March, and reprinted herein.

Morton, Paul K., 1977, Geology and Mineral Resources of Imperial County, California, County Report 7, California Division of Mines and Geology, Sacramento.

Needy, S.K., Wooden, J.L., Barth, A.P., and Jacobson, C.E., 2007, Geochronology of igneous rocks in the Chocolate Mountains region as a means to interpret the tectonic evolution of southeastern California: *Geological Society of America Abstracts with Programs*, v. 39, p. 407.

Odens, Peter, 1973, Picacho — Life and Death of a Great Gold Mining Camp; self-published.

Olmsted, F. H., O. J. Loeltz, and B. Irelna, 1973, Geohydrology of the Yuma Area, Arizona and California; *U. S. Geol. Survey Prof. Paper* 486-H, 227 p.

Reference for Business, 2012, Glamis Gold, Ltd. — Company Profile, Information, Business Description, History, Background Information on Glamis Gold, Ltd. Accessed February 19, 2012. http://www.referenceforbusiness.com/history2/30/Glamis-Gold-Ltd.html.

Richard, Stephen M., and Jon E. Spencer, 1996, Geologic Map of the Picacho Mine Area, Southeastern California, *Arizona Geological Survey Open-File Report* 96-30.

Ricketts, J.W., Girty, G.H., Sainsbury, J.S., Muela, K.K., Sutton, L.A., Biggs, M.A., and Voyles, E.M., 2011, Episodic growth of the Chocolate Mountains anticlinorium recorded by the Neogene Bear Canyon conglomerate, SE California, U.S.A.: *Journal of Sedimentary Research*, v. 81, pp. 859–873.

The San Diego Union, 1959, "Ghost Town Graveyard: The World's Richest? Gold-rich site gives Picacho its memories," by Wayne Winters; August 2, p. a–25.

Shumway, Gary L., Larry M. Vredenburgh, and Russell D. Hartill, 1980, Desert Fever — An Overview of Mining History of the California Desert Conservation Area: Bureau of Land Management, U. S. Department of the Interior, Riverside, California, Contract No. CA-060-CT7-2776, February 1980.

Van Wormer, Stephen, 2014, "Dry Diggings: A History of the Potholes Mining District;" this volume.

Barney Oldfield *(driving) and mechanic Bruno Seibel, in a modified Mercer, circa 1914. Courtesy of the San Diego History Center.*

Barney Oldfield Road

"You know me, Barney Oldfield!"

TODD A. WIRTHS
Wirths Consulting
2160 Avenida de la Playa, La Jolla, CA 92037; todd@wirths.com

ACCESS TO THE EASTERN BACKSIDE OF THE Cargo Muchacho Mountains is via Barney Oldfield Road, little more than a jeep trail. But a century ago, the name Barney Oldfield was synonymous with flamboyant, swaggering confidence. In 1902, the sport of automobile racing was in its infancy when Barney Oldfield, driving Henry Ford's 999 Racer, easily beat Alexander Winton, one of the premier racers of the era. By the time he retired in 1918 after over 2,000 race events, Oldfield was one of the most colorful and sought-after personalities of the day. During the height of his career, he set speed records (including the first to post a 100 mile-per-hour lap at the Indy 500), acted in movies, was a notorious bar-room brawler, and commanded thousands of dollars just for appearances, all the while attired in imported suits, expensive jewelry, gaudy silk neckties, topped off with his trademark cigar stub jutting from his teeth. "You know me, Barney Oldfield!" he'd say to adoring crowds as he stepped out from his limousine (Kernan, 1989; Nolan, 1959).

Beginning in 1908, automotive manufacturers, cities, and auto clubs came together to promote and stage annual overland road races from Los Angeles to Phoenix. Known as the Cactus Derby, the race crossed some of the harshest desert terrain of the American southwest. Race routes varied from year to year. Judging from a period map, portions of the 1912 and 1913 races were routed right through the Cargo Muchacho Mountains, via American Girl Canyon, on their way to the Yuma stage (American Roads, 2011a). This must have been quite a spectacle to the miners residing in the canyon and nearby Tumco. One of the race checking stations was at Black Butte, a Miocene volcano just east of American Girl Canyon. Black Butte is one of several desert varnish-coated volcanic outcrops in the area of the so-called basalt of Black Mountain, having erupted about 9.45 million years ago. These vesicular augite-olivine basalts of Black Mountain are the youngest volcanic rocks of the region (Crowe, 1978; Girty and others, 2014).

Oldfield, with his mechanic Bruno Seibel riding shotgun, entered the 1913 race in a Simplex. They didn't finish, breaking down somewhere east of Yuma. In fact, only eight of the twenty-seven entrants finished. The victors were the Olin Davis and Clarence Watts team aboard a six-cylinder Locomobile, with an elapsed time of 18 hours

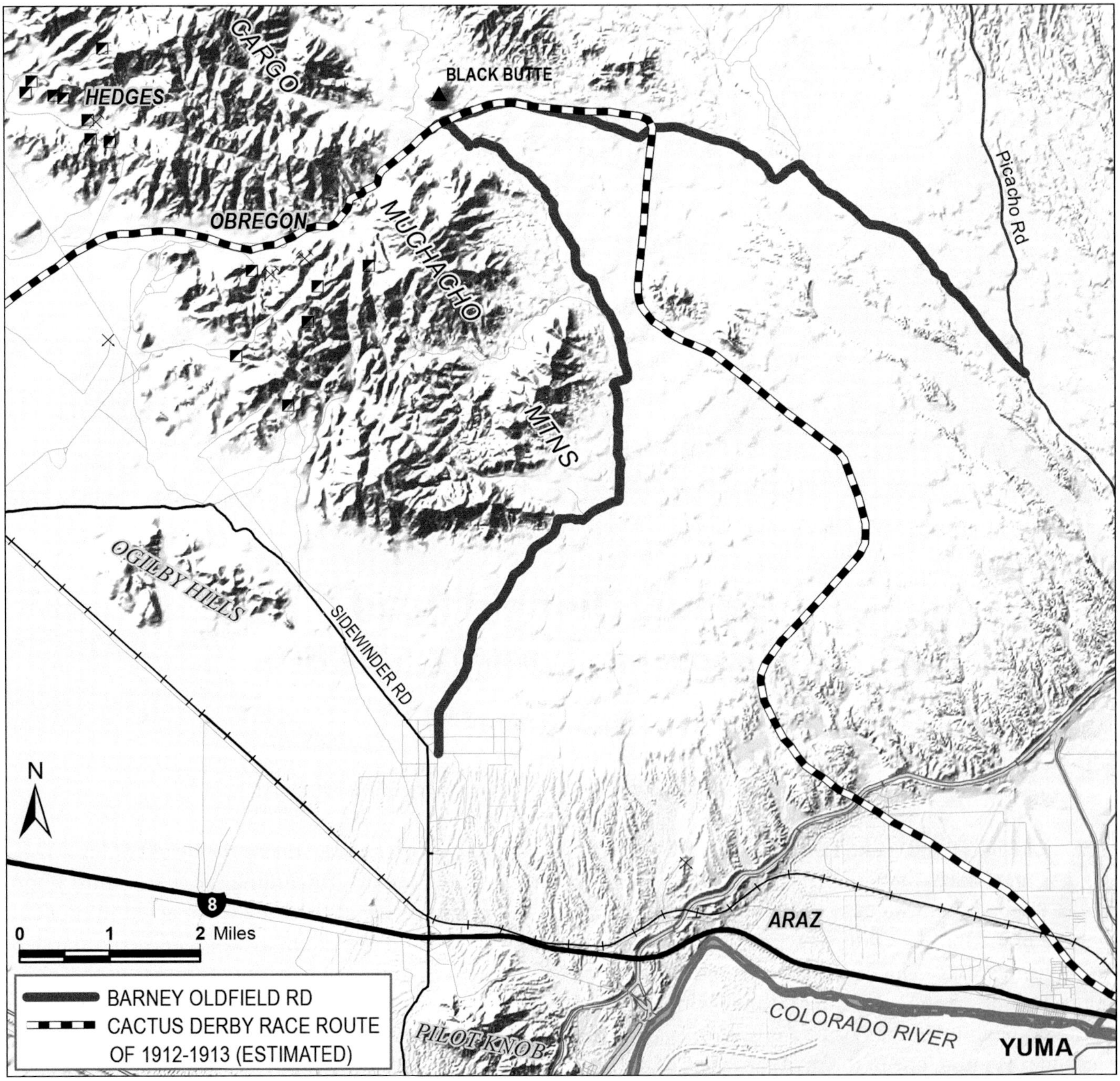

Topographic map showing Barney Oldfield Road, and approximate route of the 1912–13 Cactus Derby race course through the area. (Map by Lisa Pierce and David Bloom.)

Black Butte.

and 50 minutes (American Roads, 2011b; Smith, 1913). Oldfield won the final Cactus Derby held in 1914 racing a Stutz along a northern route through the Mojave Desert, beating other well-known racers such as Louis Chevrolet, Olin Davis, and 1909 Derby winner Joe Nikrent. At the end of the race Oldfield was awarded a medal studded with diamonds that proclaimed him "Master Driver of the World" (American Roads, 2011a; Nolan, 1959).

To traverse Barney Oldfield Road, take the Winterhaven Drive exit from Interstate 8 and head north, then turn right on Picacho Road (Imperial County Road S24). Proceed north on Picacho Road through the Fort Yuma Indian Reservation, and follow the signs to the Picacho State Recreation Area. After about 8.3 miles from Interstate 8, take the left fork from Picacho Road. Reset odometer to zero.

MILE 0.0 This is the start of Imperial County Road #986, i.e., Barney Oldfield Road. Four-wheel drive is recommended for the soft sand washes, but the passage can be accomplished with a two-wheel drive truck. At Mile 3.45, take a sharp left and drive down a steep slope of about 8 to 10 feet, where the county road designation switches to #830. After seven miles westward and a few sand washes, Black Butte will be front and center. Imperial County Road #746 circumnavigates Black Butte, and a hiking trail is said to weave its way to the top. The Cactus Derby race route was once staged right through this area.

MILE 7.75 Road #830 intersects with Imperial County Road #622. Barney Oldfield Road continues south on Road #622.

MILE 11.8 Road #622 travels under high voltage power lines and past a few intersections — keep a straight course. Approximately ¼-mile past the power lines, the desert pavement surface changes from an all locally derived angular gravel to a mix of local gravel and exotic well-rounded Colorado River "B" gravel (Metzger and others, 1973, p. G22–G23; Olmsted and others, 1973, Pl. 3). These mostly Paleozoic clasts were transported great distances by the Colorado River during Plio-Pleistocene time (Metzger and others, 1973, p. G28). This mixed-origin fan surface continues all the way to Interstate 8. Eventually, Barney Oldfield Road intersects with Sidewinder Road. Before crossing the railroad tracks, look for the California Registered Historical Landmark No. 985 monument for Camp Pilot Knob to the west, commemorating the U.S. Army's establishment of a desert training facility here:

> "Camp Pilot Knob was a unit of the Desert Training Center, established by General George S. Patton, Jr., to prepare American troops for battle during World War II. It was the largest military training ground to exist. At the peak of activity here at Pilot Knob, June–December, 1943, the 85th Infantry Division, and the 36th and 44th Reconnaissance Squadrons of the 11th (Mechanized) Cavalry trained for roles in the liberation of Europe, 1944–45."

MILE 18.75 The journey ends at the tracks of the Southern Pacific Railroad's Yuma Line with a view to the south of Pilot Knob. Pilot Knob is largely composed of the Precambrian Chuckwalla Complex, comprised mostly of quartz biotite gneiss and foliated hybrid granitic rocks, ranging in composition from gabbro to granite (Morton, 1977). A hornblende-biotite granodiorite to quartz monzonite augen gneiss at Pilot Knob yielded a U-Pb apparent age on zircons of 1.7 billion years (Dillon, 1975, in Wilkinson and Wendt, 1986). These rocks apparently underwent amphibolite-facies metamorphism prior to intrusion of Mesozoic(?) porphyritic quartz monzonite (Dillon, 1986).

Pilot Knob received its appellation due to its prominence noted by Colorado River boat navigators in the 19th and early 20th centuries. Today, the Andrade Quarry, developed in the augen gneiss for construction materials, can be seen on its north side (Wilkinson and Wendt, 1986). Currently, the quarry site is not active and no equipment is present, but past activity has left blocks and boulders lying about, providing a good 3-D view of the character of the foliation developed in the augen gneiss (W. Elliott and M. Marshall, personal communication, 2011). At the base of the peak lies the Pilot Knob Long Term Visitor Area, a snowbird retreat during the winter.

REFERENCES

American Roads, 2011a, "New Route for November Phoenix Race," reprinted from *Touring Topics*, October 1914, pp. 13–14. Accessed September 11, 2011. Includes map showing races routes for the years 1908 through 1914. http://www.americanroads.us/articles/Touring_Topics_1914_10.html.

American Roads, 2011b, "Report on the 1913 Los Angeles to Phoenix Road Race," reprinted from *Touring Topics*, November 1913, pp. 17. Accessed September 11, 2011. http://www.americanroads.us/articles/Touring_Topics_1913_11.html.

Crowe, Bruce M., 1978, Cenozoic volcanic geology and probable age of inception of basin-range faulting in the southeasternmost Chocolate Mountains, California; *Geological Society of America Bulletin* v. 89, no. 2; p. 251–264.

Dillon, John T., 1986, Summary of the Geology of the Chocolate and Cargo Muchacho Mountains, California, in *Geology of the Imperial Valley, California*, edited by Paul D. Guptill, Robert W. Ruff, and Eldon M. Gath; South Coast Geological Society, Inc.

Girty, Gary H., Krista K. Muela, Holly Olson, Lauren Sutton, Melanie Biggs, Jon Sainsbury, Jason W. Ricketts, Rachel Errthum, Thomas A. Colby, Taylor Carrasco, Erik Gordon, and Tim Pelbath, 2014, Toward a unified model for the geologic and tectonic history of the region between Indian Pass and Picacho State Recreation Area, southeastern California, in this volume.

Kernan, Michael, 1989, "Wow! A Mile A Minute!" *Smithsonian*, March.

Metzger, D. G., O.J. Loeltz, and B. Irelna, 1973, Geohydrology of the Parker-Blythe-Cibola Area, Arizona and California: *U. S. Geol. Survey Prof. Paper* 486-G, 130p.

Nolan, William F., 1959, "It's Me, Barney Oldfield!" *Sports Illustrated*, July 27.

Olmsted, F. H., O.J. Loeltz, and B. Irelna, 1973, Geohydrology of the Yuma Area, Arizona and California; *U. S. Geol. Survey Prof. Paper* 486-H, 227p.

Smith, Bert C., 1913, "Davis and Ball, Dark Horses, Take Honors," *Los Angeles Times*, Nov. 5.

Wilkinson, William H., and Clancy J. Wendt, 1986, Precious Metal Mineralization, Stratigraphy, and Tectonics in Southeastern California, in *Field Trip to Mesquite Mine, American Girl–Padre-Madre Properties, and Picacho Mine, Imperial County, California*, edited by William Wilkinson and Clancy Wendt, Arizona Geological Society Fall Field Trip, March.

2011 San Diego Association of Geologists Field Trip Road Log
Ogilby Road

BY TODD A. WIRTHS, M.S., P.G.
Wirths Consulting
2160 Avenida de la Playa, La Jolla, CA 92037; todd@wirths.com

OGILBY ROAD (IMPERIAL COUNTY ROUTE S-34) is a 25-mile long, mostly north-south highway connecting California State Route 78 and Interstate Highway 8 in southeastern Imperial County, California. The two-lane blacktop passes by several ghost towns and only one permanent settlement, Gold Rock Ranch, an RV-mobile home resort.

The rugged Cargo Muchacho Mountains lie to the east along the southern half of the road. The range is underlain by the Chocolate Mountains fault, a late Cretaceous to early Tertiary-aged subduction structure subsequently reactivated during the early Tertiary as an exhumation-related detachment fault (Owens and Hodder, 1994; Jacobson et al., 2002). Although the Chocolate Mountains fault crops out about 9.3 miles to the north and is not exposed here, the range locally contains low-angle structures interpreted as thrust faults, detachment faults, and thrust faults re-activated as detachment structures. These are occasionally cut by high-angle dextral faults, parallel to the San Andreas fault, with separation of up to one-half mile (Wilkinson and Wendt, 1986; Girty et al., this volume).

The oldest rock unit exposed in the Cargo Muchacho Mountains is the Jurassic-aged Tumco Formation, composed of quartzofeldspathic gneiss with minor hornblende schist. From unpublished data provided by Dr. Richard Tosdal of the University of British Columbia, a U-Pb Sensitive High Resolution Ion Microprobe (SHRIMP) age for the Tumco Formation is 185 ± 3 million years old. Dr. Tosdal indicated the rock is loaded with xenocrystic and inherited zircons, so the age is not as precise as it could be (Tosdal, personal communication, 2011). These rocks were intruded by four types of Jurassic granitic magmas ranging in age from 160 to 173 million years old (Tosdal, 1986; Owens and Hodder, 1994). Subsequently, all underwent amphibolite-grade (green-schist) metamorphism during the Jurassic Period (Owens and Hodder, 1994). At the Jurassic-Cretaceous transition, another round of magmatic intrusives consisting of 145 million-year-old biotite granite dikes and dike swarms, with associated aplites and pegmatites, intruded the older rocks (Dillon, 1976, cited in Wilkinson and Wendt, 1986). Still younger aplitic and pegmatitic dikes of Oligocene age intrude the west side of the range (Wilkinson and Wendt, 1986). Middle to late Tertiary vesicular basalt cones dot the landscape, and Plio-Pleistocene Colorado River sediments can be found at places along the southern and eastern sides of the range (Morton, 1977; Olmsted et al., 1973; Sterling White, BLM, personal communication, 2011; Elliott, this volume "a").

The hallmark of the Cargo Muchacho Mountains was the gold they once contained in abundance (Morton, 1977; Guthrie et al., 1987). Indeed, 19th century Mexican boys arriving into a gold placering camp with their shirts loaded with gold ore ("muchachos cargados") provided the range its namesake (Van Wormer and Newland, 1996). The gold occurs in quartz veins, veinlets, and as disseminated grains in association with the metasomatized foot walls and hanging walls of low-angle faults within the Tumco Formation and granitic intrusives (Guthrie et al., 1987, Owens and Hodder, 1994). The timing of emplacement of the ore is generally constrained to coincide with late Oligocene extension of the region, characterized in part by low-angle detachment faulting (Pridmore and Frost, 1992; Girty et al., this volume). Associated with the detachment faults are intensive areas of brecciation, where conditions are favorable for hydrothermal circulation and mobilization of metals in solution. Metals, e.g., gold, are "then deposited in breccia zones directly overlying the detachment fault and in open spaces along normal faults in the hanging wall" (Pridmore and Frost, 1992). Industrial-scale gold mining continues in the Cargo Muchacho Mountains

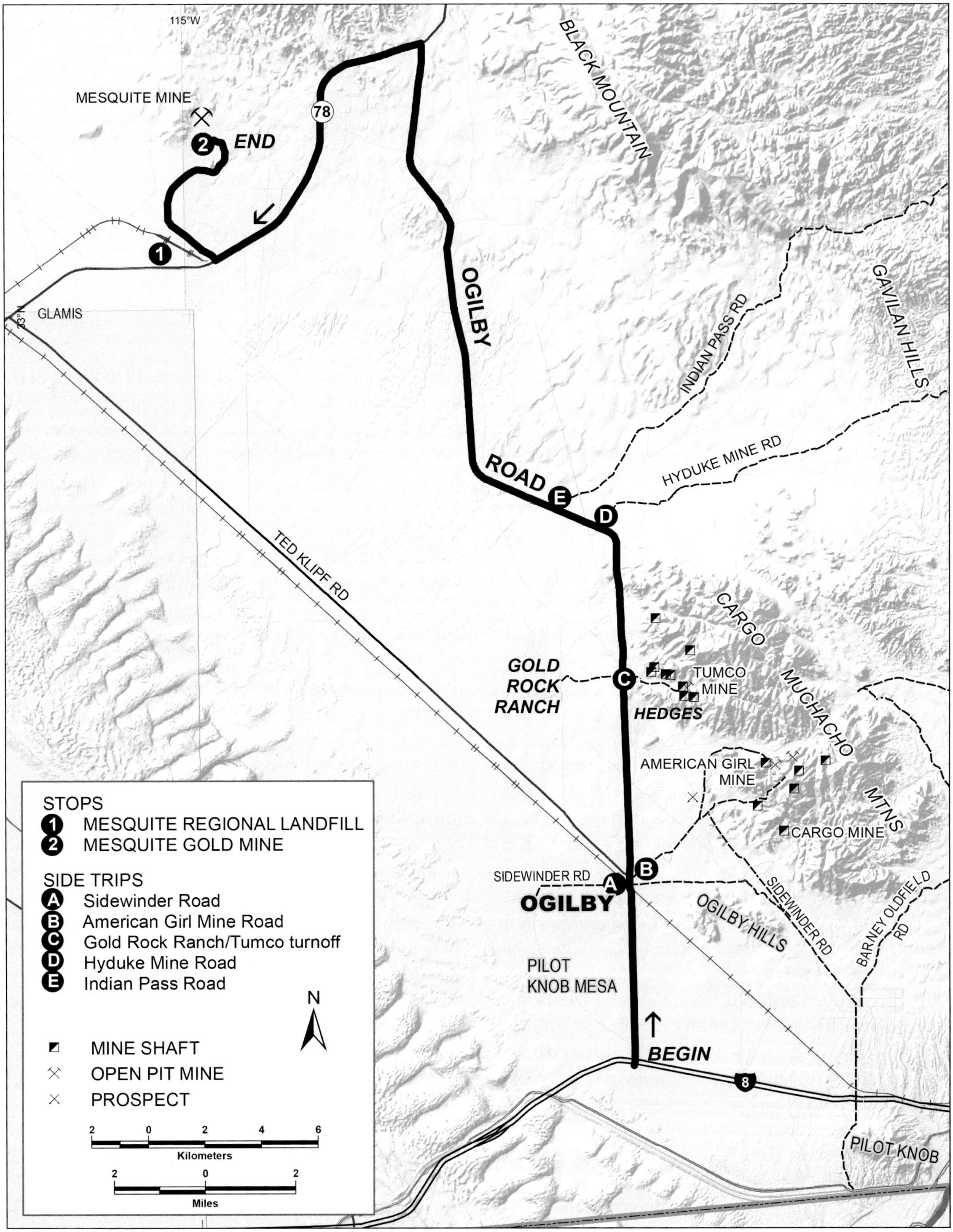

Ogilby Road *with portions of Highway 78 and Interstate 8. Map by David Bloom*

today, as the high price of gold merits mining of low-grade ore deposits that were previously considered unprofitable (Lincoln Mining Corp., 2011).

The northern half of Ogilby Road is dominated by an expansive terrain of washes and alluvial fans, with the southeastern Chocolate Mountains bestowing a distant panorama to the east. Several dirt roads provide access into the range and some of its geomorphic divisions — Gavilan Hills, Black Mountain, Peter Kane Mountain, Picacho. The major rock units comprising the range include the upper Cretaceous — lower Tertiary Orocopia Schist, Mesozoic gneisses, Jurassic Winterhaven Formation, and a suite of middle to upper Tertiary volcanics. The structure of the range is complex, and involves the aforementioned Chocolate Mountains fault, an older thrust fault reactivated as a detachment fault during the early Tertiary, and growth of the Chocolate Mountains anticlinorium, a result of multiple episodes of N-S shortening superposed on a complex early Miocene extensional fabric that occurred during the mid to late Tertiary. For a detailed account of the geologic history of this area, the reader is referred to three papers published herein, by Girty el al., Stephenson and Girty, and Marshall et al.

After exiting Ogilby Road, a short drive west on Highway 78 will take us to the pits of the Mesquite Gold Mine. The area of the Mesquite Mine was previously a hodgepodge of gold mining activity beginning over a century ago (Morton, 1977). In 1980, Gold Fields Mining Corp. began evaluating the old mining district and initiated acquisition of the lands. Exploratory drilling commenced in 1981, followed by heap leach testing in 1983. Open pit mining and pouring of gold was initiated in 1986 at the Big Chief pit, and early capital costs were paid back within one year of production from 179,000 ounces of gold at a cost of $115 per ounce. Today the mine is operated by the New Gold Division of Western Gold Fields, Inc. In 2009, New Gold produced 99,000 ounces of gold and over 7,000 ounces of silver from 14 million tons of ore, and employed over 250 employees and contractors (Western Mesquite Mines, Inc., 2010). Mining continues today, due, in part, to the high price of gold sustaining the viability of extracting relatively low-grade ore.

MILE LOG

0.0 Exit Interstate Highway 8 at the Ogilby Road, Exit 159. Proceed north on Ogilby Road. Start odometer at intersection with Center of the World Drive, the frontage road about 0.1 mile north of Interstate 8. From here to the railroad tracks at Ogilby, the road traverses Pilot Knob Mesa. This minimally dissected, relatively young, southwest-sloping alluvial fan surface is covered with locally derived angular gravel (desert pavement) (Olmsted et al., 1973).

0.4 Sign: "Welcome to Picacho Recreation Lands." The Cargo Muchacho Mountains loom ahead to the northeast.

1.0 At 2 o'clock, the Ogilby Hills, rising 420 feet above sea level, lay before the much higher Cargo Muchacho Mountains. Note the darker upper Tertiary basalts partly overlying Jurassic quartz monzonite, the bulk of the rocks composing the Ogilby Hills.

Ogilby Cemetery

3.5 Sidewinder Road. A left here would take you to the ghost town ruins of Ogilby, a former railroad station and settlement that serviced the mines, and namesake for the road. All that remains are the cemetery, with 26 unmarked graves, and the foundation of the school house. Established in 1878, the cemetery is said to contain the remains of mostly Chinese laborers who worked for the Southern Pacific Railroad (Dolan, 2005). This road leads to the "Ogilby–South Dunes" campground, part of the Imperial Sand Dunes Recreation Area. (See also Elliott, this volume "c": "Road Logs," Mile 52.1.)

3.6 Railroad track crossing. The tracks are the Southern Pacific Railroad's Yuma District "Sunset Route."

3.7 **SIDE TRIP 1:** Intersection with Sidewinder Road, a graded dirt road, on the right. Heading east on Sidewinder Road (Imperial County Road #711) will lead to the historic Cargo Muchacho Mine after about four miles (Elliott, this volume "b"). Along the way are preserved Native American geoglyphs imprinted in the desert pavement.

"Sidwinder"?

3.75 **SIDE TRIP 2:** The turn-off for American Girl Mine Road, on the right. This is a well graded dirt road (Imperial County Road #710) that leads first to the kyanite (aluminosilicate) mines of Vitrifax Hill and Blue Bird Hill. After two miles, take the left fork, and then proceed for another 0.4 mile. Look for two low-lying hills and the weathered timbers of abandoned mine structures. This is the location of the Bluebird, Drifted Snow, and Vitrifax kyanite mines that were active until 1946 (Morton, 1977). The mineral was mined mostly for the manufacturing of spark plugs and ceramic insulators. A right at the fork will lead to the Padre Y Madre open pit mine.

Travel another 1.2 miles from the left fork for the entrance to American Girl Canyon and the American Girl Gold Mine. This mine was first located in 1892 and by "1900 it had produced 30,000 tons of ore that averaged $8 per ton in gold. Inactive from 1900 until 1913, during the next 3 years the mine went on to produce 20,000 tons of ore that averaged $6.50 per ton in gold. A cloudburst during the second week of November 1914 flooded the lower workings, occasioning a 4 month delay while workers dewatered the mine and reopened the shaft. "Inactive for 20 years starting in 1916 the mine was again worked from July 1936, until 1939 and during that time delivered 150,000 tons of ore valued at $900,000. Total estimated production of the American Girl Mine is 205,000 tons of ore valued at $1,285,000. Although mined primarily for gold, other minerals found at the American Girl include silver, galena and copper. Former state governor H. H. Markham owned shares in this mine" (Shumway et al., 1980). The mine lay idle until 1987, when modern industrial mining began in earnest by Newmont Exploration, Ltd., following several years of exploratory drilling. By 1990, the new mine operators, Eastmaque Gold Mines Ltd. and Morrison Knudsen (MK) Corp., formed the American Girl Mining Joint Venture, employing 50 laborers, mechanics, and management personnel. The enterprise involved an underground mine, an open pit, and a mill. The total monthly gold production averaged about 7,500 ounces. Peripheral mining activities included the satellite ore body at the American Boy Mine and the old Padre Y Madre Mine to the south (Towner and Keifer, 1991). The following is from Towner and Keifer's (1991) mining convention pamphlet describing the gold-bearing rocks: "The ore bodies in the American Girl Canyon were formed in and around the low angle faulting. The Tumco formation is generally associated with the ore bodies of both the underground mine and open pit. The open pit ore bodies contain quartz veins in the low angle shear zones with ore disseminated into the Tumco Formation that is found between the American Girl shear and "B" shear zones. The underground ore bodies are generally deeper seated with high grade quartz veins and very little dissemination into the Tumco Formation. "The B-zone body is a true quartz vein; approximately 450,000 tons with a grade of 0.336 ounces per ton. The ore body was formed in the B-zone shear and varies from seven to thirty feet in thickness. The dip of the ore body is nineteen degrees and is located in a synclinal structure. The southwest corner of the ore body is located on the steep wing of the syncline; the dip of the ore in that section is greater than 45 degrees. However, this steeply dipping area accounts for less than 10% of the minable reserves. The hanging wall of the ore body is a sheared gneiss, very blocky and "ravelly," and is slightly mineralized for a few feet from the vein. The footwall is an altered granite, also slightly mineralized for a few feet from the vein; the contact is a detachment fault that has a very clear fault gouge seam. The detachment movement along the fault caused both block faulting and rotation of sections of ore. It shattered the quartz vein, created horses of hanging wall that displaced the ore, and caused rapid changes in height as the detachment progressed through the curvilinear fault patterns between blocks." Mining activities ceased in early 1999, due to declining gold prices (InfoMine, Inc., 2011). Reclamation activities for both the American Girl and Padre Y Madre open pits were completed in February 2000 (SMGB, 2007).

Return to Ogilby Road and turn right (north) and resume the road log at mile 3.75.

3.85 Ted Klipf Road junction is on the left. On the road shoulder to the right is the Obregon town historical monument. The monument commemorates the community established in American Girl Canyon and reads: "The historic gold mining community of Obregon lies near here in the Cargo Muchacho Mountains. As a frontier town, Obregon marks the location of several attempts to extract ore during the 1890s through the 1930s. Several different milling technologies were used during Obregon's existence. These included a 20 stamp mill, Cornish Rolls, and a Hardinge Mill, all successively employed prior to 1913. Finally, a flotation plant tried to extract gold from the meager ore in 1936 before all mining ceased in 1939. Today, modern heap leach and conventional milling operations continue the quest for gold. "In early 1989, American Girl Mining Joint Venture received the necessary governmental approvals to resume open pit and underground mining operations in American Girl Canyon. Prior to startup of these operations the remains of the original Obregon town site were thoroughly documented and 5 graves circa 1892 were relocated to the Holtville Cemetery. "Marker placed by the American Girl Mining Joint Venture in cooperation with the Bureau of Land Management and the Imperial County Historical Society. June 21, 1990."

According to Hector (1994), the five graves were those of infants and small children. Details of the archeological excavation of Obregon, including a sketch of the remains inside a coffin, can be found in Hector's article.

6.8 A pegmatite dike swarm intruding the Tumco Formation can be seen from 12:00 to 3:00.

Tumco ore cart

8.2 **SIDE TRIP 3:** Gold Rock Ranch/Tumco turnoff. Heading west on this well-graded dirt road will lead to the Gold Rock Ranch RV Resort, a full-service camping and RV facility with a store, museum, cabins, and "golf course." Mine implements, artifacts, and ore samples from Tumco festoon the grounds. This is the location of California Registered Historical Landmark No. 182 commemorating "Tumco": "Pete Walters of Ogilby discovered the first gold vein at Gold Rock on January 6, 1884. From his Little Mary Claim began a gold camp which reached its peak development between 1893 and 1899 as Hedges, with 3200 residents. Nearly closed, 1900–10, it was re-opened as Tumco, 1910–13, and worked intermittently until 1941. Tumco has long been a California ghost town."

Heading east on a rougher but passable dirt road will lead to the legendary Tumco mining district, and the ghost town of Hedges. The BLM manages the site and offers a self-guided tour of the ruins. Access to vehicles is fenced off, but the site is accessible by foot. The three principal gold mines of the Tumco mining district were the Golden Crown, Golden Cross, and Golden Queen. These deposits were discovered in 1883 and precipitated a gold rush. For an excellent history of the mine district up to 1907, see Van Wormer and Newland's two-part "The History of Hedges and the Cargo Muchacho Mining District" reprinted in this volume. In 1910, The United Mines Company took over operations and used the company acronym to rename the district "Tumco." Subsequently, several operators ran the mines sporadically until 1942. By then the mines had produced $2,863,000 in gold, about 45% of the total for the county (Morton, 1977). The district remained idle until the formation of the Oro Cruz Joint Venture in 1990, which then merged with the American Girl Mining Joint Venture in 1994. The development of the "Oro Cruz" Mine was begun in early 1995 with production commencing during the second half of 1995. Ore was trucked to the concurrent American Girl Canyon mine operations for heap leach and mill processing. Mining at Oro Cruz was anticipated to last for approximately three years, but the mine closed in September 1996, after extracting approximately 61,000 ounces of gold from open-pit and underground operations (P.M. De Dycker & Associates, Inc., 1994; Tetra Tech, Inc., 2010). Mining apparently ceased due to low gold prices, but according to the mine database website infomine.com, "The remaining Oro Cruz reserves were not economically recoverable because of high development and recovery costs. Serious production problems were encountered at Oro Cruz in fiscal year 1996 due to delays in permitting, problems with ore reserves and operational difficulties. Significant downgrading of the reserves was a result. Accelerated shutdown and alternative operating strategies for the mine were under review" (report dated November 30, 1996) (InfoMine, Inc., 2011). The Lincoln Mining Corp. of Vancouver, Canada, is conducting exploration of the "Oro Cruz Property" to resume gold mining at the site in the near future (Lincoln Mining Corp., 2011). The district is said to possess "an inferred gold resource of 341,800 ounces contained in 4,835,000 tons with an average grade of 0.070 opt (2.20 gpt) gold" (Tetra Tech, Inc. 2010).

Return to Ogilby Road and turn right (north) and resume the road log at mile 8.2.

10.2 Crossing outcrops of the Tumco Formation, described as "gray to pinkish-gray, highly indurated, massive, fine-grained arkosite and green-gray hornblende schist, undivided" (Morton, 1977, Pl. 1). In addition, the Tumco Formation possesses "Thin intervals of quartzite, marble, amphibolite, and amphibolite dikes occurring in minor amounts. The compositional laminations and bands are suggestive of sedimentary or volcanic origin. Supporting evidence for a sedimentary origin is provided by fist-sized ovoid bodies of calc-silicate minerals in the northern Cargo Muchacho Mountains which are interpreted as relict concretions. Amphibolites of metasedimentary or volcanic origin occur as thin beds with banding and gradational contacts parallel to the laminations in the surrounding quartzofeldspathic gneiss" (Sweeney and Bradley, 1981).

10.3 Black Mountain can be viewed at 1:00. Black Mountain was used as a base for General George Patton's communication systems for the World War II training maneuvers he led throughout the region. The road to the top was built by Patton (Wilkinson and Wendt, 1986). Black Mountain is composed of augite-olivine basalt flows, termed the basalt of Black Mountain. Smaller deposits of these flows occur in the vicinity, including Indian Pass, Ogilby Hills, and around the Cargo Muchachos (Crowe, 1978). The basalts of Black Mountain at Indian Pass have been dated using the $^{40}Ar/^{39}Ar$ method at 9.45 ± 0.27 million years (Girty et al., this volume).

Outcrops of the northwestern extent of the Tumco Fm. at Mile 10.5, looking west.

11.8 SIDE TRIP 4: Hyduke Mine Road (right). This is a rough 4×4 dirt road that passes over the Gavilan Hills portion of the Chocolate Mountains. Hyduke Mine Road (County Road #659) eventually intersects a jeep trail, County Open Route #875, and heads northeast, leading through Carrizo Wash, Bear Canyon, and the ultimately the Colorado River. The trail through Bear Canyon encompasses the best exposures of the intensely-studied Bear Canyon conglomerate, recording the timing of deformation of the Chocolate Mountains anticlinorium (Girty et al., this volume). A Bureau of Land Management map indicates Route #875 has an eastern route (BLM, 2011; Wilkinson and Wendt, 1986) where the path eventually intersects Picacho Road north of Picacho Mine, although communication to the author from several individuals indicates the road appears to be very difficult for travel even with four wheel drive. Heading east, Route #875 travels through White Wash, where several 23 million-year-old volcanic units are exposed, and finally around the north side of Little Picacho Peak before intersecting with Picacho Road in a broad section of coarse alluvial stream deposits.

Return to Ogilby Road and turn right (north) and resume the road log at mile 11.8.

Continuing north on Ogilby Road, Wilkinson and Wendt (1986) indicate that "gold can be panned from all the washes we are crossing from here to the intersection with State Highway 78."

13.0 SIDE TRIP 5: Indian Pass Road (right). Indian Pass Road (County Road #649) is a 15.5-mile 4×4 dirt road that passes around the Gavilan Hills and into Gavilan Wash, and is noted for outstanding exposures of the late Cretaceous to early Tertiary Orocopia Schist, Jurassic mylonitic orthogneiss, and the Jurassic Winterhaven Formation. The Winterhaven Formation has been metamorphosed to the chlorite-grade greenschist facies, and consists of a lower metavolcanic unit composed of mostly altered dacite up to 80 meters thick, overlain by metasedimentary units of mostly quartz arenite and argillic siltstone. The metasediments are approximately 360 meters thick. The structural relationships of these rock units have been the subject of several studies interpreting the tectonic history of the region (Haxel et al., 1985, 1986; Oyarzabal, et al., 1997; Jacobson et al., 2002; Girty et al., 2012;

and others). After about 8.6 miles, the ~9.45 million-year-old basalts of Black Mountain are plainly visible to the northwest and southeast. Here the road descends into Gavilan Wash (passable eastbound and down with a high-clearance 4×4, uphill westbound would be challenging with a 4x4, and impossible with a standard car; Bill Elliott, personal communication, 2011). At Mile 9.0 the track passes by the juncture of the low-angle Chocolate Mountains fault, with the Orocopia Schist encompassing the footwall and orthogneiss comprising the hanging wall, and the dextral-slip Indian Pass fault, which cross-cuts the Chocolate Mountains fault and offsets the two formations (Girty et al., this volume). Continuing west through outcrops of Winterhaven Formation metasediments and the upper Tertiary Bear Canyon conglomerate, Indian Pass Road eventually ends (when not washed out) at the Colorado River campsites of 4S Beach and Outpost, the two northwestern camp sites of Picacho State Recreation Area.

Gold placer mining was attempted in the 1960s along Indian Pass, about 7.5 miles from Ogilby Road (Morton, 1977). More recently, an industrial-scale open pit gold mine was proposed in this area in the 1990s. Known as the "Imperial Project," Glamis Gold, Inc., proposed a cyanide heap-leach operation from three open pit mines across 1,650 acres. It was estimated that up to 150 million tons of ore (with an estimated 1.5 million ounces of gold) would be mined and leached under the proposal. A plan of operations for mining was submitted to the U.S. Department of the Interior, which was subsequently not approved, as recommended by the Bureau of Land Management. The reasons were several, including claims of adverse environmental impacts and Native American concerns (U.S. Department of the Interior, 2001). However, the denial was "based on a novel legal opinion rendered by [then Interior Secretary Bruce Babbitt's] Solicitor." Therefore, later in 2001, "[new] Interior Secretary Gale Norton rescinded the Babbitt denial based on the legal opinion of her Solicitor, which held that Interior had no discretionary power to veto the mine proposal" (Crowell & Moring LLP, 2011). In response, opponents of the project persuaded California legislators to quickly pass a bill to protect cultural sites and another bill to mandate open pit backfilling and other environmental mitigation measures in gold mining operations. In the view of Glamis Gold's counsel, the bills were designed to change "the law in an unprecedented manner to prohibit any cost-effective mining operation of the Glamis mining claims." Governor Gray Davis vetoed the first bill (SB 1828) but signed the second (SB 483). As a result, SB 483, imposing mandatory backfilling requirements for proposed open pit metallic mines, did not go into effect because it was linked to SB 1828. In early 2003, the California Legislature introduced SB 22, requiring mandatory and complete backfilling of all proposed open pit metallic mines located within one mile of a Native American sacred site. Governor Davis signed the bill that year, with the aim of preventing gold mining specifically at the Glamis Imperial Project (Crowell & Moring LLP, 2006). In October 2003, Glamis Gold, a Canadian-based company, filed suit against the U.S. government under the provisions of NAFTA, asking for $49 million in damages, but lost (*Yuma Sun*, 2009). Glamis Gold claimed the complete backfilling legislation was targeted specifically at stopping the Imperial Project and rendered the property economically unfeasible. Additionally, the legislation would essentially stifle all future gold exploration in California (Crowell & Moring LLP, 2006). In late 2006, Glamis Gold Ltd. was acquired by Goldcorp, Inc. (Goldcorp, 2011).

Return to Ogilby Road and turn right (north) and resume the road log at mile 13.0.

19.0 The high peak in the distance at 11:00 is Mt. Barrow. The Mt. Barrow Granodiorite forms the bulk of the peak, topped by the Orocopia Schist. Northwestern-striking dikes cut through both these units, and appear to be comagmatic with the granodiorite. K-Ar dating suggest the age of the granodiorite and dikes to be early Miocene. The northwest strike of the dikes indicates NE-SW extension related to regional mid-Tertiary detachment systems (Frost et al., 1986).

20.6 "Low hills on right held up by moderately consolidated brown breccias and conglomerates composed of basement rock clasts of volcanic rocks. This unit is Tertiary in age and has been faulted, tilted, and locally folded" (Wilkinson and Wendt, 1986). Note the large excavation scar across the hills. This is the path of the North Baja Pipeline, designed to provide natural gas to electricity generators in Baja California, Mexico, although flow can be directed north to the U.S. (North Baja Pipeline, 2011).

23.5 Old placer workings in wash on the right.

24.25 Intersection with State Highway 78 and north end of Ogilby Road, facing the Chocolate Mountains.

A right turn here will lead to Blythe. In approximately six miles, an outcrop of gray Bouse Formation limestone can be seen about fifty yards on the right side of the highway (Metzger et al., 1973; Poulson and John, 2003). An additional four miles will bring one to the Midway Well turnoff (Elliott, this volume "c").

Turn left (west) toward Glamis. "We are driving through exposures of altered, maroon-weathering pyroxene rhyodacite to andesite lava flows of the Quechan Volcanics and partially welded to unwelded, pink, pumiceous crystal tuffs unconformably overlying the Quechan Volcanics" (Wilkinson and Wendt, 1986).

25.75 Continuing west on Highway 78, a pre-Columbian Native American trail parallels the highway just beyond the southern shoulder. A monument erected in 1965 marks the spot, and explains the trail connected the "Colorado River to Lake Coachuilla [sic] (now Imperial Valley)." The actual trail is preserved behind chain link fencing.

The remnants of the Crown Uranium Mine lie to the north. Uranium ores at the mine occur in a hydrothermally altered zone near the contacts of quartzite, sericite schist, and a basic dike. Uranium-bearing minerals, including torbernite, carnotite, pyrophylite, and rare autunite, contained an average uranium content of 0.293 percent — "quite high for a uranium ore." About 1,000 tons of ore was mined and shipped in 1959 (Morton, 1977; Mulkey, 1994).

26.6 Extensive tailings of the Mesquite Mine are in view straight ahead.

31.4 STOP 1: Turn right (north) on GFOC Mine Road, the entrance to New Gold, Inc.'s Mesquite Mine, located at the foot of the Chocolate Mountains. The Mesquite Mine is one of the largest gold mines in the U.S. A large sign will introduce you at the entrance, and a monument has been erected in 1993 to commemorate the "Mesquite Mine": "Felizario Parra discovered gold in April 1876, worked placers until 1880, and sold out for $3,000. Dry washing of low-grade ores continued intermittently for 100 years, along with many exploratory shafts. From 1957 to 1980, Dick and Anna Singer mined, recorded earlier efforts, made studies, and sold their findings to Gold Fields Mining Corporation. After investing $70 million in exploration, development, and construction, Gold Fields began full scale production in March 1986. Production reached 207,897 ounces of gold in 1992. It is expected that this rate of production will be maintained through 1999 when the mine will be exhausted. Gold Fields and its successor, Santa Fe Pacific Gold Corporation, have given steady employment to over 300 persons, contributed heavily to public and private needs, operated with full concern for environmental issues, and have made Mesquite the second largest gold producer in California."

Obviously, the authors of the commemoration did not anticipate the extended production of mining operations! Incidentally, the grave of Sr. Felizario Parra, who discovered the Mesquite placer and died in 1909, can be visited at the Picacho cemetery. It is interesting to note that Sr. Parra "sold out" for $3000, a huge sum in those days, while the discoverer of gold at the Picacho gold mine, Sr. José Mendivil, sold his claims for much less, $500 (Ernie Mendivil, personal communication, 2011).

The geology of the Mesquite Mine has been summarized by the New Gold website (New Gold, Inc., 2011):

"The Mesquite Mining District lies beneath alluvial pediment deposits at the base of the Chocolate Mountains along the San Andreas fault system. Gold mineralization occurs within complex assemblage of regional scale fault blocks that comprise distinct rock units and host the more than 3 million ounces of gold that have been produced at Mesquite since its discovery circa 1940. Currently, remaining mineral resources at Mesquite contain an estimated 4.1 million ounces of gold.

"Gold mineralization at Mesquite was emplaced during mid-Tertiary time (26–38 million years ago) within a complex system of faults and fractures that have been superimposed upon an older sequence of Jurassic age gneisses and granitic intrusives. The local stratigraphic sequence is, from bottom to top: mafic, hornblende biotite, and biotite gneiss overlain by Bay Horse Quartzite and Muscovite Schist. These units have been crosscut by a number of quartz, feldspar, biotite and muscovite rich granite sills, plutons and dikes. The gneiss units are present in a shallowly dipping stratigraphic sequence, offset by numerous high and low angle faults. This package is overlain by up to 400 feet of Tertiary lithified silts, sands, and gravels and unconsolidated Quaternary gravels.

"There are three main structural components that serve as the principal controls to gold mineralization at the Mesquite Mine. Northwest trending structures appear to have the greatest control on gold distribution, while younger northeast trending cross-faults offset mineralization in a stair step fashion. East-west striking near vertical and northwest striking low angle structures are also seen on the property. The system is

Head stone of the grave of Felizario Parra, Picacho Cemetery (Carol Teasley)

Aerial photo of the Mesquite Gold Mine, looking southwest (New Gold)

bounded by structures sub-parallel to the San Andreas fault system. Mid-Tertiary age deposition of gold was within an epithermal environment along faults and fractures, and the deposit was subsequently oxidized down to approximately 150 meters (500 feet) below current surface topography. Gold at Mesquite occurs as both submicron disseminated and coarse gold, with flakes of visible 'flour gold' occurring along fault zones throughout the mine area.

"The ores of Mesquite are proven to be readily amenable to heap leach processing, as demonstrated by the past twenty years of successful mining operations there.

"Mesquite is a run-of-mine, heap leach operation. Ore is drilled and blasted within the open pit mine, loaded on trucks and subsequently transported to the leach pad for processing. There is no requirement for crushing, grinding or milling at Mesquite.

"Mesquite's process facilities consist of a lined heap leach pad that leads to a conventional carbon-in-column recovery plant. When Mesquite was brought back into production in January 2008, its construction included expansion of the leach pad and installation of a new carbon in column plant to process the run-of- mine ore. Run-of Mine ore is stacked on the leach pad and a solution percolates through the ore leaching gold and silver from the rock to produce a "pregnant solution" which is then pumped to the carbon-in-column plant to produce gold-silver doré.

"Mesquite is expected to produce 145,000 to 155,000 ounces of gold in 2010 at a total cash cost of US$540 to $560 per ounce." (Note: Mesquite produced 169,000 ounces of gold in 2010, and 158,000 ounces in 2011. Source: New Gold website)

Welcome the Mesquite Mine. Don Wagstaff, New Gold's Chief Geologist at the Mesquite Mine, will be leading a tour of the mine. After a presentation introducing us to the details and history of the mine, we will go to two locations where we can get out and look down into the pits and do some geologic arm-waving, etc. Then we will drive by active mining areas to see equipment in action, etc. Finally we'll take a brief drive out to the leach pad and discuss the process side of things for a few minutes and then head back to the starting point. Next we will visit the nearby Mesquite Regional Landfill. The Mesquite Regional Landfill is Southern California's first operating landfill that is permitted to receive waste by rail. A 5-mile rail spur connects the landfill to the railroad Union Pacific Railroad mainline. The site is said to be well suited as a landfill because of its desert climate, distance from groundwater, distance from residential development, proximity to the railroad, and preexisting use for mining. The landfill will be designed and operated as a sanitary landfill to meet all local, state and federal requirements. Michele Ochs will lead a unique tour, where we will see a landfill site before trash from the Los Angeles area arrives.

REFERENCES

Bureau of Land Management (BLM), 2011, BLM routes of travel for eastern Imperial County, California: El Centro Field Office.

Crowe, Bruce M., 1978, Cenozoic volcanic geology and probable age of inception of basin-range faulting in the southeasternmost Chocolate Mountains, California; *Geological Society of America Bulletin* v. 89, no. 2; pp. 251–264.

Crowell & Moring LLP, 2006, In the arbitration of Chapter 11 of the North American Free Trade Agreement and the UNCITRAL arbitration rules between Glamis Gold LTD, Claimant/ Investor, and United States of America, Respondent/Party, May 5.

Crowell & Moring LLP, 2011, Law firm representing Glamis Gold. "Legislation to Protect Sacred Native American Federal Lands would Jeopardize Mining and other Land-Development Activities" by R. Timothy McCrum. Accessed October 3, 2011. http://www.crowell.com/NewsEvents/Publications/Articles/1349687.

Dolan, Christy, 2005, Ogilby: A desert station on the Southern Pacific Railroad; *Proceedings of the Society for California Archeology*, v. 18, pp. 83–89.

Elliott, William J., 2014a, "Araz Outcrop;" this volume.

Elliott, William J., 2014b, "Cargo Muchacho Mine: then and now;" this volume.

Elliott, William J., 2014c, "Desert Watering Holes in the Southern Colorado Desert Region of Southern California, Then and Now;" this volume.

Frost, Eric, Peter Drobeck, and Bud Hillemeyer, 1986, Geologic Setting of Gold and Silver Mineralization in Southeastern California and Southwestern Arizona, in *Cenozoic Stratigraphy, Structure and Mineralization in the Mojave Desert, Field Guide for the 1986 Cordilleran Section Meeting of the Geological Society of America*, March 25–28, Los Angeles, California.

Girty, Gary H., Krista K. Muela, Holly Olson, Lauren Sutton, Melanie Biggs, Jon Sainsbury, Jason W. Ricketts, Rachel Errthum, Thomas A. Colby, Taylor Carrasco, Erik Gordon, and Tim Pelbath, 2014, "Toward a unified model for the geologic and tectonic history of the region between Indian Pass and Picacho State Recreation Area, southeastern California"; this volume.

Goldcorp, 2011, Company website, accessed October 29, 2011. http://www.goldcorp.com/Home/default.aspx.

Guthrie, J.O., A.R. Cockle, and A.D. Branham, 1987, Geology of the American Girl–Padre Y Madre gold deposits, Imperial County, California, *Society of Mining Engineers Preprint* No. 87–87, January.

Haxel, G.B., R.M. Tosdal, and J.T. Dillon, 1985, Tectonic setting and lithology of the Winterhaven Formation: A new Mesozoic stratigraphic unit in southeasternmost California and southwestern Arizona: *U.S. Geological Survey Bulletin* 1599, p. 19.

Haxel, Gordon B., Richard M. Tosdal, and John T. Dillon, 1986, Field Guide to the Chocolate Mountains Thrust and Orocopia Schist, Gavilan Wash area, southeasternmost California, in *Frontiers in Geology and Ore Deposits of Arizona and the Southwest*; Arizona Geological Society Digest, v. 16.

Hector, Susan M., 1994, The Historical Archeology of American Girl Canyon, Imperial County, California: *Proceedings of the Society for California Archaeology*, Vol. 7., pp. 199–212.

InfoMine, Inc., 2011, Information on the American Girl Mine and Oro Cruz Mine. Accessed October 3, 2011. http://www.infomine.com.

Jacobson, C.E., M. Grove, M.M. Stamp, A. Vucic, F.R. Oyarzabal, G.B. Haxel, R.M. Tosdal, and D.R. Sherrod, 2002, Exhumation history of the Orocopia Schist and related rocks in the Gavilan Hills area of southeasternmost California, in Barth, A., ed., *Contributions to crustal evolution of the southwestern United States: Geological Society of America Special Paper* 365, pp. 129–154.

Lincoln Mining Corp., 2011, Company website, for proposed mining at Oro Cruz. Accessed October 3, 2011. http://www.lincolnmining.com/.

Metzger, D. G., O. J. Loeltz, and B. Irelna, 1973, Geohydrology of the Parker-Blythe-Cibola Area, Arizona and California: *U. S. Geol. Survey Prof. Paper* 486-G, 130p.

Morton, Paul K., 1977, Geology and Mineral Resources of Imperial County, California, County Report 7, California Division of Mines and Geology, Sacramento.

Mulkey, James E., 1994, "Prospecting for uranium minerals," *Rock and Gem*, March.

New Gold, Inc., 2011, Company website, accessed October 29, 2011. http://www.newgold.com/Home/default.aspx.

North Baja Pipeline, 2011, Company website, accessed October 29, 2011. http://www.northbajapipeline.com/.

Olmsted, F. H., O. J. Loeltz, and B. Irelna, 1973, Geohydrology of the Yuma Area, Arizona and California; *U. S. Geol. Survey Prof. Paper* 486-H, 227p.

Owens, E.O., and R.W. Hodder, 1994, Aluminosilicate mineral assemblages in the Cargo Muchacho Mountains, southern California: metasomatism and gold concentration associated with magmatism and deformation in mesozonal environments: *Canadian Journal of Earth Sciences* — 31(2): pp. 310–322.

Oyarzabal, F.R., C.E. Jacobson, and G.B. Haxel, 1997, Extensional reactivation of the Chocolate Mountains subduction thrust in the Gavilan Hills of southeastern California: *Tectonics*, v. 16, pp. 650–661.

P.M. De Dycker & Associates, Inc., 1994, Final Environmental Impact Statement, Oro Cruz Operation of the American Girl Mining Project, Imperial County, California; Bureau of Land Management, El Centro, California, November.

Poulson, S. R., and B.E. John, 2003, Stable Isotope and Trace Element Geochemistry of the Basal Bouse Formation Carbonate, Southwestern United States — Implications for the Pliocene Uplift History of the Colorado Plateau: Geol. Soc. America Bull. v. 115, n. 4, pp. 434–444.

Pridmore, Cynthia L., and Eric G. Frost, 1992, Detachment Faults, California's extended past, *California Geology*, California Department of Conservation, Division of Mines and Geology — 45(1): pp. 3–17.

U.S. Department of the Interior, 2001, Record of Decision for the Imperial Project Gold Mine Proposal, Imperial County, California: United States Department of the Interior, Bureau of Land Management, California State Office, California Desert District, El Centro Field Office, BLM Case File No. CA 670-41027, January 17.

State Mining and Geology Board (SMGB), 2007, Report on backfilling of open-pit metallic mines in California, Information Report 2007–02, California Department of Conservation Resources Agency, January.

Shumway, Gary L., Larry M. Vredenburgh, and Russell D. Hartill, 1980, Desert Fever — An Overview of Mining History of the California Desert Conservation Area: Bureau of Land Management, U. S. Department of the Interior, Riverside, California, Contract No. CA-060-CT7-2776, February 1980.

Sweeney, Thomas M., and Robin Bradley, 1981, Status of Mineral Resource Information for the Fort Yuma and Cocopah Indian Reservations, Arizona and California; BIA Administrative Report 85.

Tetra Tech, Inc., 2010, Oro Cruz Gold Project Resource Estimate, Imperial County, California, USA, NI 43-101 Technical Report prepared for Lincoln Mining Corporation, Vancouver, BC, Canada, September 21.

Tosdal, R.M., 1986, Gneissic host rocks to gold mineralization in the Cargo Muchacho Mountains, southeastern California, in *Field Trip to Mesquite Mine, American Girl–Padre-Madre Properties, and Picacho Mine, Imperial County, California*, edited by William Wilkinson and Clancy Wendt: Arizona Geological Society Fall Field Trip, March; and in *Cenozoic Stratigraphy, Structure and Mineralization in the Mojave Desert, Field Guide for the 1986 Cordilleran Section Meeting of the Geological Society of America*, March 25–28, Los Angeles, California.

Towner, Robert K., and Jerald W. Keifer, 1991, "American Girl Underground Mine," presented at the 97th annual Northwest Mining Association Convention, Spokane, Washington, December 4–6.

Van Wormer, Stephen R., and James D. Newland, 1996, The history of Hedges and the Cargo Muchacho Mining District, Part I, in *The Journal of San Diego History* — 42(2): San Diego Historical Society, and reprinted in this volume.

Western Mesquite Mines, Inc., 2010, New Gold 2009 Sustainability Report brochure for the Mesquite Mine; Brawley, California.

Wilkinson, William H., and Clancy J. Wendt, 1986, Precious Metal Mineralization, Stratigraphy, and Tectonics in Southeastern California, in *Field Trip to Mesquite Mine, American Girl–Padre-Madre Properties, and Picacho Mine, Imperial County, California*, edited by William Wilkinson and Clancy Wendt: Arizona Geological Society Fall Field Trip, March.

Yuma Sun, 2009, "Tribe calls mining ruling a victory," by Darin Fenger, staff writer; 15:37:26, June 9.

Desert Watering Holes in the Southern Colorado Desert Region of Southern California, Then and Now

WILLIAM J. ELLIOTT
Consulting Engineering Geologist
P. O. Box 541, Solana Beach, California 92075

PURPOSE

The purpose of this presentation is to provide an historical perspective of desert travel in southeastern California during the early years of the 20th century.

SCOPE

The scope of work for this paper consisted of reviewing several United States Geological Survey Water-Supply papers, which were researched, field-checked, and prepared during the years 1917–1922 (see references). A personal field reconnaissance by this author was made during the month of June 2011 to determine the presence or absence of mapped and/or published locations of desert watering holes.

Drenched with sweat and perhaps cursing a broken air conditioner, if the car had such a luxury, one is led to ponder the following questions: What is the same today? What is different today? Personal accounts of desert travels and experiences are included at the end to provide a perspective of typical desert experiences during the 1940s and 1950s.

INTRODUCTION

Author's note: The following text is intended to be both serious and informally fun. Portions should be taken with tongue-in-cheek, or, if you prefer, with the grain of salt needed by desert travelers to keep their electrolytes in balance. These truths and tales are timeless, whether told today or a hundred years ago.

As we embark upon our San Diego Association of Geologists 40th Anniversary Field Trip to share camaraderie with old friends, as well as the fun and adventure of making new ones, it seems fitting to look back to the uncertainty and perils facing desert travelers almost 100 years ago. Up until the time of mass-produced automobiles in the early years of the 20th Century, desert travel was largely accomplished on horse-back, on foot, or with livestock pulling wagons, buckboards, stagecoaches and the like. The objective at that time was to either carry enough water to make the planned journey from point A to point B, or to carry enough water to get from one watering hole to the next before livestock and people ran out of water. Even when traveling by automobile, making safe passage from watering hole to watering hole was a challenge (Photo 1). Roads, really not much more than wagon trails, were poorly mapped and marked. Missing a watering hole (Photo 2) or becoming stuck in sand could spell disaster for even the most experienced desert traveler.

To assist an increasing number of travelers in the deserts of the southwestern United States, the United States Geological Survey embarked upon a regional reconnaissance of water resources in these arid regions (Map 1). Of particular interest are four Water-Supply papers researched, field-checked, and prepared by four geologists (Brown, 490-A, 1920; Bryan, 490-D, 1922; Ross, 490-C, 1922; and Thompson, 490-D, 1921). An introduction to each of these papers was prepared by Meinzer (1920, 1921, 1922). Because the San Diego Association of Geologists' 2011 Field Trip takes place in the Colorado Desert Region of southeastern California (Jenkins, 1938), John S. Brown's "Routes to Desert Watering Places in the Salton Sea Region, California" is referred to and quoted extensively herein.

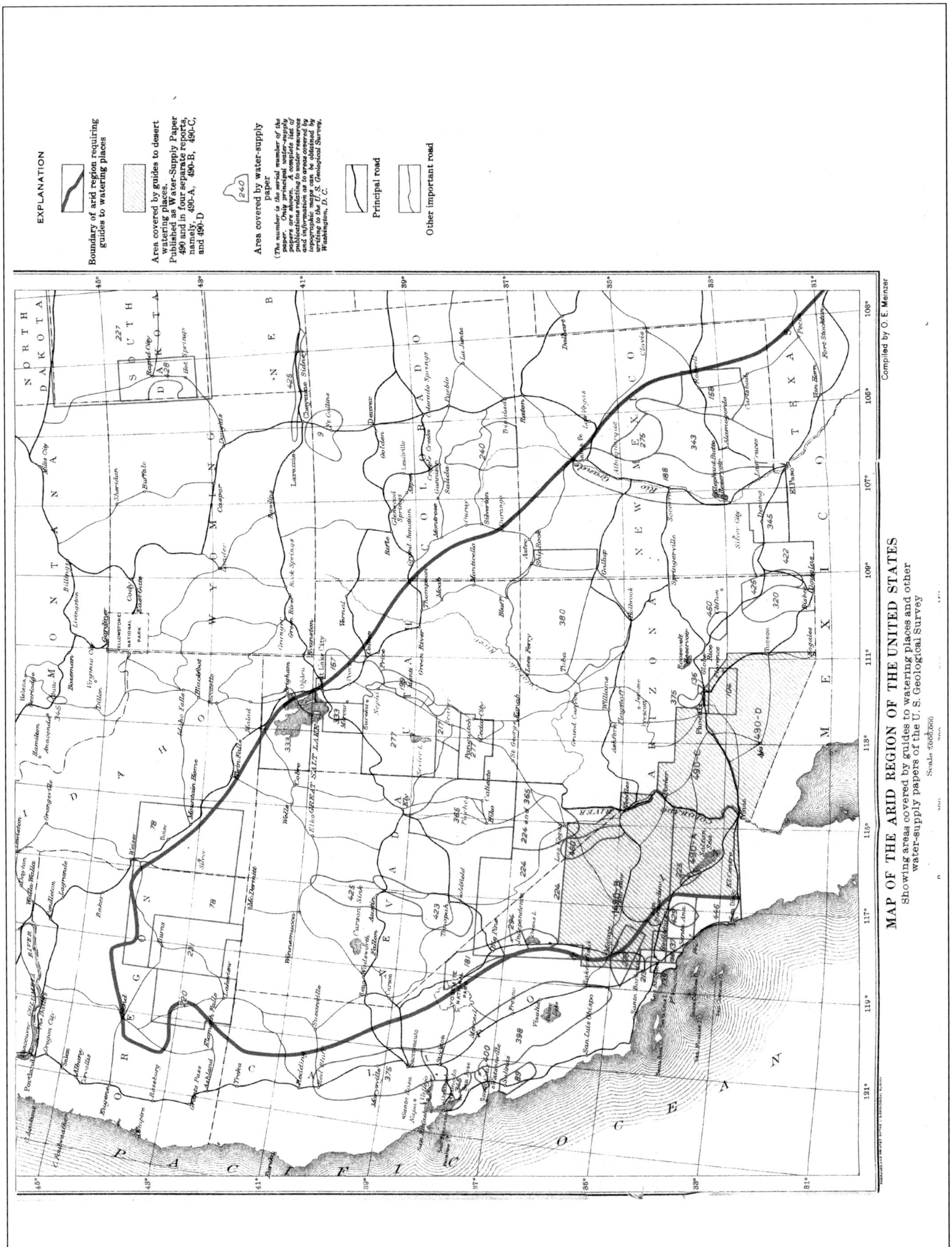
EXPLANATION
Boundary of arid region requiring guides to watering places
Area covered by guides to desert watering places. Published as Water-Supply Paper 490 and in four separate reports, namely, 490-A, 490-B, 490-C, and 490-D
Area covered by water-supply paper
(The number is the serial number of the paper. Only principal water-supply papers are shown. A complete list of publications relating to water resources and information as to areas covered by topographic maps can be obtained by writing to the U. S. Geological Survey, Washington, D. C.
Principal road
Other important road
MAP OF THE ARID REGION OF THE UNITED STATES
Showing areas covered by guides to watering places and other water-supply papers of the U. S. Geological Survey
Compiled by O. E. Meinzer

At the time these desert reconnaissance adventures occurred, during 1917 and 1918, in the "driest, hottest, and least explored part of the desert region ... in southeastern California and southwestern Arizona" (Meinzer, 1920, 1921, 1922), internal combustion engine-propelled vehicles were beginning to replace primitive human and animal powered forms of travel. With this change came efforts by automobile clubs and local and regional governments to provide, to the extent possible under the circumstances, for the health, safety, and welfare of its citizens living and traveling in this harsh environment. "Permanent" road signs (with mileages and directional arrows) began to appear at crossroads (Photos 3 and 4), and primitive sketch maps (Map 2) became available to the inquiring traveler. Besides routes of travel, these maps showed even the most out-of-the-way watering holes, lest the broken-down traveler be stuck without any idea of where to find water.

Today, with well-signed paved roads, reliable automobiles, maps, global positioning systems, air-conditioned vehicles, and relatively short distances between civilization and assistance, the need for a guide to desert watering holes would seem unnecessary. With modern communication devices, such as cell phones, amateur and civilian radios, and the like there are few remaining excuses for perishing in the desert.

But wait, what about intrepid geologists seeking to unravel the earth's secrets in remote desert areas? We still need to be fairly cognizant of most of the same things that were so important to survival 100 years ago: carry enough water, food, gas, and other supplies as may be appropriate for the journey to our destination(s) and a little more for any unexpected events along the way.

EXCERPTS FROM
Routes to Desert Watering Places in California and Arizona

Preface to Water-Supply Papers 490-A, B, C, and D by O. E. Meinzer (pages 1–4).

... Travelers in this region must depend for their existence on the desert water holes (springs, wells, or natural tanks), many of which are separated from one another by a hard day's journey with team and wagon. For most of the region the water holes have never been accurately mapped or described, no systematic provision has been made for maintaining them, and the roads leading to them have not been marked with substantial and reliable signs. Hence travel in the remote parts of the region has been a precarious and sometimes a dangerous undertaking.

... A systematic program for the survey, marking, and protection of desert watering places was authorized by an act of Congress approved August 21, 1916, which reads as follows:

> *Be it enacted by the Senate and House of Representatives of the United States of America in Congress assembled*, That the Secretary of the Interior be, and he is hereby, authorized and empowered, in his discretion, in so far as the authorization made herein will permit, to discover, develop, protect, and render more accessible for the benefit of the general public springs, streams, and water holes on arid public lands of the United States; and in connection therewith to erect and maintain suitable and durable monuments and signboards at proper places and intervals along and near the accustomed lines of travel and over the general area of said desert lands, containing information and directions as to the location and nature of said springs, streams, and water holes, to the end that the same may be more readily traced and found by persons in search or need thereof; also to provide convenient and ready means, apparatus, and appliances by which water may be brought to the earth's surface at said water holes for the use of such persons; also to prepare and distribute suitable maps, reports, and general information relating to said springs, streams, and water holes and their specific location with reference to lines of travel.

An appropriation of $10,000 [about $177,000 in 2011 dollars (United States Department of Labor, 2014)] became available for this purpose July 1, 1917. With this appropriation, supplemented with some other funds that could be used for the purpose, a survey was made in 1917 and 1918 of the driest, hottest, and least explored part of the desert region, comprising 60,000 square miles in southeastern California and southwestern Arizona... [Map 1].

Map 1 *(opposite). "Map of The Arid Region of the United States." (in Brown, 1920, Pl. I).*

Photo 1. *"Bad Road Traversed in Desert Watering Place Survey." Photo by C. P. Ross, in Brown, 1920, Plate II-B.*

Photo 2. *"Hidden Spring... in Mohave Desert, Calif. The only water within 10 miles is... [this] spring hidden under brush...." Photo by D. G. Thompson, in Brown, 1920, Plate IV-B.*

Photo 3. *"First Desert Watering Place Signpost Erected by the United States Geological Survey." Photo by C. P. Ross, in Brown, 1920, Plate II-A.*

Photo 4. *Typical Desert Watering Place Signpost Erected by the United States Geological Survey." Photo by D. G. Thompson, in Brown, 1920, Plate III.*

... The field work was done by four parties, each consisting of one geologist and one nontechnical assistant and each provided with an automobile and light camping equipment. The mapping was done using a plane table, on a scale of 1:125,000.

... Signs directing travelers to water were erected at 167 localities in California and 138 in Arizona... [Photos 3 and 4].

... Soon after the field work was completed three of the four geologists who made the survey entered the Army and the other one was assigned to work on water supplies for military establishments. Consequently, the preparation of the maps and guides to desert watering places had to be postponed until after the war [World War I].

[Author's Note: It appears that the four geologists were the authors of Water-Supply Papers 490-A, B, C, and D, see references.]

... The four abbreviated guides comprising the present series (Water-Supply Paper 490) are published in advance of the complete reports for the use of those who do not need the more general information which the complete reports contain" (Meinzer, 1920, 1921, and 1922, pp. 1–4.).

[Author's Note: The following paragraphs are quoted from Water-Supply Paper 490-A, pages 7 through 15. This interesting read describes travel in the lower Colorado Desert as it was about 90 years ago, and is in stark contrast to what we expect and experience today.]

DESCRIPTION OF THE REGION.

Physical Features. Within this region there are wide variations of physical features within short distances. From lofty mountain regions, with forests of large trees, heavy winter snows, and an annual precipitation of 30 inches or more on the west, there is a swift transition eastward to low desert ranges and valleys, nearly devoid of vegetation, where summer temperature prevails the year around and where there may be entire years practically without rain. As a whole this is the hottest and driest part of the United States.

The desert consists mainly of broad, nearly level areas with sandy or gravelly soil, separated by barren mountain ranges whose sides are commonly chocolate-colored and are deeply furrowed by dry canyons. The low flats usually constitute units to which the name valley is applied, such as Coachella Valley and Imperial Valley. They are generally covered with at least a scant growth of bushes or shrubs which are adapted to life under the prevailing conditions of heat and drought.

Settlements. Some of these valleys where a supply of water can be obtained have been developed by irrigation to a remarkable degree of productivity, and large settlements exist at these isolated places. Coachella Valley, Palo Verde Valley, Yuma Valley, and Imperial Valley are the principal centers of agricultural development.

Roads. The transcontinental line of the Southern Pacific Railroad, completed in 1879, crosses the region from southeast to northwest, and several branch lines from various places have been built in recent years to serve the irrigated areas. Consequently the centers of agricultural development are provided with good transportation facilities and constitute important supply points. Within the settled areas good roads are common. Long stretches of pavement are found on some of the chief highways. But practically all the roads through the undeveloped region are entirely unimproved. They are not laid out along definite lines, but follow in a general way the easiest natural routes available, crossing valleys and threading mountain passes, and avoiding as much sand and as many hills as possible The result is that a desert road usually consists of two wheel tracks (Photo 1), and if these become bad a new one is made somewhere in the vicinity.

Mining is practically the only business which attracts travel away from the main roads. Development at any place is subject to many uncertainties, and camps appear and disappear rapidly. Roads are made and fall into disuse. Where wagon or automobile roads are impracticable, pack animals are used and burro trails result.

Means of Travel. All the ordinary means of travel are used in this region. While teams and wagons are still used, they have been replaced to a great extent in recent years by automobiles. Trucks are coming into use on a commercial scale for freighting from mines in places away from the railways. The burro is still the faithful friend of the prospector and the horse of the 'cow man,' but automobiles are also much used by prospectors, for with them it is usually possible to get somewhere near even the most inaccessible territory. Foot travel is uncommon, although transient laborers often tramp from one valley to another, particularly during the period of seasonal occupations such as cotton picking.

SUGGESTIONS TO TRAVELERS

General Advice. To one taking the proper precautions the desert is much less to be dreaded than the average stranger imagines. Only in midsummer heat is it really a dangerous place. Nevertheless, a few short trips without mishap should not lead one to a careless contempt for

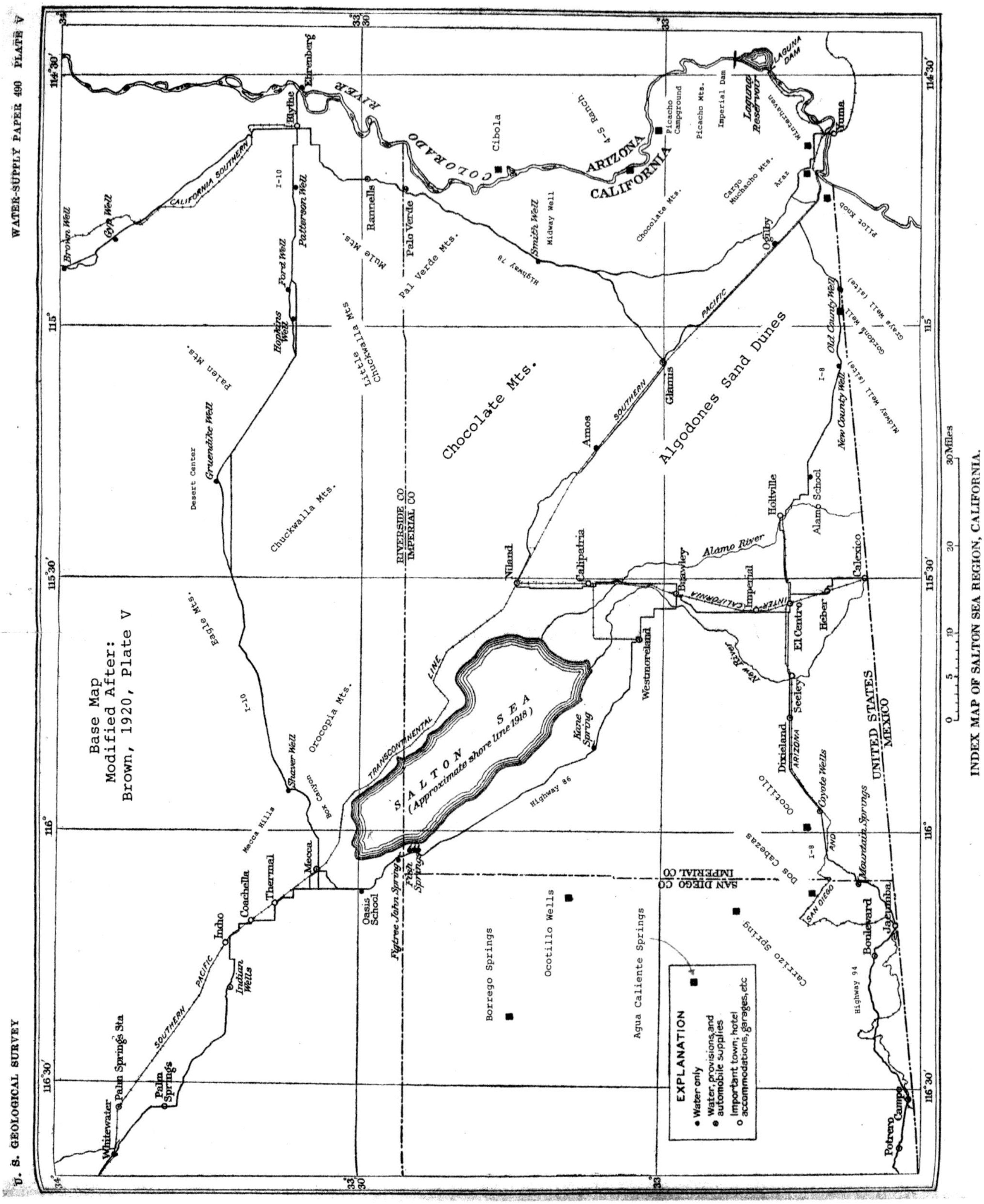
U. S. GEOLOGICAL SURVEY
WATER-SUPPLY PAPER 490 PLATE V
Base Map
Modified After:
Brown, 1920, Plate V
SALTON SEA
(Approximate shore line 1918)
Whitewater
Palm Springs Sta
Palm Springs
Indian Wells
Indio
Coachella
Thermal
Mecca
Mecca Hills
Box Canyon
Shaver Well
Orocopia Mts.
Eagle Mts.
I-10
Desert Center
Gruendike Well
Palen Mts.
Chuckwalla Mts.
Little Chuckwalla Mts
Hopkins Well
Ford Well
Patterson Well
Mule Mts.
Rannells
Palo Verde
Pal Verde Mts.
Blythe
Ehrenberg
Brown Well
Gyp Well
CALIFORNIA SOUTHERN
COLORADO RIVER
Cibola
ARIZONA
CALIFORNIA
4-S Ranch
Picacho Campground
Picacho Mts.
Imperial Dam
Laguna Reservoir
LAGUNA DAM
Yuma
Winterhaven
Araz
Cargo Muchacho Mts.
Chocolate Mts.
Smith Well
Midway Well
Highway 78
Ogilby
Pilot Knob
SOUTHERN PACIFIC
Glamis
Amos
Algodones Sand Dunes
Old County Well
Grays Well (site)
Gordons Well
Midway Well (site)
New County Well
I-8
Holtville
Alamo School
Alamo River
Calexico
RIVERSIDE CO
IMPERIAL CO
Niland
Calipatria
Brawley
Imperial
INTER-CALIFORNIA
El Centro
Heber
New River
Westmoreland
Kane Spring
Highway 86
TRANSCONTINENTAL LINE
Seeley
Dixieland
ARIZONA AND SAN DIEGO
Coyote Wells
Ocotillo
Dos Cabezas
Mountain Springs
UNITED STATES
MEXICO
SAN DIEGO CO
IMPERIAL CO
Oasis School
Figtree John Spring
Fish Springs
Borrego Springs
Ocotillo Wells
Agua Caliente Springs
Carrizo Spring
Boulevard
Jacumba
Highway 94
Campo
Potrero
EXPLANATION
• Water only
• Water, provisions and automobile supplies
o Important town; hotel accommodations, garages, etc
0 5 10 20 30 Miles
INDEX MAP OF SALTON SEA REGION, CALIFORNIA.

its very real perils. No matter in what sort of vehicle or for what length of time it is planned to enter the desert, adequate provision for possible misfortunes should be made. Probably more fatalities and hardships result at present from the failure of automobilists to know the road or to take a little food or an extra supply of water than from any other cause. Of course, with a good automobile and good luck one can travel across this region in a day; but an accident in the midst of a lonely waste may leave a party stranded without help for days, or until it is possible to walk to civilization.

It is necessary, first, to learn definitely about the road. Procure the best maps available. The one in this guidebook is believed to be the best general map of the region yet published, but for numerous byways and burro trails it is not accurate in detail. Moreover, desert roads are frequently changed with the exigencies of travel, and what is a plain road one day may be obliterated by a sandstorm and abandoned the next. Still the main routes are controlled by surface features and remain much the same. Numerous private maps of local districts are valuable, and the automobile clubs and touring bureaus supply good maps of the chief automobile roads, which, however, are not drawn accurately to scale.

Inquiries should be made of local garage men and hotel keepers about the roads and watering places. Garage men usually have the latest information as to the condition of automobile roads. Inquiry about watering places and their condition at the time should be made, as a broken pump may render a good well entirely useless for a water supply.

Some food should be taken even for an automobile trip of only a few hours. A surplus of water over probable needs of men and automobiles should be provided. Oil and gasoline more than enough for probable needs should be taken, and it should be remembered that desert roads may require twice as much per mile as pavement. Supplies for horses should be ample for all possible contingencies.

Teams, Hay, Grain. The following excellent suggestions to desert travelers are taken from a paper by Mendenhall [1909, p. 21] and are especially pertinent for persons traveling with teams [of livestock]:

> Where teams are used animals accustomed to the desert should be procured if possible, for horses or mules that are unused to desert condition fret on the sandy roads, and rapidly weaken from drinking the saline waters. They are also in danger of pneumonia from the cold of the winter nights and the wide extremes of temperature. During winter journeys blankets should be provided to protect the animals at night.
>
> Travel in the desert far from the railroads and from food supplies is, of course, more expensive than in other regions. A party leaving a supply station to go 100 miles or more into an uninhabited part of the desert must take along everything needed, even to the most minute detail. This means that if the trip is to last for two weeks enough hay and grain for each animal and enough provisions to last each man that length of time must be taken. For four horses, drawing a wagon that carries four persons and bedding, provisions, and tools, another team of four horses must also be taken to haul sufficient hay and grain to feed the eight horses for two weeks. There are but few places in the desert, away from the railroads, where grain or hay of any kind can be procured. As the teams are rarely able to travel faster than a walk, heavy horses that are good walkers should be selected. The tires should be as wide as can be procured. Desirable widths of tires for freight wagons are 6 to 9 inches; for light wagons 3 inches * * * For packing trips experienced prospectors select burros on account of their endurance of heat and thirst, their foraging abilities, and the slight amount of care they require. They are slow and too light to carry heavy packs, so that on expeditions where speed is essential, or heavy freight is to be moved and feed is available, horses or mules are to be preferred.'

Map 2 *(opposite). "Index Map of Salton Sea Region, California." (in Brown, 1920, Pl. V).*

Special Suggestions to Motorists. More people travel the desert now in automobiles than in any other way, although horses are not unknown and even foot travelers are sometimes seen. Low-geared trucks with large tires have an advantage in freighting or traveling very sandy roads. With an experienced desert driver the average car can travel almost any road that is passable for wagons. Without careful driving it may fail to get any where on a comparatively good road. Automobile parties should always carry a supply of spare tires and tubes. A vulcanizing outfit for making patches is especially desirable. A tire gage [*sic*] is very useful, and an air pump and a jack are necessary.

Sand is the worst obstacle in this part of the desert. Fortunately it is less prevalent than popular fancy imagines. The average road consists of a pair of wheel ruts (Photo 1); and in sandy places it is essential to stay in these ruts. Leave them only to pass another vehicle and then keep two wheels of the car in a rut if the sand is bad. Parties attempting to pass on a sandy road can usually do so by helping push the autos if other means fail. Wheel ruts, if fresh, are easily traversed even in deep sand, but old ruts or wagon tracks make very difficult traveling for automobiles [Photo 1]. On such roads if a car gets stuck it is often possible to back up and by getting a fresh start in one's own tracks break the road ahead through bad sand. A shovel is sometimes useful in short stretches for cleaning out covered ruts.

It is common practice in case of trouble in sand to deflate the tires. This gives the tire a greater bearing surface by allowing it to flatten out and increases the effectiveness of a car's gearing by reducing the diameter of the wheel. There is danger of rim cutting by having the tires too soft; so that no more air should be allowed to escape than is absolutely necessary. No fixed rule is known, but for Ford cars a pressure of 35 or even 30 pounds was found safe and always gave good results. Tires are not damaged by running "soft" in sand, but they should be immediately pumped up when hard ground is reached, or they will suffer rim cuts, stone bruises, or blow-outs. The tire gage [sic] is a necessity for judging the safe reduction of air pressure.

One great trouble in soft sand is that the wheels lose traction and spin, digging down and down into the sand. This is frequently brought about by attempting to start too suddenly. On the other hand , going too slowly when moving induces the wheels to spin. After a wheel has 'dug in' it has to be 'dug out' with a shovel, jacked up, and the hole surfaced with brush, canvas, or stones to give a bearing. Very effective use can be made of two strips of heavy canvas, say 30 feet long and 18 inches wide, for such difficulties. The strips must be thrust under the rear wheel, then laid lengthwise ahead in the ruts, and it is necessary to lift the front wheels and set them on the canvas to hold it down while the rear wheels pull. Otherwise the canvas is chewed up and 'spit' out in the rear by the spinning wheels. Canvas solved the trouble of the worst sand for the Survey party without much recourse to brush or shoveling. Progress is slow, but nearly any bad place may be crossed in this manner. The use of canvas for occasional trips on well-traveled roads is seldom necessary. Most travelers, instead of using canvas, fill the ruts with broken twigs, brush, stones, or anything else available when they get stuck, but unfortunately the brush is usually thinnest where the sand is thickest. There are various devices on the market for pulling out automobiles which get stuck, and one of these may be a valuable part of the equipment. On some roads in this region there are steep hills and sandy grades which automobiles can not ascend but which they can descend with ease. Such roads can be used only in one direction. To reach certain places it is necessary to choose circuitous routes in order to obtain feasible grades, and it may be necessary to follow a different route on the return trip from that taken in approaching.

Provisions. In the larger agricultural settlements in this region there are towns which have hotels and restaurants and at which food, clothing, and mechanical equipment can be purchased. Persons planning to take trips of a day's duration or longer should take plenty of food. This will necessarily be in condensed form, chiefly canned goods and staples such as flour, sugar, coffee, and bacon. For short trips it was found very easy to carry fresh bread in paper seals, especially by using a tight can for a container. Crackers can be used for longer journeys or bread can be made in camp. Butter in tight containers will keep for some time in winter. Cans with tightly fitting lids are desirable containers for such things as flour, coffee, etc. As much of vegetables, fruit, and canned milk should be taken as space will permit, because these things add variety to the rather scanty list of staples. Above all do not forget essential things which will be unobtainable in camp.

Clothing and Bedding. Persons camping in this region should remember that the nights are cool. Warm blankets and plenty of bedding are essential to comfort and to health. Bedding is usually made into a roll for the day, and a large sheet of canvas is convenient to protect it by day or night. There are no particular fashions or inherent necessities in clothing for the desert and one may choose here practically as he would anywhere else of the standard kinds of goods, remembering, as with bedding, that something extra may be needed for warmth at night.

Fuel. Fuel can be found at most places in the region, though it may be necessary sometimes to camp at a completely barren spot. Dead wood decays very slowly in this climate and is abundant at most places where small brush grows freely. Roots or branches burn readily with great heat. Most of the desert plants bear thorns to some extent and heavy gloves are useful protection in gathering firewood. An ax or hatchet may be used but is rarely necessary. Fuel is often stripped away in the vicinity of water holes, so that it may be necessary to go several hundred feet away from these common camping grounds to get a good supply.

Some kind of camp light is necessary, ordinary kerosene lanterns being most used. Carbide or electric lamps are sometimes satisfactory substitutes.

WATER.

Methods of Carrying Water. The most serious problem in desert travel is to obtain enough good water for drinking and camp uses and for teams or automobiles. It is necessary to have canteens, kegs, or water bags of sufficient capacity for all needs. Even tourists who do not plan to stop at all should cry a few canteens of water. Good canteens can be purchased at stores in all the towns. Water bags are excellent for keeping drinking water cool but waste water by evaporation. For cooking, 5-gallon kegs or cans are satisfactory containers. A day's supply is usually sufficient to carry, as there are very few places where one can not travel from one watering place to the next in a day's journey, even on foot.

Kinds of Watering Places. The usual watering places are springs or wells. Springs are frequently clogged with gravel or rubbish or sometimes even with the bodies of dead animals, and it may be necessary to clean them out. For this work a shovel is needed. Wells may or may not be equipped with pumps. Frequently the pumps are broken and useless, and a rope and bucket are then necessary to obtain water. Most of the wells in this region are less than 100 feet deep, but some are deeper, and 100 feet of rope is not too much to provide. As a rule the rope and bucket at a well, if they were ever provided, soon disappear, and one should never trust the chance of finding them there. Open wells are sometimes contaminated in the same way as springs and need to be cleaned out, particularly in little-frequented places where they are unused for months at a time.

Indications of Water. Water usually occurs in the lowest parts of the broad valleys or in canyons and gulches in the mountains. Plants are sometimes a useful indication of its presence. For instance, in this part of California a clump of palm trees is a reliable sign of water, though sometimes the water may be bad or it may be necessary to dig a foot or two for it at the base of the palm trees. Certain mesquite trees, particularly the creeping variety, are often indicators of water, especially where an isolated clump appears along a stream course. Tules (rushes) are also a reliable indication of water. There are, however, many desert plants which depend entirely on the scanty rainfall and do not indicate permanent water.

Numerous agencies, private and official, have erected signposts at different places in this region which call attention to watering places. A particular effort was made to place Geological Survey signs near water holes which were so situated as to [not] be easily overlooked... [Photo 2].

GETTING LOST.

The main roads are usually very plain and are adequately sign-posted by public and private agencies, so that there is little danger of getting lost while traveling on them. However, new roads are often made and old ones are abandoned, causing uncertainty at some places. Also a branch road to an active mine may be plainer than the main road. Some of the little-used roads are so indistinct that it is often extremely difficult to follow them. To be able to get back upon the right road if one has gone astray one should form the habit of careful observation, noting the prominent landmarks, such as mountains and peaks, and the branch roads, signposts, and minor details of the road. One should remember, however, that the aspect of mountains is very different from different points of view. More-over, all objects in desert regions, as a rule, appear much nearer to unaccustomed observers than they actually are. It is well to scale the distance on the map if possible before attempting to walk to any particular place. One should be able to tell directions by the stars, and it is well to carry a compass. Persons using the compass should remember that it is affected by iron or steel objects and electric currents at close distances.

If it is necessary to walk a long distance one should take a canteen of water and a pocket full of food. If the weather is warm it is better to walk by night than by day. Follow a road if possible, as the greatest danger of perishing is in wandering about where there is no road. If it is necessary to travel where there is no road take a straight course toward some star or distant landmark. Distances can be estimated by pacing, allowing about 2,000 paces to the mile.

If one meets with an accident or gets lost far out in the desert it is important not to get excited, nor to rush wildly about wasting energy and ruining self-control. Any bad situation should be thought over carefully, and a sensible solution arrived at by studying the available maps, estimating probable distances to the most certain points of relief, and then making plans to walk, wait, or signal. It should be remembered that there are usually several travelers passing every day on the main automobile roads and on these roads help may be obtained most easily by waiting. On less frequented roads waiting is inadvisable.

MAIN ROADS ACROSS THE REGION.

Several roads or combinations of roads are available for tourists passing across this region [Map 2]. A traveler starting at Los Angeles generally goes eastward through Banning and the San Gorgonio Pass, then southeastward through Coachella Valley. At the south end of this valley he may turn east at Mecca and cross the uninhabited desert to Blythe, or he may continue southeastward into Imperial Valley. From Blythe a road leads eastward across Colorado River at Ehrenberg Ferry and on by several slightly divergent routes into Phoenix. Another goes south to Yuma, from which Phoenix is reached by a well-established road. From Imperial Valley two roads are available to Yuma. One, by way of Niland, follows closely the Southern Pacific railroad; the other, by way of Holtville, crosses the Sand Hills and is the more direct route. From San Diego there is practically but one road as far as Imperial Valley, from which the same choice of roads as before, by Niland or Holtville, is available to Yuma. There is also an important road to Phoenix that goes from Los Angeles through San Bernardino and by way of Cajon Pass across San Bernardino County. This crosses [the] Colorado River at Parker. Although it is the longest road to Phoenix, it has natural advantages which afford compensation for the greater distance. This road lies north of the region here considered.

The shortest route from Los Angeles to Phoenix, by way of Blythe and Ehrenberg, is about 425 miles long. The distance by way of Imperial Valley is 492 miles. The road through Cajon Pass is about 530 miles long."

[Author's Note: At this point the author presents 72 pages of detailed road logs for the southeastern California Desert Region. One of these, found on pages 23–24, is a detailed description of dirt and sand roads from Niland to Yuma. It was chosen to illustrate a typical description of a route not far from our Picacho Area field trip. Bold text in the road log that follows highlights places where water could be found. A "plain" route or track is interpreted to be the "main" road or track.]

ROAD LOGS...

Niland [California] to Yuma [Arizona] (73 miles).

0.0 Niland. Corner by post office, one block south of railway station. Geological Survey sign on corner. Go east.

0.2 Cross Southern Pacific Railroad and turn southeast along north side of track.

0.5 Auto club sign. Plain road leads north. Continue along track.

3.7 Turn east, away from track, crossing East Highline canal of Imperial Irrigation District, about 100 feet from the railway. A dim road leads southeast (right) along track to Flowing Well, a railway station half a mile away. Go straight ahead, up hill that marks old beach [Lake Cahuilla shoreline]. This is very sandy. It may be necessary to deflate tires to about 35-pound pressure here.

4.5 Top of hill. Follow main road.

6.8 Plain trail turns to left up large wash. There are numerous tracks in next 3 miles that turn to the left toward the Chocolate Mountains. Some of these go to Salvation Spring.... Ignore them and continue southeast on main traveled road.

8.9 Another very plain trail leads to left toward mountains. Continue straight ahead. There are bad sandy spots in next 3 miles.

14.4 Auto club sign. Cross trail running northeast and southwest. Go straight ahead.

15.5 Auto club sign. Faint trail continues straight ahead but main road turns to right, toward railway.

16.2 Railway. Cross and go southeast along track. A dim branch road comes down railway from Mammoth Wash.

19.3 Amos railway station. Water obtainable from cistern supplied by railway company. Charge of 5 cents per head for stock. Continue along south side of track.

32.7 Pass Glamis railway station, which is on north side of railway, and continue southeast 0.1 mile.

32.8 Cross to north side of track and turn back northwest to railway station.

32.9 Glamis railway station. Water can be had from railway cistern. Charge for stock. Gasoline and meals are obtainable at store. To leave Glamis go northeast, away from railway, on main road, which is well signed.

33.1 Road forks. Auto club sign. Left-hand road continues northeast to Blythe. Right-hand road turns southeast to Yuma. Road parallels railway but keeps at a distance.
37.9 Turn south toward railway.
38.9 Cross track to south side of railway. A dim branch road comes in down railway. Turn southeast beside track.
52.1 Ogilby railway station. Water is obtainable from cistern kept by railway company. Gasoline can be purchased. Continue along railway.

[Author's Note: Water was an important commodity along early railroad mainlines. Wood and/or coal burning steam engines of the day needed regular replenishment with water, as well as fuel. Hence, the many way-stations and water tanks (with interesting names like Amos, Glamis, Ogilby and Araz) along main-stem rail lines are only recalled today by seemingly out-of-place and lonely sign-posts. Today, all that remains at Ogilby is a small cemetery (Photo 5) on the west side of Ogilby Road (County Road S-34) and an at-grade Southern Pacific Railroad crossing guard with modern electro-mechanical crossing arms.]

56.0 Branch road comes in from southwest. This is the Sand Hills or Holtville route to Yuma. Continue along railway.
58.8 Road forks. Goodrich sign. Right-hand road continues along railway into Yuma, passing close to Colorado River for last 5 miles. This road is very hilly and sandy and consequently less traveled. Left-hand road crosses railway and goes north.
Left-hand or 'mesa' road.
58.8 Cross railway and go north.
59.8 Dim road goes left. Continue to right.
61.9 Geological Survey sign. Branch road goes northwest to Blythe. Turn southeast here on main road.
65.0 Descend hill and follow sandy wash 1½ miles, passing through cut in river terrace, 50 feet high.
66.5 Branch road goes southwest. Turn left (east) along irrigation ditch.
67.4 Turn south (right).
67.9 Touch railway and turn east (left). Branch road crosses railway to south.
70.1 Cross bridge over very large canal.
70.2 Bridge over large canal.
70.9 Geological Survey sign at cross-roads. Turn south to Yuma.
72.4 Cross a railway track and turn east, uphill to bridge over Colorado River. River road (right-hand branch) joins here. Fort Yuma Indian school on hill.
72.5 California side of bridge over Colorado River. Follow principal street into Yuma.
72.9 Yuma, Ariz., at post office or railway station.
Right-hand or river road.
58.8 Take right-hand road, continuing along track.
63.2 Turn south (right), away from railway, along levee.
64.2 Turn east, along levee. Exact course of road not known here but it goes nearly east into Yuma. Follow main road.
67.3 (Approximate distance.) Winterhaven. Small village. Water, store and supplies.
68.2 Pass under railway bridge and turn north, uphill.
68.3 Join left-hand road and turn east uphill to bridge over Colorado River. Fort Yuma Indian School on hill.
68.4 California side of bridge. Follow principal street into Yuma.
68.8 Yuma, Ariz. at post office or railway station.

Photo 5. Old Historical Ogilby Cemetery. View to northeast. Southern Pacific Railroad crossing guard in right-middle ground (Highway S-34). Cargo Muchacho Mountains in distance. Photo by W. J. Elliott, 2011.

ANOTHER VIEW OF DESERT TRAVEL

A memorable account of a two-month journey across the Colorado and Sonoran deserts in mid-summer 1922 can be found in Dickey (2012). The story unfolds with Dad, a recent widower, setting off from San Diego with his two pre-pubescent sons to visit Grandma in Massachusetts. When their Chevrolet broke down in La Mesa, California, a scant few miles into the trip, on the advice of a mechanic, he bought a Model T Ford to finish the journey. If you've already read the preceding pages, you'll have no trouble identifying with the washboard and rutted roads, soft hot sand and frequent tire changes in Dickey's article.

Photo 6. *Gordon's Well on north side of Interstate-8, east of El Centro. View to east. Photo by W. J. Elliott, 2011.*

IN SEARCH OF WATER WELLS—JUNE 2011

Many watering holes (springs and wells) relied upon by early-20th Century desert travelers have long since dried up and/or been abandoned. About the only record of these necessities for early desert travel are old auto club and United States Geological Survey Topographic maps. Trying to find some of these watering holes proved fruitless in June of 2011, with about the only clue to the location of an abandoned watering hole being a clump of tamarisk trees or a grove of mesquite bushes.

Two well sites were found. One, Gordon's Well, located on the north side of Interstate-8 (Map 2, Photo 6), is approximately 5 miles west of Imperial Sand Dunes Recreation Lands Rest Stop (U. S. Geological Survey, 1964–79). This relatively newer well is not listed by Brown (1920, p. 29) in his road log between El Centro, California and Yuma, Arizona. The wellhead at this RV-park and recreation vehicle storage yard was easily viewed from the Interstate-8 frontage road in June of 2011. Lush green grass in front of the well suggests that it is functioning properly.

The other, Midway Well, previously known as Smith Well (Brown, 1920, p. 59), is located about 1/2 mile west of State Highway 78 (United States Geological Survey, 1988), approximately 10 miles north-northeasterly of the State Highway 78 and Ogilby Road Intersection (Map 2, Photo 7). A partially disassembled windmill and a dry masonry cistern (pila) attest to the presence of this one-time desert watering hole. An old cabin, furnished with a potbelly stove, and deteriorating remains of corrals suggest this was once a welcome stopover along the unimproved rocky and sandy track between Blythe and El Centro.

Three other older wells were searched for along the Interstate-8 Corridor, but not found.

Photo 7. *Midway Well, west side of State Highway 78, northeast of Glamis. This was known as Smith Well in the late 19-teens (Brown, 1920, p. 59). View to west. Photo by W. J. Elliott, 2011.*

Coyote Wells Site (Brown, 1920, p. 27; U. S. Geological Survey, 1957) is located approximately 1¾ miles east of Ocotillo at a San Diego and Arizona Eastern Railroad siding (Map 2). According to Brown (1920, p. 27), a "Store, post office, and railway station…" were present at the time of his reconnaissance. All that remains today are a few phreatophyte tamarisk (salt cedar) trees (Photo 8), a common indicator of free water not too far below the ground surface.

Photo 8. *Coyote Wells Site. View to northeast. San Diego and Arizona Eastern Railroad Siding, an Interstate-8 overpass, and a grove of tamarisk trees are all that remain. Photo by W. J. Elliott, 2011.*

Photo 10. *Old Plank Road. Historical monument, south side of Interstate-8, west side of Algodones Sand Dunes. Photo by W. J. Elliott, 2011.*

Photo 9. *Midway Well Site, on North Side of Interstate-8 at intersection with State Highway 98. View to west. This was known as New County Well in the late 19-teens. Old U.S. Highway 80 pavement and scattered creosote bushes are all that remain. Photo by W. J. Elliott, 2011.*

Photo 11. *Old Plank Road. View to east. South side of Interstate-8, west side of Algodones Sand Dunes. Photo by W. J. Elliott, 2011.*

Midway Well (Site) located approximately 28½ miles east-southeast of El Centro at the intersection of Interstate-8 and State Highway 98 (U. S. Geological Survey, 1954) was known as New County Well in the late 19-teens (Map 2). A windmill pump system was reported to produce good water. Brown (1920, p. 29) states that the "Next 10 miles very sandy." Today, there is no evidence to suggest the presence of a formerly productive water well. All that remains are a bleak, dry, desert landscape of sand, creosote bushes and paved highways (Photo 9).

Grays Well Site is located approximately 38 miles east-southeast of El Centro at the western terminus of the Old Plank Road (U. S. Geological Survey, 1964–79). It was known as Old County Well in the late 19-teens (Map 2). Good water was reported, as well as the presence of a roadmen's house. The old planked road extended 6.8 miles further east, across the Algodones Sand Dunes (Brown, 1920, p. 29). Today, an historical marker (Photo 10) stands near the Old Plank Road (Photo 11) and a grove of tamarisk trees (Photo 12) suggest the presence of shallow ground water (Grays Well Site). It is interesting to reflect on how important these stopping places must have been to the survival of early desert travelers; whether on foot, with livestock, or in early 20th Century automobiles. Clearly being prepared would have been a key ingredient to survival in the harsh and unforgiving desert landscape.

Lastly, let me mention a modern-day system of desert watering places. Half-drums, filled with up to 6 one-gallon plastic jugs of drinking water, are located in the Yuha Desert along State Highway 98 (U. S. Geological Survey, 1957). Loose, easily removable wooden lids protect and shade the water. Placed approximately ⅓-mile apart along the

Photo 12. *Grays Well Site. View to east. South side of Interstate-8, west side of Algodones Sand Dunes. Photo by W. J. Elliott, 2011.*

Photo 13. *Cistern (Agua Cache) Southeast of Ocotillo. View to northwest in Yuha Desert on south side of State Highway 98. Photo by W. J. Elliott, 2011.*

Photo 14. *Imperial Diversion Dam. Completed in 1938, impounds Colorado River water for delivery to All American Canal. It is an 85' high, 3,475' long concrete slab and buttress structure located 18 miles northeast of Yuma, Arizona (U. S. Dept. Interior, Bureau of Reclamation, 2009). View to northeast. Photo by W. J. Elliott, 2011.*

south side of the highway by humanitarian groups, blue and orange flags fly on top of approximately 18-feet high metal poles (Photo 13) to draw the attention of thirsty foot travelers (Sifuentes, 2012).

WHAT IS THE SAME TODAY?

1. The climate is about the same. It is still hot and dry in summer, save an occasional monsoon thunder storm (Wikipedia, 2011a), and cold and dry in winter, save a few scattered inches of rain dropped by low pressure systems originating in the Gulf of Alaska (Heidorn, 2005). Wind, sand, and dust storms still accompany springtime storms, to which any knowledgeable and sandblasted desert rat, will readily attest.

2. Temperature extremes are the norm in the Colorado Desert. Daytime summer temperatures typically range from 90°F to 120°F (The Weather Channel, 2014). July, August, and September monsoon moisture, originating in the Gulf of California, can make late summer temperatures even more uncomfortable, not to mention the chance of flash flooding arising from intense thunder storms.

3. Winter temperatures can fall into the 30°F to 50°F (The Weather Channel, 2014) range at night, a bit chilly for one caught without proper clothing to mitigate hypothermia. Hence, the need to be prepared for most any emergency when traveling in the desert remains important today.

4. What rainfall that does occur falls mainly in the winter months and is scarcely enough to keep highly adaptable desert community plants and hardy critters alive from season to season.

6. The Colorado and Gila Rivers still flow through desert valleys and canyons, though modified in today's world by dams and weirs that divert water for human benefit (Dowd, 1960, p. 2; Photo 14). Smaller ephemeral rivers and streams continue to appear/disappear with seasonal cycles.

5. Water supplies are still scarce; the exception being along rivers, canals, and areas developed with agriculture, cities, towns, and designated recreation sites. Wells and natural springs are sparse to non-existent in most remote and uninhabited regions, especially away from well-traveled highways and byways.

7. Enormous numbers of square miles of the desert landscape still have sparse or no permanent inhabitants and consequently few if any human services or necessities, a condition that does not appear likely to change any time soon.

8. Dehydration is still, and will always continue to be an ever-present danger in the dry desert environment.

Photo 15. *Cistern on the up-hill side of Mountain Springs Grade, Interstate-8, Southwest of Ocotillo. Photo by W. J. Elliott, 2011.*

Photo 16. *Desert Varnish (Patina) on well-rounded Colorado River "B" gravel near the Cargo Muchacho Gold Mine, Cargo Muchacho Mountains, Imperial County, CA. Photo by W. J. Elliott, 2011.*

Photo 17. *Modern Paved Two Lane Road, State Highway 78, at turnoff to Midway (Smith) Well northeast of Glamis. Left to right, Little Chuckwalla Mountains and Palo Verde Mountains in the distance. View to north. Photo by W. J. Elliott, 2011.*

Photo 18. *Wide and modern 4-lane highway, Interstate-8, and sign showing 10 miles to Winterhaven, California, and points east. Pilot Knob in distance, under sign. Photo by W. J. Elliott, 2011.*

9. Small cisterns containing radiator water still exist on the 11½-miles up-hill side (elevation gain: 2,640′) of the I-8 Mountain Springs Grade (Photo 15; U. S. Geological Survey, 1959–75). Newer and older automobiles alike are not immune to over-heating on hot summer days — especially while driving up the long grade at warp-speed with the air conditioner cranking!

10. Critters adapted to harsh desert conditions, such as burros, sheep, deer, coyotes, foxes, tortoise, horned toads, lizards, a variety of snakes and the like, continue to abound.

11. Desert varnish, or patina ("sunburn" if you will), still cover individual rocks and colors entire mountain ranges with a ubiquitous mineral coating which produces distinctive shades of a dark orange- to red- to black-sheen depending upon the length of exposure to the sun and the elements. Longer exposures produce darker colors (Photo 16).

12. Ventifacts still litter abandoned stream terraces and fan deposits. Windborne sand, an effective "sandpaper", shapes, smooths, and polishes these rocks into seemingly whimsical shapes.

13. Oh, and the geology of the desert rocks and dirt don't change, but our interpretations certainly do….

WHAT IS DIFFERENT TODAY?

So, what is different in the desert environment today? After all, it is still the same Colorado Desert.

1. Many roads, once no more than washboard dirt and/or rutted sandy tracks, are now paved (compare Photo 1 with Photo 17), or have become "improved" dirt roads suitable for most modern passenger vehicles. Divided super highways (Photo 18) allow safe travel at speeds unheard of in the early 1900s.

2. Road signage is much better in today's world (Photo 18), with more and better information available to the

Photo 19. *Dark Gray "Smoke Signal" on Horizon. Old Plank Road on sand dunes in middle-ground. View to east-southeast. Photo by W. J. Elliott, 2011.*

Photo 20. *Car-Camping at Agua Caliente Springs, San Diego County, California. November, 1940. View northerly of the author and his folks' 1936 Chevrolet parked among mesquite bushes. Vallecito Mountains and Whale Peak rise on the skyline. W. T. Elliott family photo, 1940.*

traveling public. Many display boards are larger, better located, more informative, and in some cases, lighted along major travel routes.

3. Whereas livestock likely outnumbered people during the early years of the 20th century, the reverse may well be true today. Sprawling population centers, such as in the Imperial, Coachella, and Yuma valleys have increased the need for stable and sustainable domestic, as well as agricultural water supplies.

4. Well-water aside, since 1942, when the All American Canal was completed (as a part of the 1928 Boulder Canyon Project—Boulder Dam, now officially Hoover Dam), most water is obtained from the Colorado River (Billington and others, 2005, p. 129–142; Rowley, 2006, p. 319–328). A spiderweb of water distribution canals supply farms, ranches and city dwellers with life-giving water. Similarly, a web of waste-drains carry excess and used water to the Salton Sea.

5. Never mind smoke signals sent up by the burning of fallow fields in agricultural areas (Photo 19), communications in today's world far surpass that of 100 years ago. The infancy of radio, telegraph and telephone at the turn of the last century are dwarfed by modern cell phones and satellite communications, and are taken for granted by today's public. Even amateur (ham) radio has its place in the world of emergency communications.

6. Mail, package and freight delivery back then was slow at best (via wagon, sailing ship and rail). Now-a-days we expect overnight delivery of even the largest of our belongings.

The desert can still be a foreboding place to the unwary traveler. But, with a tincture of common sense and a spirit of Boy Scout preparation, life and travel in the desert can be many times safer and enjoyable than my grandparents experienced 100 years ago.

PERSONAL EXPERIENCES AND GENERAL RECOLLECTIONS

My first experience in the desert was in November 1940, car-camping with my folks at Agua Caliente Springs, in east-central San Diego County, California (Photo 20; U. S. Geological Survey, 1959). This desert watering hole was listed on J. S. Brown's (1920, p. 40) road log for desert travel from Julian to El Centro (via Vallecito and Carrizo) and notes Agua Caliente Springs as a place to obtain Good Water. As a child, I was told that Agua Caliente, "Hot Springs," was an alternate stopping place for the Butterfield Overland Stage (mail) Route during its operation from 1857–1861 (Wikipedia, 2011b).

During the 3¾ years of World War II, I recall meandering around southern portions of the Colorado Desert with my paternal grandfather (Pepa), my dad (Bill, Senior), and my younger brother, Bob. Traveling mostly along dirt roads and faint tracks in the sand, we collected seashells in the Yuha Desert (United States Geological Survey, 1957), drove over portions of the Old Plank Road (Photo 11), and visited submerged mud volcanoes near Mullet Island (Dowd, 1960, p. 30; Jenkins, 1938; Lynch and Hudnut, 2006) at the southeastern end of the Salton Sea.

East of the Salton Sea, near the base of the Chocolate Mountains, we bathed in mud-lined warm water pools (Photo 21) located down-stream from a 174°F artesian water well (Photo 22) drilled in the fall of 1938 by the builders of the Coachella Branch of the All American Canal (Seddon, 1989). It is located approximately 16 miles northwest of Niland, California at the north end of Hot Mineral Spa Road (U. S. Geological Survey, 1992b). Early improvements to this desert watering hole were affectionately referred to as "Old Spa." With time came capital site improvements as well as name and ownership changes, first to Imperial Spa and in 2007 to Glamis North KOA (http://www.glamisnorthkoa.com/). With abandonment and capping of the old original well (Photo 23) came a new nearby well to supply hot "medicinal" mineral water to the swimming pools and tubs that have attracted people to the area since the heyday years of "Old Spa" in the 1950s (Seddon, 1989).

Approximately 0.3-mile east of the old original "Old Spa" wellhead, a gravel bench/beach (at an approximate elevation of 40 feet above mean sea level) marks the ancient Lake Cahuilla Shoreline (Brown, 1923, p. 32, 33, 39; Jefferson and Lindsay, 2008; Testa, 2008; p. 5–7; U. S. Geological Survey, 1908, Van De Kamp, 1973, p. 832). Both the old un-lined, and new concrete-lined Coachella Canal lie approximately 0.1 mile east of the old shoreline (San Diego County Water Authority, 2008). This area of hot-water springs lies along the northeastern edge of the Salton Trough transform/pull-apart boundary between the North American and Pacific plates (Irwin, 1990; Jennings, 1994).

An analysis (by Union Oil Company, circa 1960s) of well water from adjacent Bashford's Hot Mineral Spa was reported as follows (Seddon, 1989, p. 50) (Table 1). Anions and cations are reported in parts per million (ppm).

Frink Spring, a desert watering hole along the old dirt track between Mecca and Niland (Brown, 1920, p. 55–56) is located approximately 4½ miles southeast of "Old Spa," about one mile northeast of the State Highway 111 and Frink Road Intersection (U. S. Geological Survey, 1992a).

Photo 21. *"Mud Baths at Old Spa. Rattle Snake Sam in foreground." (Photo credit: Marion Seddon [©1989, Mike Stanley], p. 16). This image is reminiscent of the author's experience during the years of World War II.*

Photo 22. *"Original mineral fountain at Old Spa." (Photo credit: Marion Seddon [©1989, Mike Stanley], p. 4). This 1938 well supplied 174° F water to builders of the first Coachella Canal, and subsequently to Old Spa which later became Imperial Spa.*

Photo 23. *Old Spa (formerly Imperial Spa, now Glamis North KOA). View to east. Capped original well drilled in fall 1938 in lower left- hand corner between witness posts (this is the well the author remembers from the World War II years). White water from the new well splashes out of a large-diameter pipe at the southeast edge of the large blue cooling tank. Photo by W. J. Elliott, February, 2006.*

Table 1. *Analysis of well water from Bashford's Hot Mineral Spa (Seddon, 1989, p. 50). Anions and cations in parts per million (ppm).*

pH	7.14
Total Dissolved Solids	415 mg/L
Settled brine	0.0
Chlorides	18,875
Sulfate	5,750
Arsenic	0.0
Boron	33
Calcium	638
Cobalt	1
Fluoride	5
Iron	0.0
Lithium	15.8
Magnesium	734
Manganese	0.0
Potassium	40
Silica	625
Sodium	14,000

Photo 24. *Agua Caliente Springs, San Diego County, California, April, 1952. Northwesterly view of the campground. Mom marked our camping spot with a white X over my folks' 1947 DeSoto Suburban 9-passenger sedan. Tierra Blanca Mountains in the left-center foreground; Vallecito Mountains in the right background. W. T. Elliott family photo, 1952.*

Photo 25. *Desert Water Bag. Ubiquitous and sometimes infamous artifact of the mid-20th Century. The author is pointing to a desert water bag. Photo by Wendy Elliott, October 2011, at the Owens Valley Cruisers Fall Colors Car Show, Bishop, California.*

In April, 1952, we camped again at Agua Caliente Springs, this time traveling over the still unpaved State Highway S-2 in our 1947 DeSoto Suburban 9-passenger sedan (Photo 24) to reach our campsite. Little did I know at the time, but our hot springs camp was situated atop the Elsinore Fault Zone (Jennings, 1994; Todd, 1983). Hmmm, perhaps this was an omen for a future career path! To boot, the first annual San Diego State University Geology Department Alumni Camping Trip was held at Agua Caliente Springs in late winter 1987.

And, how could I leave my favorite Colorado Desert environment without mentioning a few memories and musings? I have fond memories of warm summer vacations to our western National Parks, with the ubiquitous standard driver irritants — "he's touching me!" — "she's looking at me!" Never mind "are we there yet?" Expressing one's thirst usually brought a cup of warm, if not hot, water from mom. Complaints about the water temperature brought the standard response, "yes, but its wet!" Doubtless most of us can fill in the rest of that conversation....

And, who could forget the ubiquitous Desert Water Bag? These one gallon canvas bags (Photo 25) were usually hung near the front of the car (off a head light, hood ornament, or front bumper). Porous canvas would sweat and a breeze created by the vehicle's forward motion would cool the water by evaporation.

As a last resort, I have always been told that the flesh inside a barrel cactus could be sucked on to stave off dehydration. Fortunately, I've never needed that service. I'm not sure I'd be able to get past the thorns, much less the reported bitter taste. A better choice is probably to be prepared, and travel with plenty of water.

In days before modern refrigeration, most folks drove across the summertime desert at night to minimize creature discomforts and to try to keep the motor from overheating. I remember farmers in Yuma, Arizona, plowing their fields at night, with headlights on their tractors, to beat the blistering summer heat.

A luxury "deluxe" car in the 1940s and 1950s came with an analogue clock, an AM radio, and a heater. There was no factory-installed refrigerated air conditioning. A

Photo 26. *Swamp Cooler. This torpedo-shaped device collected bugs at highway speeds, as well as provided about 20° F of humid cooling for desert travelers in the mid-20th Century. For not so obscure reasons it was always mounted at the top of the passenger side window (forget the "hairdo"). Photo by W. J. Elliott, October 2009, at the Owens Valley Cruisers Fall Colors Car Show, Bishop, California.*

Photo 27. *Red Rubber Inner-Tube and Whitewall Spare Tire. In the mid-20th Century, these, along with tire-irons and a hand-operated air-pump, were among the necessities of road travel, desert or not. A well-stocked picnic basket, a thermos bottle full of coffee, and a glass Mason jar filled with drinking water were a passable substitute for today's quick stop at a fast-food restaurant. Photo by W. J. Elliott, October 2009, at the Owens Valley Cruisers Fall Colors Car Show, Bishop, California.*

"poor-man's" air-conditioning usually consisted of opening hood- and/or side-vents, as well as rotating wind-wings to blow outside air directly on the driver and passenger, which at highway speeds rarely exceeded 55 miles per hour. If one had to cross the desert on a hot summer day, and could afford the luxury, an after-market passenger-window swamp-cooler was "top-drawer." These torpedo-shaped devices (Photo 26) gave travelers about 20 degrees of cooling. But don't jump too quickly, at 120°F. outside, one had to be happy with about 100°F inside and a relative humidity near 100%. How spoiled we have all become with factory-installed air conditioned refrigeration, standard in most of today's vehicles.

As a side note to traveling American highways and byways in mid-20th Century, one could expect at least one blown tire and/or at least one flat-tire per day. Hence the need for "tire-irons" to pry a destroyed tire from the rim and coax a new one on. Can't do that with today's tubeless tires! Does anyone remember when tires came with inner-tubes (Photo 27)? And, don't forget the messy tube-patching "first aid" kits!

ACKNOWLEDGEMENTS

A special thanks goes to Todd Wirths for encouraging this paper — an effort to provide some historical perspective to our otherwise jargon-filled and perfunctory academic efforts to understand the earth upon which humanity lives and depends for its existence. Carole Ziegler's encouragement and careful, critical manuscript review resulted in a much improved presentation. Additionally, I would be remiss if I failed to thank my most helpful and understanding wife, Wendy. Without her computer tech support, patience, and frequent offerings of something to eat and drink, this effort would not have been such a fun and interesting adventure!

REFERENCES

Billington, D. P., Jackson, D. C., Melosi, M. V., 2005, The History of Large Federal Dams — Planning, Design, and Construction in the Era of Big Dams: U.S. Dept. Interior, Bureau of Reclamation, Denver, Colorado, 605p.

Brown, J. S., 1920, Routes to Desert Watering Places in the Salton Sea Region, California: U.S. Geol. Survey Water-Supply Paper 490-A, 86p.

Brown, J. S., 1923, The Salton Sea Region, California: U.S. Geol. Survey Water-Supply Paper 497, 292p.

Bryan, K., 1922, Routes to Desert Watering Places in the Papago Country Arizona: U.S. Geol. Survey Water-Supply Paper 490-D, p. 317–429.

Dickey, F., 2012, Going on a Road trip Lacked One Thing — A Road: San Diego Union-Tribune, July 8, 2012, pp. B-1 and B-6.

Dorsey, R., 2006, Stratigraphy, Tectonics, and Basin Evolution in the Anza-Borrego Desert Region, in Jefferson, G. T., and Lindsay, L., eds., Fossil Treasures of the Anza-Borrego Desert: California State Parks, Colorado Desert District, Anza-Borrego Desert State Park, The Anza-Borrego Foundation and Institute; published by Sunbelt Publications, San Diego, CA, p. 89–104.

Dowd, M. J., 1960, Historic Salton Sea — History of the Imperial Irrigation District: Imperial Irrigation District, El Centro, California, 32p.

Heidorn, K. C., 2005, For Strong Winds — Gulf of Alaska Storms: World Wide Web, "The Weather Doctor," 2011.

Irwin, W. P., 1990, Geology and Plate-Tectonic Development, in The San Andreas Fault System, California: U. S. Geol. Survey Prof. Paper 1515, p. 61–82.

Jefferson, G. T., and Lindsay, L., 2008, The Lake Cahuilla High Shoreline and a Stable Overflow Path Elevation, Southeastern California and Northeastern Baja California, in Reynolds, R. E., ed., Trough to Trough — The Colorado River and the Salton Sea: Calif. State Univ., Desert Studies Consortium and LSA Assoc., Inc., Field Trip Guide and 2008 Desert Symposium Proceedings, p. 107–109.

Jenkins, O. P., 1938, Geomorphic Map of California: Calif. Div. Mines (Bulletin 158, Plate 2), scale 1:2,000,642, contour interval 2,000′

Jennings, C. W., 1994, Fault Activity Map of California and Adjacent Areas: Calif. Div. Mines and Geology; Calif. Geol. Data Map Series, Map No. 6, scale 1:750,000, contour interval 500′.

Lynch, D. K., and Hudnut, K., W., 2006, Geothermal Structures Southeast of the Salton Sea, in Reynolds, R. E., ed., Trough to Trough — The Colorado River and the Salton Sea: Calif. State Univ., Desert Studies Consortium and LSA Assoc., Inc., Field Trip Guide and 2008 Desert Symposium Proceedings, p. 96–101.

Meinzer, O. E., 1920, 1921, and 1922, Routes to Desert watering Places in California and Arizona — Preface: U.S. Geol. Survey Water-Supply Papers 490-A, -B, -C, and -D, 4p.

Mendenhall, W. C., 1909, Some Desert Watering Places in Southeastern California and Southwestern Nevada: U.S. Geol. Survey Water-Supply Paper 224, 98p.

Ross, C. P., 1922, Routes to Desert Watering Places in the Lower Gila Region, Arizona: U.S. Geol. Survey Water-Supply Paper 490-C, p. 271–315.

Rowley, W. D., 2006, The Bureau of Reclamation, Origins and Growth to 1945, Volume 1: Bureau of Reclamation, U.S. Dept. Interior, Denver, Colorado, 512p.

Seddon, Marian, 1989, What Ever Happened to the Chocolate Mountain Gang? Self-published — purchased at Imperial Spa store (now Glamis North KOA spa store), [©1989, Mike Stanley]. 71p.

Testa, S. M., 2008, Early Geological Excursions and Observations of the Colorado Desert Region by William Phipps Blake, 1853 and 1906, in Reynolds, R. E., ed., Trough to Trough — The Colorado River and the Salton Sea: Calif. State Univ., Desert Studies Consortium and LSA Assoc., Inc., Field Trip Guide and 2008 Desert Symposium Proceedings, p. 72–83.

San Diego County Water Authority, 2008, http://www.sdcwa.org/water-authority-receives-two-awards-coachella-canal-lining-project.

Sifuentes, E., 2012, Group Needs Help Preventing Desert Deaths: North County Times – San Diego County, California, March 25, 2012, pp. B–1 and B–4.

Thompson, D. G., 1921, Routes to Desert Watering Places in the Mohave Desert Region, California: U.S. Geol. Survey Water-Supply Paper 490-B, p. 87–269.

Todd, V. R., 1983, The Elsinore Fault Zone South of Julian, in Steller, D. L., Bryant, M., Gath, E., eds., Geology of the Northern Elsinore Trough: South Coast Geological Society Annual Field Trip, 1983, p. 57–84.

U.S. Department of the Interior, Bureau of Reclamation, 2009, Imperial Diversion Dam: from the internet via Wikipedia, 2011.

U.S. Dept. of Labor, 2014, "Consumer Price Index, Inflation Calculator," Bureau of Labor Statistics. http://www.bls.gov/data/inflation_calculator.htm.

U.S. Geological Survey, 1908, Reconnaissance Map of the Salton Sink, California, scale 1:500,000, contour interval 50′.

U.S. Geological Survey, 1954, Midway Well, California, 15' topographic quadrangle map, scale 1:62,500, contour interval 10′.

U.S. Geological Survey, 1957, Coyote Wells, California, 15′ topographic quadrangle map, scale 1:62,500, contour interval 40′.

U.S. Geological Survey, 1959, Agua Caliente Springs, California, 7½′ topographic quadrangle map, scale 1:24,000, contour interval 40′.

U.S. Geological Survey, 1959–75, In-Ko-Pah Gorge, California, 7½′ topographic quadrangle map, scale 1:24,000, contour interval 40′.

U.S. Geological Survey, 1964, Grays Well, California, 7½′ topographic quadrangle map, scale 1:24,000, contour interval 20′.

U.S. Geological Survey, 1988, Buzzards Peak, California, 7½′ topographic quadrangle map, scale 1:24,000, contour interval 40′.

U.S. Geological Survey, 1992a, Frink, California, 7½′ topographic quadrangle map, scale 1:24,000, contour interval 10′.

U.S. Geological Survey, 1992b, Frink NW, California, 7½′ topographic quadrangle map, scale 1:24,000, contour interval 40'.

Van De Kamp, P. C., 1973, Holocene Continental Sedimentation in the Salton Basin, California — A Reconnaissance: Geol. Soc. America Bull., v. 84, p. 827–848.

The Weather Channel, 2014, http://www.weather.com/weather/wxclimatology/monthly/graph/USAZ0275.

Wikipedia, 2011a, "North American Monsoon." http://en.wikipedia.org/wiki/North_American_Monsoon.

Wikipedia, 2011b, "Butterfield Overland Mail in California." http://en.wikipedia.org/wiki/Butterfield_Overland_Mail_in_California.

OVER THE LINE — The Chocolate Mountains in Arizona as viewed from California's Picacho State Park.

A Place of Enchantment on the Lower Colorado

JOHN V. YOUNG

Editor's note: Noted author and journalist John V. Young (1909–1999) contributed this early look of the recently inaugurated Picacho State Park to the "Travel and Resorts" section of the New York Times *on April 27, 1969. Young authored several regional guidebooks and histories of California and the southwest.*

WINTERHAVEN, CALIF. — WE CAME TO Picacho State Park at early dusk, just as the jagged peaks along the western skyline were beginning to loom like blue-black cardboard cutouts stark against the brilliant evening sky. Their long maroon shadows were creeping down to the edge of the Colorado River, there to reappear as reflections on the glassy water.

As we entered this quiet campground in the southwest corner of California, we were greeted by a chorus of bird notes: chuckles and little singing queries from coveys of Gambel's quail parading out from the thickets to seek their supper, soft whirrs from shiny black phainopeplas (silky flycatchers), raucous salutes from a solitary golden, flicker and the lovely, plaintive melody of white-crowned sparrows. Later, we saw cactus wrens, towhees, juncos, roadrunners and innumerable unidentifiable L.G.B.'s (the ornithologist's shorthand for little gray birds).

Siege of Herons

From the dense reeds bordering the blue Colorado came the hoarse croaking of a siege of herons and egrets as they busied themselves catching frogs and preparing their rookeries. It was spring on the lower Colorado, and all the birds knew it.

Across the river, in Imperial National Wildlife Refuge, we could see many ducks swimming and diving. Behind the river rose the beautiful and weathered ramparts of the Chocolate Mountains in Arizona, turning to deep crimson under the last rays of the sun. On all sides, highly colorful rock and clay formations glowed as if with fire, as if we were in a miniature Death Valley with a wide river running through it. Scattered through the spacious campground, half-hidden by huge mesquite and ironwood and palo verde trees, with one or two tall palms standing aloof above them, were perhaps a half-dozen camping vehicles. Smoke rose from campfires, and lights gleamed in the peacefully gathering gloom. We had come to this place of enchantment almost by happenstance, having noticed on a road map a little red triangle denoting a campground 25 miles north of Yuma, Ariz., but on the California side of the Colorado River.

Conflicting Stories

The guidebooks told conflicting stories: that the road was passable only by four-wheel-drive vehicles, that it had a commercial trailer park and marina, that it was an old mining town inhabited by squatters, that it was a new and still-undeveloped California state park with no camping facilities. Some did not mention the place at all, or referred only to the abandoned mines.

But the road sign on U.S. 80 between Yuma and Winterhaven, just north of the Colorado River bridge, simply said, "Picacho, 22 Miles," and pointed northward up a paved secondary road. So we tried it. The pavement ran out after four or five miles, and the road degenerated into a dusty washboard of gravel. It was slow to travel, but entirely passable by almost any vehicle except perhaps a large trailer. We encountered no high centers, no boulders and no sharp dips, but plenty of twists and turns, in the ensuing 20 miles of gravel.

TWISTS AND TURNS — The road into Picacho is slow to travel, but entirely passable by almost any vehicle.

Out-of-Date Guidebooks

We found the distance to the park headquarters to be 24 miles from the highway. We also found that the guidebooks were out of date. The road had been realigned and graded, with ample evidence that it has to be regraded after each summer's floods in some of the sections traversing arroyos. There used to be a couple of commercial trailer parks at the end of the road, but they are no longer in evidence. All of the occupants were bought out by the state and their mobile shacks removed. This has been a state park since 1961, but only in the last year or so have all the title disputes been settled and the area cleared for public use. Picacho was once a mining town, and a very lively one at that, but there is little to show for it. Nothing at all remains of the town itself, only a fenced cemetery or two and the ruins of a big ore stamp mill.

There are a couple of places where boats can be launched in the river, but there is no established marina. A small store near the river has food and camping supplies, butane, gasoline and oil, ice, firewood (there is none in the campground) and boats to rent (sans motors).

The public campground (the fee is $1 a night, 30-day limit), offers piped water, picnic tables and firepits, toilets and cold showers. Old mine roads and burro trails wander all over the wild, 4,800-acre park, but they are unmarked. All but the campground area has been left primitive, and probably will stay that way. The park is open all year, but its use is largely confined to the period between October and May. It gets pretty hot here in the summer, and they say the mosquitos are bad then. Heaviest use last year was during Easter week and at Thanksgiving, when all 42 of the established campsites were full and 64 additional families were set up in primitive camps elsewhere.

Popular With Campers

Park officials expect this pattern to continue at least until a projected paved road is brought in from the West. However, despite its comparative seclusion and lack of publicity to date, the park drew 16,800 visitors last year, most of them campers.

Picacho takes its name from the Spanish word for peak, referring in this case to a 1,947-foot volcanic [sic] spire rising a few miles south of the campground. Gold was discovered in the vicinity about 1860, and the first claims were staked there in 1862 by Mexican prospectors who had not heard, apparently, that the region no longer belonged to them. American developers opened up new ore bodies in the 1880's, built a narrow-gauge railroad to the river and erected the stamp mill there. A town sprang up, complete with stores and saloons, and supported by a payroll of $40,000 a month.

Although the Picacho mines were reputed to be among the richest in California, they played out not long after the turn of the century. Soon thereafter, the place was deserted, save for some desultory, small-scale mining operations that continued into World War I. Aside from the ruins of the old mill, some rusty mining machinery and the remains of the old railroad embankment are all there is to show for the mining days.

The road to the old mines, turning west from the main road just south of the park boundary, is closed and posted against trespass, with warnings that deep, unguarded shafts await the unwary intruder. The mines and the cemeteries, including a small one at the foot of Picacho Peak, comprise a state historical landmark.

Perhaps the most poignant of the relics of more boisterous times at Picacho are two rock-lined graves beside the road near the mines. One grave is marked with a white cross bearing an inscription in Chinese (the other cross has stolen).

Buried Where They Fell

According to local legend, the graves contain the mortal remains of two Chinese miners who got into a row over a mine and took to knives to settle the matter. They wounded each other mortally and were buried where they fell. Later, friends persuaded a Chinese merchant in Yuma to provide the inscriptions.

Near the campground is a large graveyard with many marked and unmarked graves. It is labeled as dating from 1875, but it actually is more recent. The cemetery was moved from its original site near the river, a site that was flooded by the rising waters of a lake behind Imperial Dam, near Yuma. The lake has formed several side arms that are popular with bass and catfish fishermen and with wildfowl hunters. These somewhat ghostly bodies of still, blue water, surrounded by impenetrable thickets of willow, reeds and tules, are host to myriads of birds, frogs and small animals.

Surrealistic Art Forms

They are called "stump" lakes because the water has drowned large stands of desert trees whose stumps and branches, bleached white, rise from the surface like

ORE CHUTES — Old stamp mill in Picacho State Park, right, is one of the relics of the area's boisterous gold-mining days.

surrealistic art forms. They make fine perches for egrets and herons. Many of the campers we talked with at Picacho had come here, they said, to fish, gather minerals or hunt. But most of them had found it an ideal place to do nothing but sit in the shade and listen to the birds — and the silence. There were people here from several states and one couple from Canada, and some had been here a month. Nearly all of them, it seems, had stumbled over the place just as we did — astonished and delighted to find such an unspoiled and hospitable desert retreat so close to civilization, and wondering how long it would last.

A MOUNTAIN OF GOLD ORE.

Senators Morgan and Jones and Mr. Pettigrew Working Large Deposit Near Yuma.

Special to The New York Times.

SAN FRANCISCO, Cal., April 28.—Senators Morgan of Alabama and Jones of Nevada, and ex-Senator Pettigrew of South Dakota, are interested with ex-Senator Dorsey in a mountain of low grade gold ore at Picacho, Cal., on the west bank of the Colorado River, twenty-six miles above Yuma. The preliminary work of development is now being done, and the plans are for a mill with 280 stamps, with a cyanide plant larger than any in the country.

The mountain of ore is inexhaustible, but it would be a waste of money to put in a small mill, for the ore is not rich enough to pay without a mill which will crush hundreds of tons a day. A branch railroad will carry the ore to Yuma.

Senator Jones and ex-Senator Dorsey were at Picacho last week and gave further instruction for the work. The road from the Southern Pacific will be the first work done.

From New York Times, *April 29, 1901.*

SECTION 2

The PICACHO REGION of the SOUTHEASTERN CHOCOLATE MOUNTAINS

MIGHTY CRASH OF FLY WHEEL.

CURIOUS ACCIDENT AT MINE AT PICACHO.

With Governor of Engine Put Out of Operation, Machinery Gains Terrific Momentum and Monster Wheel Goes to Pieces, One of Big Fragments Being Thrown Half a Mile.

[SPECIAL CORRESPONDENCE OF THE TIMES.]

SAN DIEGO, Aug. 22.—News has been received here of the closing of the California Gold King Company mine at Picacho. Two men employed in the mill and mine are laid off. The governor belt broke, and the fly wheel, weighing twenty-one tons, driven by the ungoverned engine, attained a tremendous speed and finally burst into fragments, many of them going through the roof. One piece weighing several hundred pounds was found half a mile away. Fortunately no one was hurt. It is expected that the mine and mill will be in operation by October 1.

From Los Angeles Times, *August 23, 1904.*

Toward a unified model for the geologic and tectonic history of the region between Indian Pass and Picacho State Recreation Area, southeastern California

GARY H. GIRTY, KRISTA K. MUELA, HOLLY OLSON, LAUREN SUTTON, MELANIE BIGGS, JON SAINSBURY, JASON W. RICKETTS, RACHEL ERRTHUM, THOMAS A. COLBY, TAYLOR CARRASCO, ERIK GORDON, TIM PELBATH

Department of Geological Sciences
San Diego State University, San Diego, California 92182

ABSTRACT

BETWEEN INDIAN PASS AND PICACHO STATE Recreation Area, exposures of an approximately 1.7 km thick lower Miocene volcanic and epiclastic unit are wrapped about the Chocolate Mountains anticlinorium. Lying unconformably beneath the lower Miocene volcanic and epiclastic rocks and forming the core of the anticlinorium is a regionally extensive metamorphic basement complex. Rocks of the lower Miocene volcanic and epiclastic unit are variably altered by hydrothermal metasomatism, and as a result, they cannot be classified using the total-alkali and silica classification scheme recommended by the IUGS. Hence, we used the Zr/Ti versus Nb/Y magma-series discrimination diagram as it utilizes elements that are commonly immobile.

Given that the rocks in the lower Miocene volcanic and epiclastic unit are metasomatically altered, any proposed stratigraphic terminology should not utilize an igneous-rock classification term such as dacite, rhyodacite, and rhyolite. Following this guideline we erected a working lithostratigraphic framework for the lower Miocene volcanic and epiclastic unit that includes from lowest to highest: (1) the 23.4 ± 0.4 Ma Quechan volcanics, (2) the Marcus Wash unit and its apparent stratigraphic equivalent the Rojo Grande unit, (3) the White Wash unit, (4) the 23.5 ± 1.0 Ma Taylor volcanics, and (5) the 23.2 ± 0.2 Ma ignimbrite of Ferguson Wash. Published regional paleotectonic patterns and the geochemical character of samples analyzed from the lower Miocene volcanic and epiclastic unit suggest eruption within a continental-margin island arc setting.

The Neogene Bear Canyon conglomerate unconformably overlies the lower Miocene volcanic and epiclastic unit. It can be subdivided into three unconformably bounded sequences that are designated from oldest to youngest I, II, and III. Though the exact timing of formation of the three sequences in the Bear Canyon conglomerate is difficult to constrain, stratigraphic and our new $^{40}Ar/^{39}Ar$ geochronologic data when viewed in the context of regional tectonic patterns suggest that sequence I probably was deposited after ~17 Ma and prior to ~ 9.45 Ma. In contrast, sequence II was folded prior to or after ~9.45 Ma while sequence III likely was folded after ~9.45 Ma and prior to deposition of flat-lying Quaternary alluvial sediments.

The east west trending Copper Basin reverse fault cuts Sequence II. In addition, key dextral components of a conjugate set of NW and NE trending dextral and sinistral strike-slip faults, respectively, transect the Copper Basin fault, the axis of the Chocolate Mountains anticlinorium, and the ~9.45 ± 0.27 Ma basalts of Black Mountain. However, neither the Copper Basin nor the conjugate faults disrupt flat-lying Quaternary alluvial sediments.

The episodic growth of the EW trending Chocolate Mountains anticlinorium during deposition of the Neogene Bear Canyon conglomerate, and subsequent formation of the EW trending Copper Basin fault and the conjugate set of dextral and sinistral faults all record transient NS shortening within the margin of the North America plate. These transient periods of NS shortening followed extensional deformation of early Miocene continental arc volcanic rocks, and are strain compatible with the 12–10 Ma to 6 Ma development of the Eastern California Shear Zone.

Numerous other examples of late Cenozoic EW trending folds and reverse faults are reported in the literature to have formed during development of this 100–110 km wide zone of dextral shear. We therefore conclude that the geology between Indian Pass and Picacho State Recreation Area reflects changing continental margin strain patterns as the boundary between the Farallon and North American plates was transformed from a Paleogene convergent margin to a wholly Neogene transform setting.

INTRODUCTION

Lying within southeastern California, near the boundary with Arizona, are wide spread exposures of the upper Cretaceous–lower Tertiary Orocopia Schist. In map view, the Orocopia Schist commonly forms discontinuous elongated dome-like structures that mark the trace of the Chocolate Mountains anticlinorium, a widely recognized feature that extends from the San Andreas fault on the west eastward into south-central Arizona (Fig. 1). Between Indian Pass and Picacho State Recreation Area exposures of an ~1.7 km thick lower Miocene volcanic and epiclastic unit are wrapped about the anticlinorium (Figs. 1, 2) (Crowe, 1973, 1978; Girty et al., 2006; Ricketts, 2010; Sainsbury, 2010; Sutton, 2010, Muela, 2011; Ricketts et al., 2011). Though Crowe (1973, 1978) recognized that the lower Miocene volcanic and epiclastic unit was tilted, he did not generally appreciate the complex structural and tectonic history of the area in and around Picacho State Recreation Area, and its potential effects on the lithostratigraphy of the region. For example, Girty et al. (2006), Ricketts (2010), Sainsbury (2010), Sutton (2010), Muela (2011), and Ricketts et al. (2011) showed that the lower Miocene volcanic and epiclastic unit was involved in two phases of early Miocene extensional faulting, and then was folded about the Chocolate Mountains anticlinorium sometime after about 23 Ma. Though the exact timing of the post-23 Ma history of the anticlinorium is difficult to constrain, between ~17 Ma and the deposition

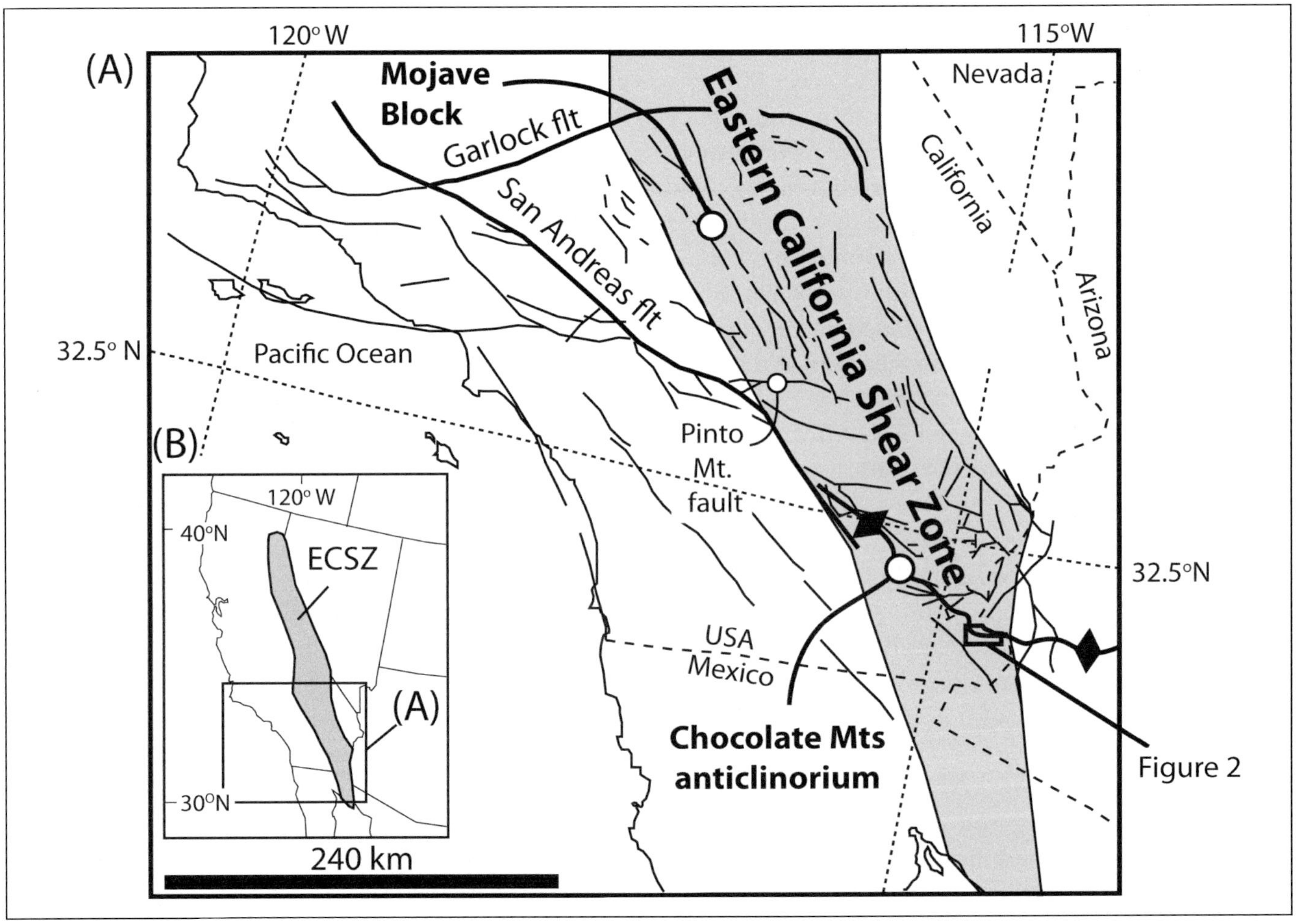

Figure 1. *(A) Map of western United States highlighting location of the Eastern California Shear Zone. (B) Close-up of southern California and Arizona showing location of the region between Indian Pass and Picacho State Recreation Area as Figure 2.*

of undeformed Quaternary sediments, the anticlinorium, as attested to by the three unconformably bounded units making up the ~1.8 km thick Bear Canyon conglomerate, grew episodically (Ricketts et al., 2011). Moreover, the Copper Basin reverse fault and a set of conjugate NW and NE trending dextral and sinistral faults, respectively, transect the anticlinorium (Sainsbury, 2010; Muela, 2011). As discussed later in this paper, these structures are constrained to have formed sometime after ~9.45 Ma and prior to the deposition of flat lying Quaternary gravels (Muela, 2011; Rickets et al., 2011).

Ricketts et al. (2011) argue that the above complex structural history reflects the changing configuration of the Paleogene to Neogene plate boundary between the North American, Farallon, and Pacific plates. In such a scenario, about 28–26 Ma, the area between Indian Pass and Picacho was located within the margin of the North American plate several hundred kilometers eastward of the point of first contact of the Pacific and North American plates (e.g., Atwater, 1989; Dickinson, 2002). As the plate boundary continued to evolve, a slab-window formed and enlarged beneath the North America plate as the Mendocino and Riviera triple junctions migrated northwestward and southeastward transforming the converging margin into a transform setting (e.g., Atwater, 1989; Dickinson, 2002). Though Atwater (1989) argued that the slab-window may have had a significant influence on volcanic activity, Dickinson (2002) suggested that such activity was restricted to the immediately adjacent plate boundary, and according to Glazner (2008) there is no geochemical evidence supporting the idea that the slab-window had any significant effect on Cenozoic volcanic patterns in the western Cordillera. In short, early Miocene volcanism in the region between Indian Pass and Picacho State Recreation Area occurred within a continental-margin island arc located inboard (i.e., continent ward) and away from the influences of the slab-window (Olson, 2010).

Below we first present a general overview of the geology between Indian Pass and Picacho State Recreation Area focusing much of our discussion on how our mapping,

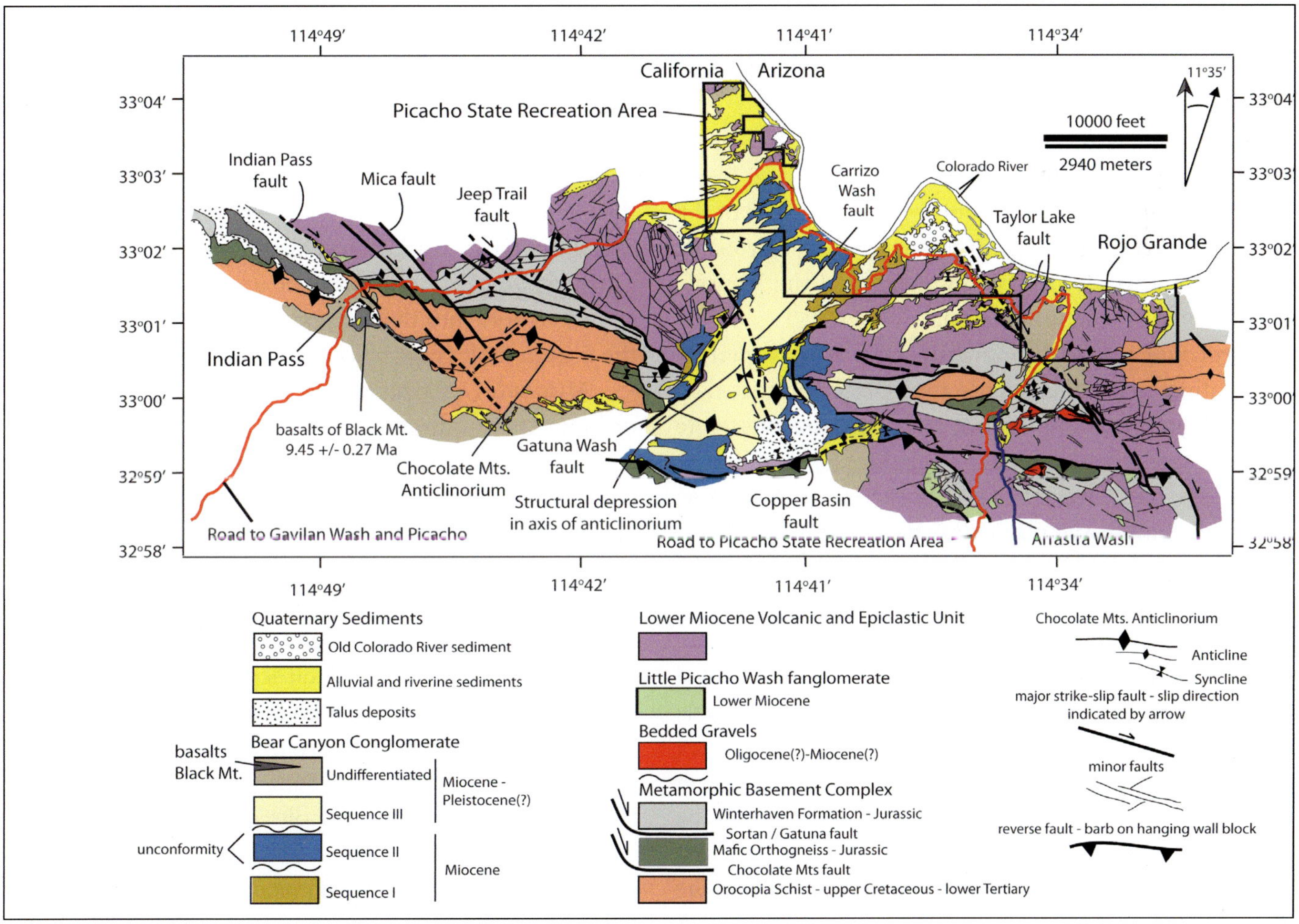

Figure 2. *Simplified geology map of the region between Indian Pass and Picacho State Recreation Area. Key geographical features and structures are labeled.*

petrology, and geochemical work relates to published accounts of the lithostratigraphy of the lower Miocene volcanic and epiclastic unit. In this early part of our paper, we integrate the results of 99 new whole rock XRF and ICP-MS analyses of samples collected in and around Picacho State Recreation Area with the results of geochemical data provided in an earlier paper (Girty et al., 2006). A detailed description of our analytical methods and a tabulation of our new geochemical data are provided in an accompanying appendix. We then describe the major components of the structural fabric between Indian Pass and Picacho State Recreation Area, and end our paper with a summary of the tectonic implications of our work.

GENERAL GEOLOGY

Major geologic units at Picacho include a Mesozoic to Paleogene metamorphic basement complex, Oligocene (?)–Miocene (?) bedded gravels, the lower Miocene Little Picacho Wash fanglomerate, the lower Miocene volcanic and epiclastic unit that is the focus of this report, and the Neogene Bear Canyon conglomerate (Figs. 2, 3). The most salient aspects of each of these units are provided in the following sections.

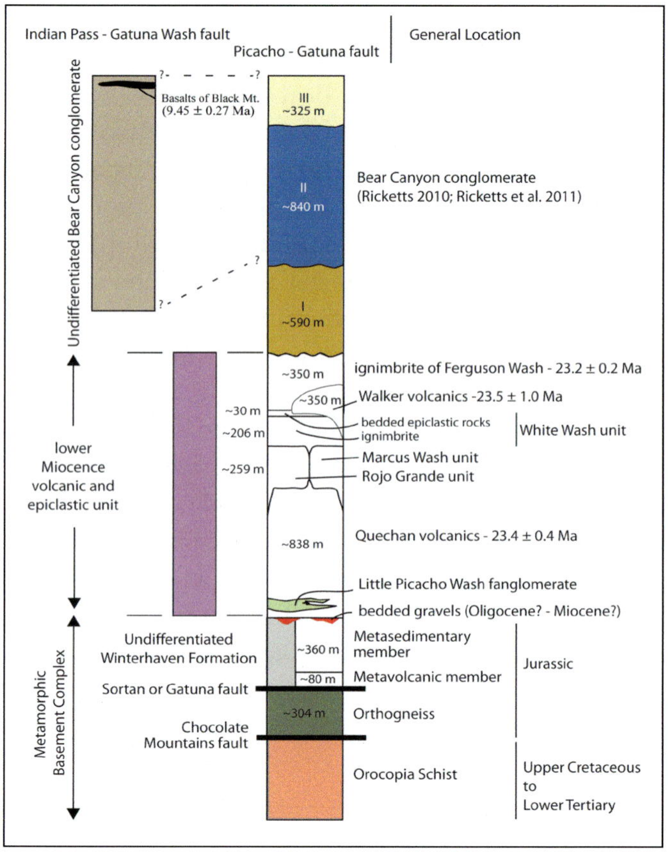

Figure 3. *Working stratigraphy of the region between Indian Pass and Picacho State Recreation Area. Color-code follows that shown in Figure 2.*

Metamorphic Basement Complex

The metamorphic basement complex consists of the Upper Cretaceous–Paleogene Orocopia Schist, a probable Jurassic mafic mylonitic orthogneiss, and the Jurassic Winterhaven Formation (Fig. 2).

Orocopia Schist. The Orocopia Schist is a quartzofeldspathic, amphibolite-grade assemblage dominated by quartz, albite, oligoclase, biotite, and phengitic muscovite (Haxel, 1977; Jacobson et al., 2002; Stephenson, 2006; Stephenson and Girty, 2006). Thin section studies show that locally biotite and plagioclase have been partially replaced by chlorite and epidote, respectively. Separating the Orocopia Schist from the overlying mylonitic orthogneiss is the Chocolate Mountains fault (Figs. 2, 3). Originally a thrust, the Chocolate Mountains fault was later reactivated as an extensional detachment (Dillon, 1976; Haxel, 1977; Dillon et al., 1990; Simpson, 1990; Jacobson et al., 2002).

$^{40}Ar/^{39}Ar$ thermochronologic studies by Jacobson et al. (2002, 2007) indicate two periods of cooling within the Orocopia Schist; ~52–50 Ma and ~28–~24 Ma. The first cooling period is associated with normal slip along the Chocolate Mountains fault, and the second with slip on the structurally overlying Gatuna fault. Movement along the Gatuna fault may have continued through ~23 Ma as Sainsbury (2010) and Losh et al. (2005) show that the ~23 Ma lower volcanic and epiclastic unit is cut by faults that appear to sole into the Gatuna or its eastern correlative, the Sortan fault (Figs. 2, 3).

Mylonitic Orthogneiss. The probable Jurassic mylonitic orthogneiss (minimum age ~144 Ma, unpublished zircon age; Haxel et al., 1985; Haxel et al., 2002; Haxel personal communication to G. H. Girty, 2003), is metamorphosed to middle-upper amphibolite facies (Oyarzabal et al., 1997). The orthogneiss is characterized by alternating bands of coarse-grained felsic and mafic materials and has a structural thickness of ~304 m (~1000 feet) (Sutton, 2010; Muela, 2011). At Indian Pass, the Gatuna fault separates the orthogneiss from the overlying Winterhaven Formation, and at Picacho State Recreation Area, the Sortan fault separates the same units (Figs. 2, 3). Because of their similar structural position, the Sortan and Gatuna faults are likely the same fault.

Winterhaven Formation. The probable Jurassic Winterhaven Formation has been metamorphosed to the chlorite-grade greenschist facies and can be divided into a lower metavolcanic member and an upper metasedimentary member (Haxel et al., 1985). The lower metavolcanic member is approximately 80 m (262 feet) thick (Fig. 3). At Picacho, it is characterized by highly fractured and jointed dark gray to black aphyric and plagioclase-phyric flows (Haxel et al., 1985; Marsh, 2003; Gordon, 2011). The overlying metasedimentary unit is approximately 360 m (1181 feet) thick (Fig. 3). It consists of metamorphosed quartz arenite, foliated argillite, and stretched-pebble conglomerate (Haxel et al., 1985). Quartz, albite, white mica, chlorite, epidote, and calcite represent a common metamorphic assemblage (Marsh, 2003; Girty et al., 2006; Gordon, 2011).

The results of our analysis of 20 samples collected from the metavolcanic member of the Winterhaven Formation are plotted on the Zr/Ti versus Nb/Y magma-series discrimination diagram of Winchester and Floyd (1977), as modified by Pearce (1996) in Figure 4a. Note that most samples plot in the basalt and andesite + basaltic andesite fields. REE-data are shown in Figure 4b. The light rare earth element (LREE) enriched character and the small to no Eu-anomaly are general characteristics of continental-margin andesitic lavas (Taylor and McLennan, 1985). Thus, our geochemical data are consistent with the paleotectonic and paleogeographic model of Busby-Spera (1988) which portrays the Jurassic continental-margin arc system of the SW Cordillera as being dominated by an arc-graben depression. In such a model, the metavolcanic member of the Winterhaven Formation would form part of the volcanic floor or wall of the depression while the metasedimentary member would represent sediments shed into the graben system (Marsh, 2003; Gordon, 2011).

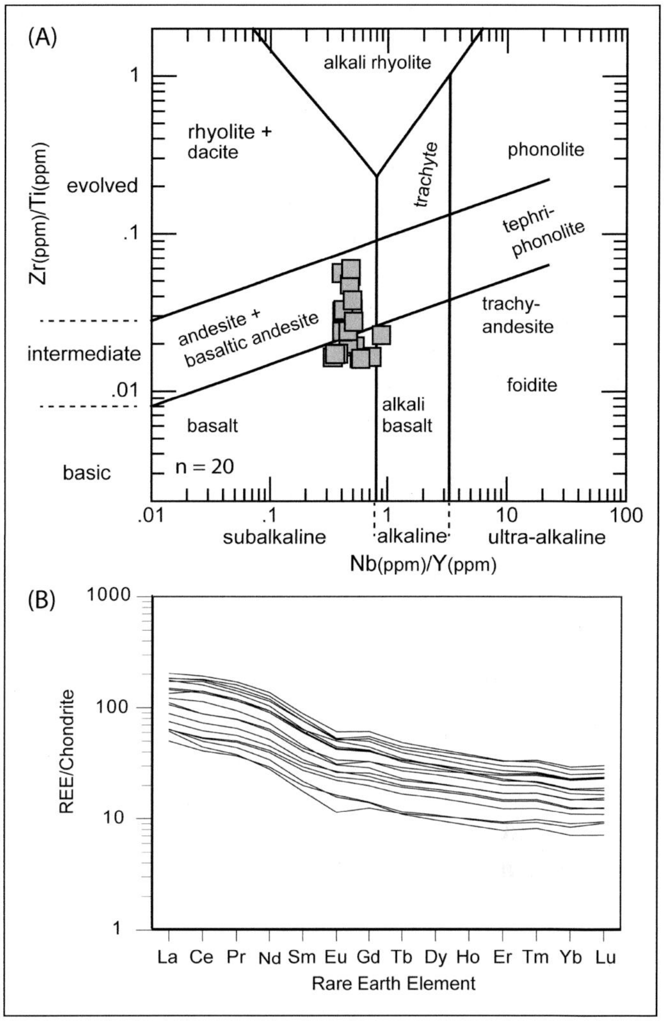

Figure 4. *(A) Chemical data derived from the metavolcanic member of the Winterhaven Formation plotted on the Zr/Ti versus Nb/Y magma-series discrimination diagram of Winchester and Floyd (1977), as revised by Pearce (1996). (B) Chondrite-normalized REE distributions. Chondrite values are those listed in Taylor and McLennan (1985).*

Bedded Gravels and the Little Picacho Wash Fanglomerate

Locally unconformably overlying the metamorphic basement complex and lying below the lower Miocene Quechan volcanics is a discontinuous section of unmetamorphosed Oligocene (?)–Miocene (?) bedded gravels (Figs. 2, 3). In areas where the bedded gravels are absent, the Quechan volcanics lie unconformably above the Winterhaven Formation. Locally, in the lower portion of the bedded gravels are 1 to 10 cm thick beds of white, tuffaceous coarse-grained sandstone composed of poorly sorted and angular quartz, feldspar and volcanic clasts. Calcarenite beds are also present in the lower portion, and greenstone (metamorphosed mafic volcanics), granite, and tonalite boulders are prominent features in conglomeratic beds (Hartung, 2005, Campbell, 2006).

The Little Picacho Wash fanglomerate is as thick as ~35 m (114 feet) and lies stratigraphically below the Quechan volcanics and above the Winterhaven Formation (Figs. 2, 3). We have yet to observe sections that contain both the bedded gravels described above and the Little Picacho Wash fanglomerates. However, locally the fanglomerate is interstratified with the lower part of the Quechan volcanics, and thus, as shown in Figures 2 and 3, may lie stratigraphically above the bedded gravels. The Little Picacho Wash fanglomerate is composed mostly of interstratified 0.5 m to 3 m thick beds consisting of normal and reversely graded gravel clasts derived mostly from the underlying Winterhaven Formation. However, boulders of mylonitic gneiss and leucogranite locally are present.

Lower Miocene Volcanic and Epiclastic Unit

Though much work remains to be done, our ongoing thin section studies indicate that rocks of the lower

Miocene volcanic and epiclastic unit are variably altered by hydrothermal metasomatism. For example, amygdules and cracks are commonly lined with quartz or chalcedony while veins and veinlets of quartz or calcite also are relatively common (e.g., Pelbath, 2011; Muela, 2011). Moreover, the centers of some quartz-lined amygdules were later infilled by calcite, and the microcrystalline groundmass in many porphyritic samples is variably altered to quartz ± calcite ± chlorite ± white mica. This alteration assemblage also affects feldspar and mafic phenocrysts to variable degrees, as well as the microcrystalline matrix of pyroclastic rocks containing ghosts of shards and fragments of pumice.

The above observations clearly indicate that the classification of rocks within the lower Miocene volcanic and epiclastic unit must be done with care, and that the total-alkali and silica classification scheme recommended by the International Union of Geological Sciences (IUGS) (e.g., Le Bas et al., 1986), or for that matter, any alternative scheme that utilizes elements that were involved in the metasomatic events producing the above alteration products, should not be used. Hence, in the following sections we will use the Zr/Ti versus Nb/Y magma-series discrimination diagram of Winchester and Floyd (1977), as modified by Pearce (1996), to gain some insight into original igneous compositions.

Given that the rocks in the lower Miocene volcanic and epiclastic unit are metasomatically altered, any stratigraphic terminology that we use should not utilize an igneous-rock classification term such as dacite, rhyodacite, and rhyolite, as such terminology implies a degree of certainty that simply does not exist for the group of rocks under discussion here. We therefore recommend that such terms be dropped, and replaced by more general field terms such as volcanics or unit. In following such a procedure, the Quechan volcanics of Crowe (1978), and the Walker volcanics of Ricketts (2010) and Sainsbury (2010) remain unchanged while the White Wash trachyte or rhyolite, the Marcus Wash trachyte or rhyolite, and the Rojo Grande trachyte or rhyolite (Girty et al., 2006; Ricketts, 2010) all become simply the White Wash, Marcus Wash, and Rojo Grande units respectively (Fig. 3). Finally, for historical reasons and to pay homage to the early work of Crowe (1973, 1978) the name of the ignimbrite of Ferguson Wash remains unchanged.

Quechan Volcanics. The Quechan volcanics have been dated by the laser-ablation U-Pb zircon technique to be 23.4 ± 0.4 Ma (Needy et al., 2007). In the field, the Quechan volcanics consist of a series of lava flows, flow breccias, and lesser amounts of laharic breccia. On weathered surfaces, colors typically range from grayish red (5R 4/2) to dark greenish gray (5GY 4/1). Moderate brown (5YR 3/4) iron-stained liesegang bands also are a characteristic feature of this unit.

Though Crowe (1973, 1978) suggested that the Quechan volcanics were about 300 m (~984 feet) thick, our fieldwork and carefully constructed cross-sections suggest a significantly greater thickness (e.g., Muela, 2011). For example, based on work by Muela (2011), the Quechan volcanics must be at least 838 m (2750 feet) thick.

Within the Quechan, individual flow thicknesses are variable, but typically range from ~3 m to ~10 m in the lower and upper parts to ~10 m to over ~20 m in the middle part (Olson, 2010). In general, thinner flows characteristic of the lower and upper parts of the Quechan weather more readily than do the thicker flows characteristic of the middle part. As a result of this difference in weathering characteristic, the thicker flows of the middle part tend to occupy higher topographic areas than do the lower and upper parts. Moreover, flows in the lower and upper parts are typically weakly phyric with augite, olivine, and/or plagioclase phenocrysts while thick plagioclase-phyric flows dominate the middle part. Large plagioclase phenocrysts reaching ~0.1 mm to ~0.6 mm in size are characteristic of flows in the middle part. In contrast, olivine, augite, and plagioclase phenocrysts in the lower and upper parts generally range from ~0.1 mm to ~0.3 mm in size (Olson, 2010). A final important difference between the upper and lower Quechan volcanics is that the former commonly includes lahar and volcanic breccia while similar features are uncommon in the lower part.

On the Zr/Ti versus Nb/Y magma-series discrimination diagram of Winchester and Floyd (1977), as revised by Pearce (1996), samples from the Quechan volcanics plot primarily within the alkali basalt field (Fig. 5a) making them the most mafic and primitive lavas in the study area. Chondrite-normalized REE patterns show LREE enrichment (La-Sm) with respect to the HREE (Gd-Lu) (Figs. 6a-6c), and resemble patterns commonly derived from ocean island basalts, a characteristic that is consistent with the results derived from the magma-series discrimination diagram (Olson, 2010).

As noted by Olson (2010), all three members of the Quechan volcanics display parallel steeply inclined LREE enriched patterns (Fig. 6a-6c) that vary little with Mg# and Zr/TiO_2 values. These characteristics suggest that the chondrite-normalized patterns are not influenced by variations in the degree of crystal fractionation. Instead, the steepness of these trends is consistent with small degrees of partial melting within the garnet stability field (Winter, 2010).

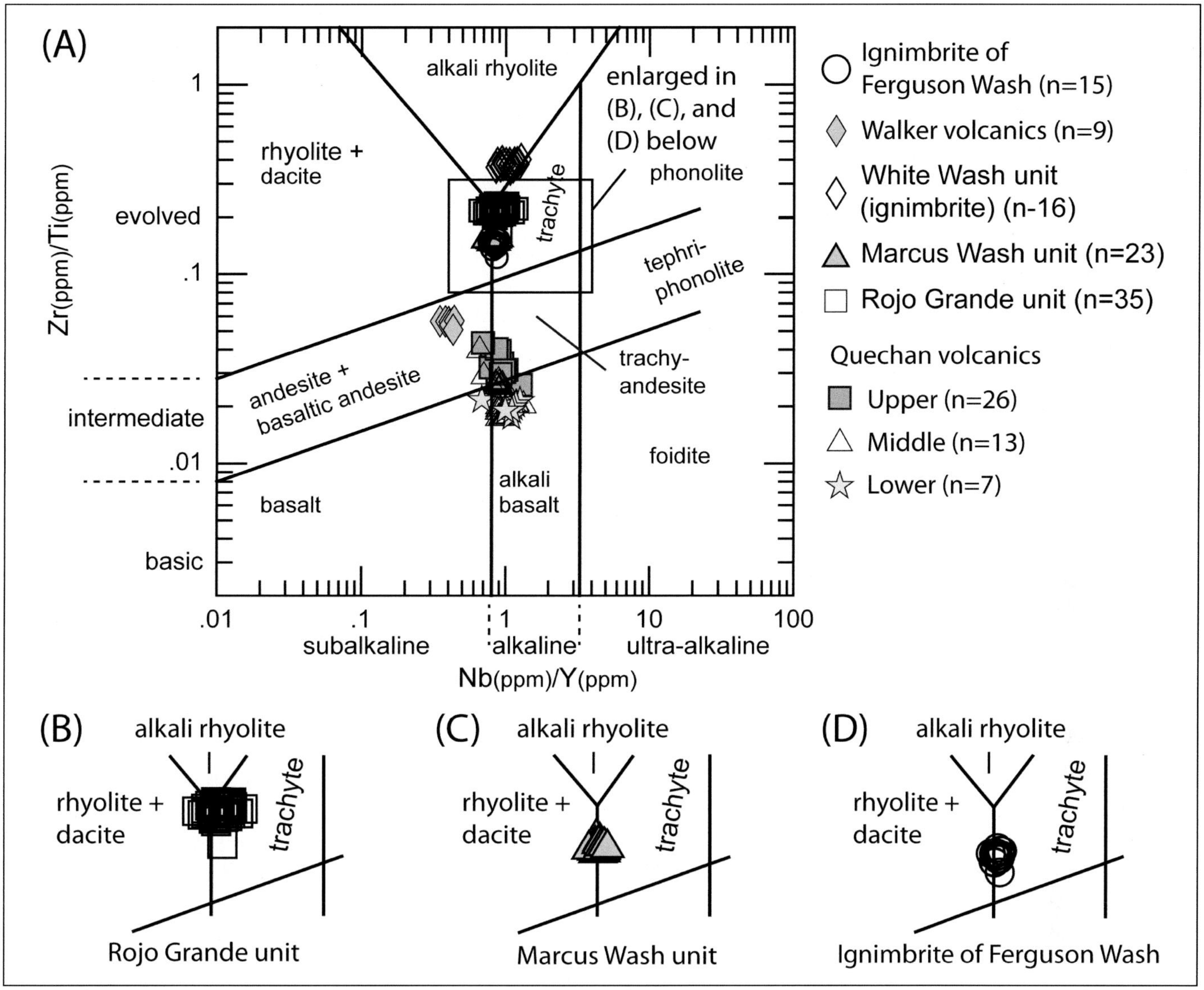

Figure 5. *(A) Chemical data derived from the Quechan volcanics, Rojo Grande unit, Marcus Wash unit, White Wash unit, Walker volcanics, and the ignimbrite of Ferguson Wash plotted on the Zr/Ti versus Nb/Y magma-series discrimination diagram of Winchester and Floyd (1977), as revised by Pearce (1996). (B) Enlargement of (A) showing distribution of samples from the Marcus Wash unit. (C) Enlargement of (A) showing distribution of samples from the Rojo Grande unit. (D) Enlargement of (A) showing distribution of samples from the ignimbrite of Ferguson Wash. Samples from the Marcus Wash unit are from Lovering (2005) and Smith (2005). Some samples from the Rojo Grande unit are from Campbell (2004). Samples from the ignimbrite of Ferguson Wash are from Biggs (2006). Samples from the Walker volcanics are from Nielson (2004) and Biggs (2008).*

In addition, the LREE's are also susceptible to transfer in aqueous solutions, and, as a result, their steepness may be the result of derivation from a metasomatised mantle.

According to Coney and Harms (1984), crustal thicknesses within the western Cordillera following the late Cretaceous to Paleogene Laramide orogeny were about 30–40 km in the general area of Picacho State Recreation Area. In general, the top of the garnet stability field lies at about ~60–80 km depth (e.g., McKenzie and O'Nions, 1991; Su, 2008; Kay et al., 2009; Winter, 2010). Hence, the initial magma or magmas that ultimately fed the Quechan volcanics may have originated at these or greater depths, and if the continental thicknesses provided in Coney and Harms (1984) are reasonable, then the initial melts were developed within the sub-continental lithospheric mantle of the North American plate.

Marcus Wash and Rojo Grande Units. Overlying the Quechan volcanics are the Marcus Wash and Rojo Grande volcanic units (Fig. 3). The former unit is well-exposed south of Picacho State Recreation Area and east of the Taylor Lake fault (Fig. 2). In contrast, as its name implies, the Rojo Grande unit crops out in and about Rojo Grande, a prominent red hill lying in the SE part of Picacho State Recreation Area (Fig. 2).

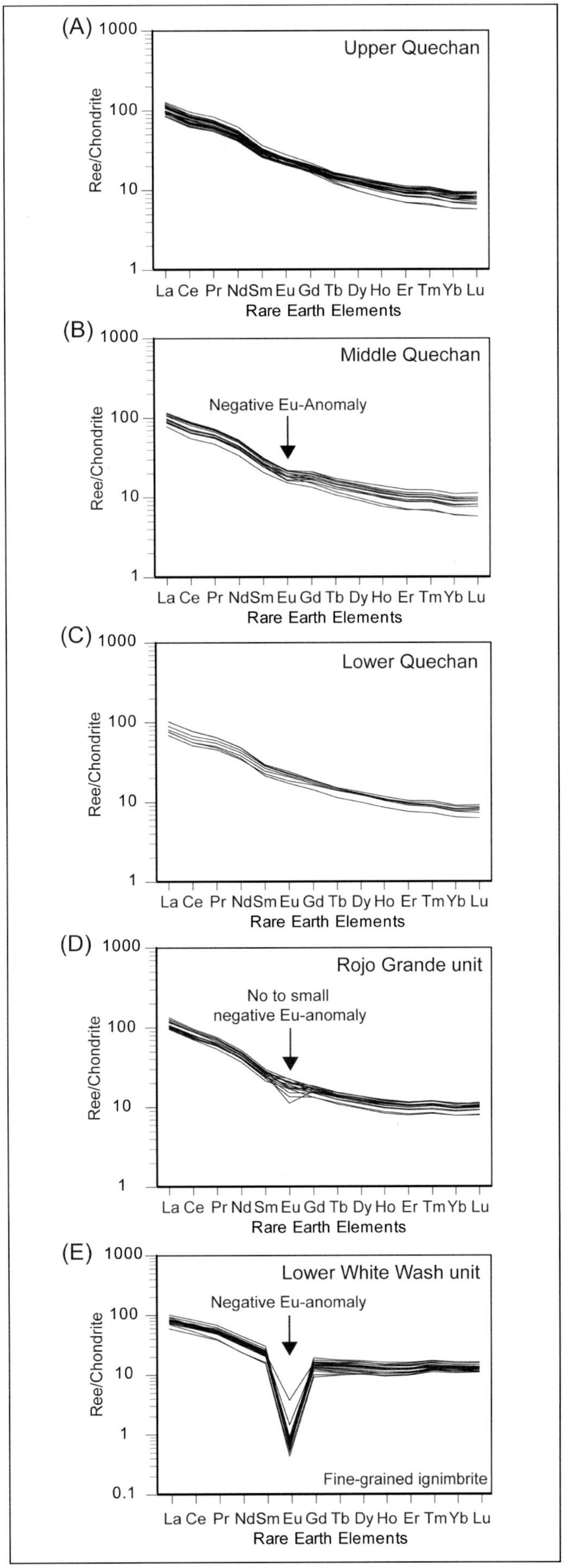

Figure 6. *Chondrite-normalized data derived from (A) the upper Quechan volcanics, (B) the middle Quechan volcanics, (C) the lower Quechan volcanics, (D) the Rojo Grande unit, and (E) the white to tan find-grained ignimbrite making up the lower portion of the White Wash unit.*

In and around Rojo Grande (Fig. 2), the Rojo Grande unit lies stratigraphically between the Quechan volcanics and the White Wash unit (Fig. 3). Similarly, south of the Picacho State Recreation Area, the Marcus Wash unit lies in an identical stratigraphic position. At Rojo Grande, work by Muela (2011) suggests that the Rojo Grande unit is ~259 m (850 feet) thick (Fig. 3). Based on field relationships projected into cross sections, we estimated a similar thickness for the Marcus Wash unit.

According to Lovering (2005), the Marcus Wash unit contains about 3% to 14% phenocrysts while the matrix is a variably altered microcrystalline material. Our field and thin section estimates of these features in the Rojo Grande unit are similar (Pelbath, 2011). Moreover, both the Marcus Wash and Rojo Grande units range in color from pale red purple (5RP 6/2) to very dusky red purple (5RP 2/2), and both contain prominent chalky white feldspar phenocrysts, and embayed to doubly terminated quartz phenocrysts. In thin section, feldspar phenocrysts are euhedral to subhedral, and display either polysynthetic or Carlsbad twins. Relict amphibole phenocrysts are uncommon. South of the Picacho State Recreation Area, pink, meter-thick, spherulitic lava locally makes up the uppermost portion of the Marcus Wash unit.

The most striking field difference between the Marcus Wash and Rojo Grande units are the characteristics of the flow banding. In the Rojo Grande unit, flow banding is characterized by alternating deep purple to light purple banding while in the Marcus Wash unit flow banding is lighter in color and is paralleled by the long dimension of small cracks lined by quartz (Campbell, 2004; Lovering, 2005; Smith, 2005; Pelbath, 2011). Additionally, hydrothermal solutions have produced large meter-scale white and/or red irregular-shaped bleached regions throughout the Rojo Grande unit. Such irregular-shaped bleached areas are not present in the Marcus Wash unit. In addition, on the Zr/Ti versus Nb/Y magma-series discrimination diagram, samples from the Marcus Wash unit cluster tightly and lie about the boundary separating the trachyte and rhyolite + dacite fields while data from the Rojo Grande unit cluster primarily just inside the trachyte field (Fig. 5b, 5c). On chondrite-normalized REE diagrams (Fig. 6d), samples from the Rojo Grande unit are LREE enriched with relatively flat HREE distributions that cluster around

~10× chondrite values. A small negative Eu-anomaly is evident in some samples from the Rojo Grande unit.

The lack of any distinctive and recognizable stratigraphic horizons within either the Marcus Wash or Rojo Grande units, suggests that they are largely the result of a single outpouring of probably highly viscous relatively high silica lava. Though the internal flow geometry of both units has not been studied in detail, we tentatively interpret both to represent extrusive dome complexes.

White Wash Unit. Previous investigators mapped the White Wash unit as a pyroclastic rhyolite or trachyte that intruded or was deposited above the Rojo Grande unit and above an interval of bedded pyroclastic and epiclastic rocks (Crowe, 1973, 1978; Girty et al., 2006). However, detailed mapping by Muela (2011) in and around Rojo Grande indicates that the White Wash unit is a fine-grained ignimbrite that in fact makes up the lower portion of the interval of bedded pyroclastic and epiclastic rocks that overlies the Rojo Grande unit; hence, both the fine-grained ignimbrite and the bedded pyroclastic and epiclastic interval are herein included within the White Wash unit (Fig. 3).

At Rojo Grande, a thin 1-m to 2-m-thick discontinuous section of epidotized bedded epiclastic rocks separates the White Wash unit from the underlying Rojo Grande unit. Above this thin discontinuous epiclastic unit is about ~206 m (~675 feet) of tan to white fine-grained ignimbrite (Fig. 3). In thin section, ghosts of shards and pumice are evident in the ignimbrite. In both thin section and outcrop, fiamme are locally prevalent, and thus suggest at least weak to moderate welding (Muela, 2011).

Along the north side of Rojo Grande, the tan to white fine-grained ignimbrite is overlain by an ~30 m (~100 feet) section of pebble to boulder bedded epiclastic rocks (Fig. 3). South of Picacho State Recreation Area exposures of the tan to white ignimbrite are too poorly preserved and sparse to allow separation of the pyroclastic from the epiclastic section. Hence, in this area these units were mapped by Muela (2011) as undifferentiated.

On the Zr/Ti versus Nb/Y magma-series discrimination diagram, 16 samples from the tan to white fine-grained ignimbrite forming the lower part of the White Wash unit, plot in the alkali rhyolite field (Fig. 5a). On chondrite-normalized REE diagrams (Fig. 6e), samples of the tan to white fine-grained ignimbrite display large negative Eu anomalies, enriched LREE patterns, and flat HREE patterns at ~10X chondrite values. These data suggest that feldspar either (1) was left as a residue or (2) fractionated from the parental melt that eventually erupted to yield the fine-grained ignimbrite.

Walker Volcanics. At Picacho State Recreation Area, Girty et al. (2006) mapped a WNW trending dome complex located near Walker Mountain that intruded the White Wash unit. Biggs (2008) provided a laser ablation U-Pb zircon age of 23.5 ± 1.0 Ma for the dome complex and along with Nielson (2004) provided the geochemical data plotted in Figure 5a for the Walker volcanics. Rocks making up the extrusive dome complex include flow banded feldspar porphyry. Vitrophyre occurs along the margins and as large irregular shaped masses within sub-dome interiors (Biggs, 2008).

West of Picacho State Recreation Area, Sainsbury (2010) mapped a compositionally similar suite of volcanic rocks that also include ignimbrite, lahar, and flow breccia. This group of rocks also intrudes or overlies the White Wash unit. As shown in Figure 5a, all of the samples that we have collected and analyzed from the Walker volcanics at Picacho State Recreation Area plot in the andesite + basaltic andesite compositional field.

Ignimbrite of Ferguson Wash. The ignimbrite of Ferguson Wash is a multiple-flow, single-cooling unit with zonal variations typical of pyroclastic sheet flows having a thickness in excess of 350 m (~1,148 feet) (Crowe, 1973, 1978; Biggs, 2006). Laser ablation U-Pb zircon data suggest an eruption date of about 23.2 ± 0.2 Ma (Needy et al., 2007). Texturally, the unit is a non-welded to welded vitric lapilli tuff with common fiamme (Biggs, 2006). The welded portion commonly forms resistant ridges throughout the Indian Pass to Picacho region (Sainsbury, 2010). The matrix of the ignimbrite ranges in color from light brownish gray (5YR 6/1) to grayish orange pink (5YR 7/2). In hand sample, small crystals of biotite and plagioclase are evident, as are small pumice and devitrified glass fragments (Biggs, 2006). On the Zr/Ti versus Nb/Y magma-series discrimination diagram, samples from the ignimbrite of Ferguson Wash plot along the boundary separating rhyolite + dacite from trachyte (Fig. 5d).

Bear Canyon Conglomerate

Between Indian Pass and Picacho State Recreation Area, the axis of the doubly plunging Chocolate Mountains anticlinorium forms a structural depression (Fig. 2). Within the depression, the ~1.76 km thick Bear Canyon conglomerate has been studied extensively by Ricketts (2010) and Ricketts et al. (2011). It can be subdivided into three unconformably bounded units referred to as sequences I, II, and III (Figs. 2, 7). Sequence I, composed of paraconglomerates and orthoconglomerates containing entirely volcanic clasts, has a total exposed thickness of

~590 m (Ricketts, 2010). In sequence I, paraconglomerates are medium to thickly bedded and poorly sorted. They contain sub-angular clasts ranging in size from pebbles to boulders, and commonly are inversely graded. In contrast, orthoconglomerates are thinly bedded, contain clasts no larger than pebbles, and are massive or display normal grading. The interface between adjacent beds is typically planar, except along the tops of paraconglomerates, where gravel clasts protrude into overlying beds.

Based on the above characteristics, Ricketts et al. (2011) interpreted paraconglomerates and orthoconglomerates in sequence I to be derived from debris flows and sheet floods respectively.

Sequence II has a maximum exposed thickness of ~840 m (Ricketts, 2010). It overlies sequence I along an angular unconformity (Fig. 7) (Girty et al., 2006; Ricketts et al., 2011). The stratigraphically lowest ~200 m of sequence II contains thinly to very thickly bedded polymictic paraconglomerates and orthoconglomerates. Orthoconglomerates form tabular beds, and also make up a nested succession of narrow (~5–8 m wide and ~1–2 m deep) asymmetric alluvial channels. Clasts within lenticular beds and channels range in size from pebbles to cobbles, with boulders making up a subordinate percentage of the total clast population. The clast population in paraconglomerates is poorly sorted and sub-angular, whereas clasts in orthoconglomerates are well sorted and sub-rounded. Clasts in both types of conglomerates were derived from volcanic, conglomerate, schist, metavolcanic, metasedimentary, and metaplutonic rocks.

The sub-angular nature of clasts and their poor sorting suggested to Ricketts et al. (2011) that the polymictic paraconglomerates within the lower ~200 m of sequence II were derived from debris flows. In contrast, orthoconglomerates were interpreted to represent the deposits of channelized-stream-flow and associated overbank sheet-floods.

Overlying the lower ~200 m of sequence II is ~200 m of thinly bedded generally tabular orthoconglomeratic beds interstratified with medium to thick beds containing pebble- to cobble-sized clasts (Ricketts, 2010). Thin orthoconglomeratic beds contain a clast population that is well sorted and sub-rounded to rounded, while clasts within medium to thick beds are typically poorly sorted and sub-angular. Clasts within both types of beds are predominantly volcanic, although fragments of older conglomerate, metaplutonic rocks, and schist are evident.

Ricketts et al. (2011) interpreted the fine-grained nature of the thin tabular orthoconglomeratic beds in the ~200-m-thick section overlying the lower part of sequence II as the deposits of overbank sheet-floods. In contrast, the thicker poorly sorted beds containing sub-angular clasts were interpreted to have been derived from debris flows.

The upper part of sequence II consists of ~440 m of thickly to very thickly bedded paraconglomerates (Ricketts, 2010). Such beds contain angular clasts of schist, and typically display coarse-tail reverse grading with clasts in individual beds ranging in size, from bottom to top, from ~1–5 cm to ~50–60 cm, respectively. In some

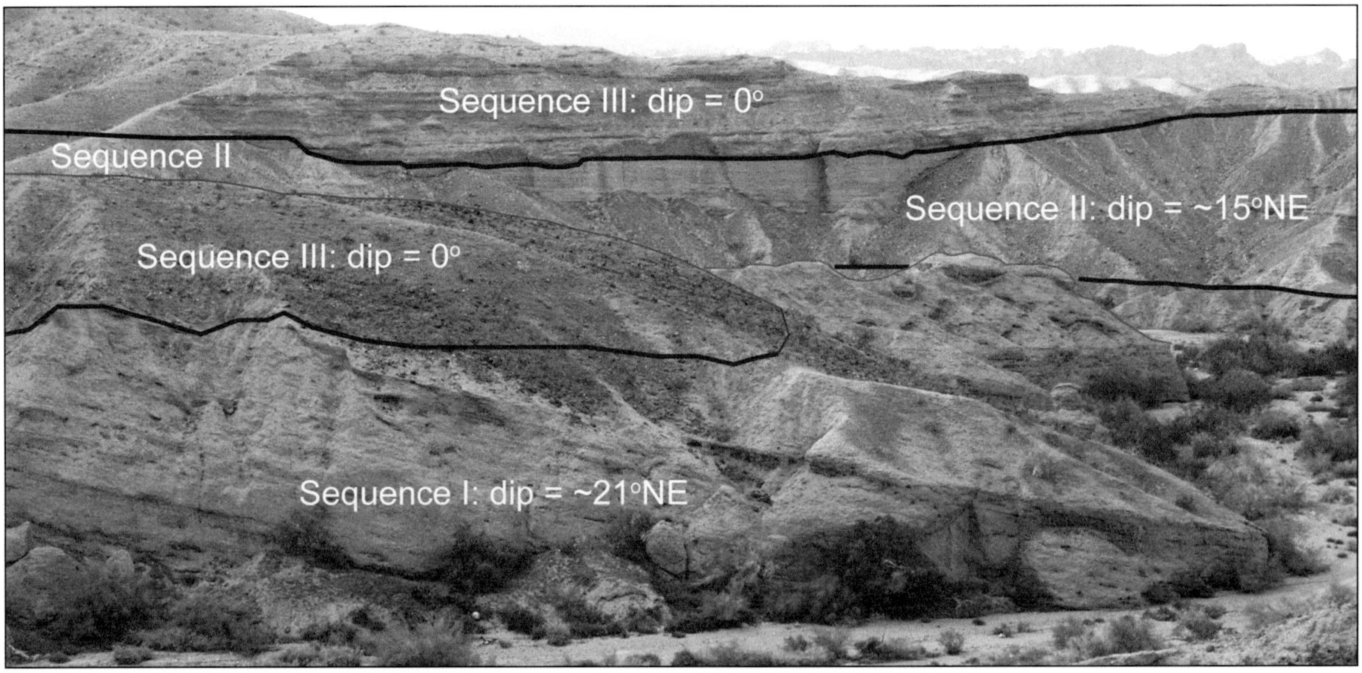

Figure 7. *Photo looking NW across Bear Canyon at the three sequences of the Bear Canyon conglomerate.*

paraconglomerates, clasts protrude into overlying layers. In contrast to beds of paraconglomerate, thickly to very thickly bedded lenticular sandy orthoconglomerates (~0.5 m to ~2.0 m in thickness) represent a minor proportion of the upper part of sequence II. They commonly contain schist fragments and pebble-sized volcanic clasts of diverse composition and texture.

The wide range in clast size and abundance of reverse grading displayed by monomictic paraconglomerate beds, and the thick to very thick beds characteristic of poorly sorted sandy orthoconglomerates in the upper part of sequence II all suggested to Ricketts et al. (2011) deposition primarily by debris flows.

Sequence III, with an exposed thickness of ~325 m, overlies both sequences I and II along an angular unconformity (Fig. 7) (Girty et al., 2006; Ricketts et al., 2011). The lower ~115 m of sequence III is composed mostly of thinly to thickly bedded, cobble to pebble, lenticular orthoconglomerate beds. In a few exposures, orthoconglomerates represent the infill of obvious channels. The clast population in orthoconglomerates is poorly sorted and sub-angular. Medium to thickly bedded paraconglomerates also are present but make up a minor proportion of the lower part of sequence III. They are characterized by a poorly sorted sub-angular clast population. Clasts in both types of conglomerates were derived from older conglomeratic, volcanic, schistose, and metaplutonic sources.

The poor sorting and sub-angular nature of clasts within beds of the lower ~115 m of sequence III suggests a debris-flow origin (Ricketts et al., 2011). In contrast, the infill of channels by orthoconglomerates implies channelized-stream flow.

Lying above the lower ~115 m of sequence III is a ~210 m succession of very thickly bedded boulder paraconglomeratic beds with a very poorly sorted clast population, ranging in size from ~5 cm to over 3 m (Ricketts, 2010). Locally interstratified with the boulder paraconglomerates are thin to medium bedded orthoconglomerate layers. Clasts in the boulder and pebbly beds were derived from dark-red older well-cemented conglomerate.

Ricketts et al. (2011) interpreted the ~210 m section making up the upper part of sequence III to have been derived primarily from debris-flows and minor intermittent overbank sheet-floods.

Based on a rigorously defined and carried out geochemical and clast-count procedure Ricketts (2010) and Ricketts et al. (2011) concluded that the sources of gravels in the Bear Canyon conglomerate were local, and that alluvial material was funneled into and down the NW to NE trending depression within the axial trace of the Chocolate Mountains anticlinorium (Fig. 2). As noted by these authors, within this setting deposition was dominated by debris flows and sheet floods. However, depression-parallel stream flow is clearly indicated by the nested channels in sequence II. In general, though deposition was episodically interrupted by periods of fold growth, these characteristics suggest deposition within middle to outer alluvial fan settings (e.g., Miall, 1996; Collinson and Thompson, 1982).

STRUCTURAL GEOLOGY

The structural fabric of the region lying between Indian Pass and Picacho State Recreation Area is dominated by the following: (1) the Chocolate Mountains fault, (2) the Sortan and Gatuna faults, (3) the doubly plunging Chocolate Mountains anticlinorium and associated EW trending folds, (4) the EW trending Copper Basin fault, and (5) a set of NW and NE trending conjugate dextral and sinistral strike-slip faults (Fig. 2). The Chocolate Mountains fault separates the Orocopia Schist from the mafic orthogneiss and $^{40}Ar/^{39}Ar$ thermochronology suggests that it last moved as a detachment fault ~52–50 Ma (Jacobson et al., 2002, 2007). The Gatuna and Sortan faults separate the Winterhaven Formation from the Orocopia Schist and the mafic orthogneiss (Figs. 2, 3). $^{40}Ar/^{39}Ar$ thermochronology suggests that the Gatuna, and by correlation, the Sortan fault moved about 28–24 Ma (Jacobson et al., 2002, 2007). However, Sainsbury (2010) has documented clear evidence that the ~23 Ma, lower Miocene volcanic and epiclastic unit is cut by a series of normal faults that appear to sole into the Gatuna fault. Hence, the structures mapped and described by Sainsbury (2010) may suggest reactivation of the Gatuna fault after ~23 Ma or possibly continuing movement during eruption and deposition of the lower Miocene volcanic and epiclastic unit. Both the Gatuna and Sortan faults along with the entire lower Miocene volcanic and epiclastic unit are folded about the Chocolate Mountains anticlinorium (Fig. 2). These relationships suggest that folding generally post-dates eruption of the lower Miocene volcanic and epiclastic unit, but does not rule out the possibility that the trace of the anticlinorium was initially a series of megamullion culminations, turtlebacks, or large-scale corrugations in the detachment surface of the Sortan and Gatuna faults. In fact, we favor the idea that following slip on the Sortan and Gatuna faults, such structures likely existed and were cored by the Orocopia Schist. However, subsequently, the anticlinorium was reactivated and tightened following deposition of sequence I, sequence II, and sequence III of the Bear Canyon conglomerate (Ricketts, 2010; Ricketts et al., 2011). Unfortunately, the timing of the deposition and folding of each of these unconformably bounded units is problematical. For example, as shown in Figure 3, the basalts of Black Mountain may be interstratified with conglomerates in sequence III. Alternatively, they may be part of sequence II. If the former interpretation is correct, then folding of sequence II occurred prior to eruption of the basalts during deposition of sequence III, ~9.45 Ma. Alternatively, if the latter interpretation is correct, then both the folding of sequences II and III occurred after ~9.45 Ma. We are continuing to work on this important timing relationship and hope to have more to report in 2012.

The Copper Basin fault trends ~EW, and dips 56° to the south. In exposures east of Arrastra Wash, it places amphibolites-grade mafic orthogneiss above Quechan volcanics, and in exposures west of the road leading into Picacho State Recreation Area it places the same gneiss above sequence II of the Bear Canyon conglomerate (Fig. 2). These relationships are consistent with the idea that the fault records significant reverse-sense slip. Hence, the parallelism between the Copper Basin fault and the trace of the Chocolate Mountains anticlinorium suggests that these two structures are related and are the record of a major NS shortening event or events. Unfortunately, because of the problematical relationship between the eruption of the basalts of Black Mountain and the deposition of sequences II and III, we again do not know the exact timing of the Copper Basin fault. However, as noted below its geometry suggests that it may have formed contemporaneously with the conjugate set of faults after ~9.45 Ma.

As shown in Figures 2 and 8, in addition to the EW-trending Chocolate Mountains anticlinorium and Copper Basin fault, a prominent set of conjugate strike-slip faults also occur between Indian Pass and Picacho State Recreation Area. The major NW components of the conjugate set of strike-slip faults include the Indian Pass, Mica, Jeep Trail, and Taylor Lake faults. Two major sinistral faults also are shown in the above illustrations, and numerous dextral and sinistral faults with too small of a slip to be shown at the scale of Figures 2 and 8 also have been mapped. Though oblique-slip is likely in many cases, the dominance of strike-slip motion is indicated by sub-horizontal slickenlines and offset fold axes.

West of Picacho State Recreation Area, the Indian Pass fault, a key dextral component of the conjugate set of faults, displaces the trace of the Chocolate Mountains anticlinorium and the ~9.45 Ma basalts of Black Mountain (Figs. 2, 8). South of the eastern boundary of Picacho State Recreation Area, the Taylor Lake fault, another key dextral component of the conjugate set of faults, displaces the Copper Basin fault which in turn transects sequence II of the Bear Canyon conglomerate. If the conjugate set of faults formed more or less contemporaneously with the Copper Basin fault, then the above data suggest that the area between Indian Pass and Picacho State Recreation Area was subjected to NS shortening after ~9.45 Ma.

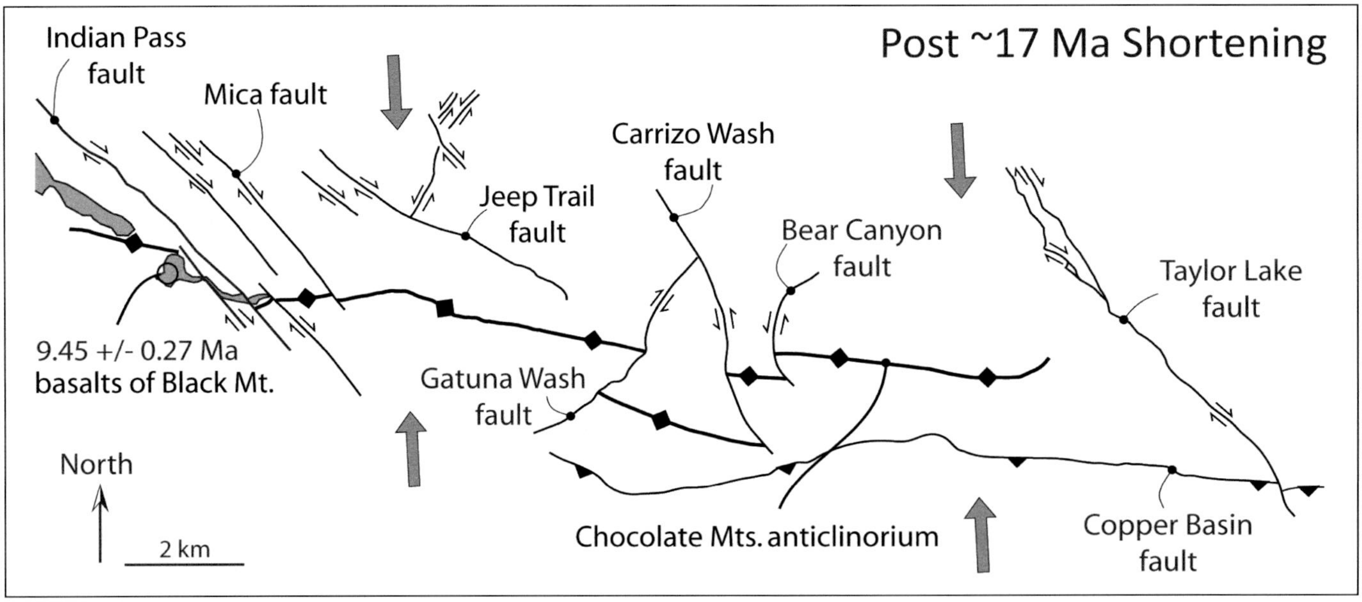

Figure 8. *Simplified map showing the distribution of major structures lying between Indian Pass and Picacho State Recreation Area. Most of the illustrated structures suggest NS shortening.*

The Gatuna Wash, Carrizo Wash, and Bear Canyon faults disrupt the above structural pattern and displace the main axis of the anticline to the SW (Fig. 8). Though difficult to interpret, we favor the idea that as the axial depression grew a space problem was created, and as a result, a large segment was transported out of the zone of high strain.

In summary, though the timing of structural forming events is complex, numerous lines of evidence converge on the idea that significant NS shortening affected the region between Indian Pass and Picacho State Recreation Area after ~9.45 Ma and prior to the deposition of flat-lying Quaternary alluvium. As discussed in Ricketts et al. (2011), the timing of the various structural forming events outlined above, was likely controlled by Tertiary changes in the plate boundary configuration between the Farallon, Pacific, and North American plates. Below we utilize our current understanding of the plate tectonic history of the SW Cordillera to gain insight into the timing and significance of the structural history of the region between Indian Pass and Picacho State Recreation Area. In providing this summary, we rely heavily on a similar discussion provided in Ricketts et al. (2011).

DISCUSSION AND CONCLUSIONS

During the Late Cretaceous to Paleogene Laramide orogeny, the Farallon plate subducted eastward beneath the North American plate along a low-angle trajectory, and as a result, continental-margin arc volcanism migrated eastward to as far as Colorado (Coney and Reynolds, 1977; Dickinson and Snyder, 1978; Atwater, 1989; Dickinson, 2002). The subducting plate dragged with it remnants of the subduction complex and possibly parts of the forearc basin (i.e., the Orocopia Schist), which underplated the North American plate (e.g., Burchfiel and Davis, 1981; Crowell, 1981; Hamilton, 1987; Jacobson et al., 1996, 2002, 2007; Saleeby, 2003; Stephenson and Girty, 2006; Saleeby et al., 2007). These combined effects resulted in over-thickened continental crust (Coney and Harms, 1984; Coney, 1987; Spencer and Reynolds, 1990; Dickinson, 2002). Initial juxtaposition of the orthogneiss above the Orocopia Schist along the Chocolate Mountains fault may have occurred during underplating, but gravitational collapse of the over-thickened crust transformed the earlier thrust into an extensional detachment about 52–50 Ma (Dillon et al., 1990; Jacobson, 1990; Simpson, 1990; Robinson and Frost, 1996; Yin, 2002; Jacobson et al., 1996, 2002, 2007). Eventually, slab rollback of the Farallon plate resulted in the leading edge of arc volcanism north of the Colorado Plateau to sweep southwestward while a southern branch swept NW back toward the trench (e.g., Atwater, 1989; Dickinson, 2002; and references therein). By about 23 Ma, the NW sweeping branch of the leading edge of arc volcanism had reached the latitude of the Indian Pass–Picacho area and resulted in eruption and deposition of the lower Miocene volcanic and epiclastic unit discussed in this paper. Data presented in Olson (2010) suggests that magma feeding the Quechan volcanics was derived from the subcontinental lithosphere, and that trace and REE distribution patterns are consistent with a metasomatised mantle source much like that currently underlying the central flat segment of the Andes of South America.

Further to the west of the island arc, the spreading center separating the Pacific and Farallon plates first contacted the North American plate about 28–26 Ma (Atwater, 1989). Following initial contact, the Mendocino and Riviera triple junctions formed and migrated NW and SE, respectively, from a point located near what is now Los Angeles (Stock and Hodges, 1989; Atwater, 1989), transforming the convergent margin into a transform setting (Dickinson, 2002). The volcanic arc lying within the margin of the North American plate responded to this local transtensional regime through formation of the Oligocene-Miocene Extensional Belt of Dickinson (2002). The Indian Pass–Picacho area lies within this belt, and the initial 28–24 Ma and subsequent post-23 Ma movement along the Gatuna-Sortan faults discussed in this paper likely formed during this period. Moreover, during this local transtensional period, the Chocolate Mountains anticlinorium may have originated as a series of megamullion culminations, turtlebacks, or large-scale corrugations in the detachment surface (e.g., Davis and Coney, 1979; Spencer and Reynolds, 1991; Holm et al., 1994; Frost et al., 1996; Spencer, 2010) with the Orocopia Schist in their cores.

Extension in SE California and SW Arizona generally ended about 17 Ma (e.g., Eberly and Stanley, 1978; Spencer et al., 1995; Dickinson, 2002), and was replaced by NW-SE dextral shear as the Eastern California Shear Zone began to form within the North American plate margin about 12–10 Ma (Dokka and Travis, 1990a, 1990b). The ECSZ is a precursor to the formation of the San Andreas fault system, and represents an ~100–110 km wide zone of dextral shear that initiated along the North American plate margin around 12–10 Ma. It currently takes up as much as 20% to 25% of the total displacement between the Pacific and North American plates (Richard, 1989, 1993; Dokka and Travis, 1990a, 1990b; McQuarrie and Wernicke, 2005; Frankel et al., 2008; Lease et al., 2009; Plattner, 2009; Liu et al., 2010). Strain associated with the development of the Eastern California Shear Zone can be subdivided into components of ~NS shortening and ~EW extension (e.g., Bartley et al., 1990; Dolan et al., 2007). Such strain can be accommodated by EW-trending folds and reverse faults, NS-trending extensional faults, and/or NW- and NE-trending dextral and sinistral faults, respectively (e.g., Wright, 1976; Bartley et al., 1990; Dolan et al., 2007; Twiss and Moores, 2007). This point is significant, as the folding of the Chocolate Mountains, Gatuna, and Sortan faults along with the ~23 Ma early Miocene volcanic and epiclastic unit about an EW trending axis is reflective of NS shortening, and is therefore only strain compatible with the latter regime. If such an interpretation is correct, then the initial megamullion, turtleback, or large-scale corrugation structures that formed during movement on the Chocolate Mountains and Gatuna/Sortan faults were subsequently telescoped and enhanced after ~17 Ma by the subsequent succession of Neogene NW shortening events.

Although high-angle normal faults with minor displacements (< 1 m) occur locally within the Bear Canyon conglomerate, our work suggests that the Bear Canyon was not significantly affected by development of widespread extensional faults, and it overlies the metamorphic basement complex and the lower Miocene volcanic and epiclastic unit along a pronounced angular unconformity. This observation suggests that sequence I of the Bear Canyon conglomerate was deposited after initial formation of the anticlinorium as a fold and prior to eruption of the basalts of Black Mountain about 9.45 Ma. Unfortunately, timing of deposition and folding of sequence II is not well constrained, and is dependent upon whether or not the basalts of Black Mountain are interstratified with sequence II or III. If the former is the case, then sequence II was deposited and folded after sequence I sometime between ~9.45 Ma and sometime within or prior to the Quaternary. If, on the other hand, the basalts of Black Mountain are interstratified with sequence III, then sequence II was deposited and folded sometime between ~17 Ma and prior to ~9.45 Ma. However, given that folding of sequence II requires NS shortening, and that such shortening is strain compatible with initial development of the Eastern California Shear Zone, ~12–10 Ma, both sequences I and II likely were deposited and folded closer to ~9.45 Ma than to ~17 Ma. Alternatively, successive deposition and folding of sequences I and II simply could indicate early, ~17 Ma to ~9.45 Ma, NS shortening prior to the development of strike-slip faults in the Eastern California Shear Zone. Regardless of the problematical timing of sequences I and II, deposition and folding of sequence III, and formation of the Gatuna Wash, Carrizo Wash, Bear Canyon, Copper Basin, and conjugate set of faults developed after eruption of the basalts of Black Mountain, ~9.45 Ma, and prior to deposition of flat lying Quaternary deposits.

EW trending folds and reverse faults like the Chocolate Mountains anticlinorium and the Copper Basin fault, respectively, along with development of the conjugate NW and NE dextral and sinistral faults all reflect NS shortening (Fig. 8). Such shortening is not unique to the Indian Pass and Picacho region. For example, Wright (1976), Glazner and Bartley (1984), and Wernicke et al. (1988) proposed that eastern California has undergone significant NS shortening in the Neogene, while Bartley et al. (1990) provided a map showing the distribution of EW trending folds and reverse faults that occur within the Mojave Block, i.e., the region bounded on the west by the big bend in the San Andreas fault, and lying between the Pinto Mountain fault to the south and the Garlock fault to the north (Fig. 1). Significantly, in a manner similar to that recognized between Indian Pass and Picacho, in the eastern Mojave Block, the set of EW trending structures are superposed on an early Miocene (22–18.5 Ma) extensional fabric (Dokka, 1983; 1989; Bartley et al., 1990; Lease et al., 2009).

Though details are lacking, the formation of EW trending structures within the Mojave Block appears to have varied in both space and time, and may have accompanied dextral slip and/or developed intermittently between periods of dextral slip on major strike-slip faults within the Eastern California Shear Zone (e.g., Bartley et al., 1990; Glazner et al., 2002). For example, some EW trending structures may have formed as early as ~19 Ma, while others formed after ~13 Ma, and still others in the Quaternary (Bartley et al., 1990). Such observations suggest that within the Eastern California Shear Zone structures that accommodated NS shortening developed sporadically in both a spatial and temporal sense.

If the EW trending fabric at Picacho was produced by NS shortening during the Neogene development of the Eastern California Shear Zone, as the above timing relationships suggest, then at Picacho NS shortening occurred without the development of major large-slip (e.g., > 5 km) dextral faults. Apparently, the general dextral shear pattern and NS shortening characteristic of the Mojave Block during the Neogene was taken up further to the south at the latitude of Indian Pass and Picacho by *continued growth* of the Chocolate Mountains anticlinorium and formation of structures like the Copper Basin and conjugate fault system identified during this study. A similar more modern pattern is developed where dextral slip on the faults of the southern

San Andreas system is taken up in the Los Angeles area by EW trending folds and reverse faults and a conjugate set of NW and NE trending dextral and sinistral faults (Dolan et al., 2007). Hence, we interpret the structures identified in this study to form part of a regionally developed but discontinuous network of EW trending folds and reverse faults, and conjugate strike-slip faults that accommodated transient and spatially variable NS shortening in an otherwise dextral shear environment within the margin of the North American plate as the Eastern California Shear Zone evolved after ~12–10 Ma. As noted in Ricketts et al. (2011), such an interpretation can be tested by further detailed mapping and geochronological work.

ACKNOWLEDGEMENTS

Work reported here was funded by NSF EAR 091163 awarded to G.H. Girty.

REFERENCES CITED

Atwater, T., 1989, Plate tectonic history of the northeast Pacific and western North America, *in* Winterer, E. L., Hussong, D. M., Decker, R. W., eds., The geology of North America, v. N: The Eastern Pacific Ocean and Hawaii: Boulder, CO, Geological Society of America, p. 21–72.

Bartley, J.M., Glazner, A.F., and Schermer, E.R., 1990, North-south contraction of the Mojave block and strike-slip tectonics in southern California: Science, v. 248, p. 1398–1401.

Biggs, M.A., 2006, The ignimbrite of Ferguson Wash and the Taylor Lake fault system: Key temporal markers in the tectonic evolution of the lower Colorado River region [senior thesis]: San Diego State University, San Diego, California, 26 p.

Biggs, M.A., 2008, Miocene volcanic rocks and conglomerates, SE California: Evidence for Neogene reactivation of the Chocolate Mountains anticlinorium [MS thesis]: San Diego State University, San Diego, California, 30 p.

Burchfiel, B. C., and Davis, G. A., 1981, Mojave Desert and environs, *in* Ernst, W. G., ed., The geotectonic development of California (Rubey Volume I): Englewood Cliffs, New Jersey, Prentice-Hall, p. 217–252.

Busby-Spera, C. J., 1988, Speculative tectonic model for the early Mesozoic arc of the southwest Cordilleran United States: Geology, v. 6, p. 1121–1125.

Campbell, C., 2006, Provenance and tectonic implications of greenstone clasts in Eocene (?) gravels, Picacho State Recreation Area [senior thesis]: San Diego State University, San Diego, California, 31 p.

Campbell, K., 2004, The Rojo Grande trachyte, Picacho State Recreation Area, SE California: Its chemistry, petrology, and structure [senior thesis]: San Diego State University, San Diego, California, 20 p.

Collinson, J.D., and Thompson, D.B., 1982, Sedimentary structures: London, UK, George Allen and Unwin, 194 p.

Coney, P.J., 1987, The regional tectonic setting and possible causes of Cenozoic extension in the North America Cordillera, *in* Continental Extensional Tectonics, eds., Coward, M.P., Dewey, J.F., and Hancock, P.L., Geological Society of London Special Publication 28, p. 177–186.

Coney, P.J., and Harms, T.A., 1984, cordilleran metamorphic core complexes: Cenozoic extensional relics of Mesozoic compression: Geology, v. 12, p. 550–554.

Coney, P. J., and Reynolds, S. J., 1977, Cordilleran Benioff zones: Nature, v. 270, p. 403–406.

Crowe, B.M., 1973, Cenozoic volcanic geology of the southeasternmost Chocolate Mountains, California [Ph.D. thesis]: Santa Barbara, University of California, 117 p.

Crowe, B.M., 1978, Cenozoic volcanic geology and probable age of inception of basin-range faulting in the southeasternmost Chocolate Mountains, California: Geological Society of America Bulletin, v. 89, p. 251–264.

Crowell, J. C., 1981, An outline of the tectonic history of southeastern California, *in* Ernst, W. G., ed., The geotectonic development of California (Rubey Volume I): Englewood Cliffs, New Jersey, Prentice-Hall, p. 583–600.

Davis, G.H., and Coney, P.J., 1979, Geologic development of the Cordilleran metamorphic core complexes: Geology, v. 7, p. 120–124.

Dickinson, W.R., 2002, The Basin and Range province as a composite extensional regime: International Geology Review, v. 44, p. 1–38.

Dickinson, W.R., and Snyder, W.S., 1978, Plate tectonics of the Laramide orogeny, *in* Matthews V., ed., Laramide folding associated with basement block faulting in the western United States: Geological Society of America Memoir 151, p. 355–366.

Dillon, J.T., 1976, Geology of the Chocolate and Cargo Muchacho Moutains, southeasternmost California [Ph.D. thesis]: Santa Barbara, University of California, 405 p.

Dillon, J.T., Haxel, G.B., and Tosdal, R.M., 1990, Structural evidence for northeastward movement on the Chocolate Mountains thrust, southeasternmost California: Journal of Geophysical Research, v. 95, p. 19953–19971.

Dokka, R.K., 1983, Displacements on late Cenozoic strike-slip faults of the central Mojave Desert, California: Geology, v. 11, p. 305–308.

Dokka, R.K., 1989, The Mojave Extensional Belt of southern California: Tectonics, v. 8, p. 363–390.

Dokka, R.K. and Travis, C.J., 1990a, Late Cenozoic strike-slip faulting in the Mojave Desert, California: Tectonics, v. 9, p. 311–340.

Dokka, R.K. and Travis, C.J., 1990b, Role of the Eastern California shear zone in accommodating Pacific-North American plate motion: Geophysical Research Letters, v. 17, p. 1323–1326.

Dolan, J.F., Bowman, D.D., and Sammis, C.G., 2007, Long-range and long-term fault interactions in southern California: Geology,

v. 35, p. 855–858.

Eberly, L.D. and Stanley, T.B., 1978, Cenozoic stratigraphy and geologic history of southwestern Arizona: Geological Society of America Bulletin, v. 89, p. 921–940.

Frankel, K.L., Glazner, A.F., Kirby, E., Monastero, F.C., Strane, M.D., Oskin, M.E., Unruh, J.R., Walker, J.D., Anandakrishman, S., Bartley, J.M., Coleman, D.S., Dolan, J., and Finkel, R.C., 2008, Active Tectonics of the eastern California zone, *in* Duebendorfer, E.M., and Smith, E.I., eds., Field Guide to plutons, volcanoes, faults, reefs, dinosaurs, and possible glaciations in selected areas of Arizona, California, and Nevada: Geological Society of America Field Guide 11, p. 43–81.

Frost, E.G., Fattahipour, M.J., and Robinson, K.L., 1996, Neogene detachment and strike-slip faulting in the Salton trough region and their geometric and genetic interrelationships, *in* Abbott, P.L. and Cooper, J.D., eds., Field Conference Guide 1996, Pacific Section American Association of Petroleum Geologists, GB 73, Pacific Section Society of Economic Paleontologists and Mineralogists, Book 80, p. 263–276.

Girty, G.H., Stephenson, D., Canfield, A., Marsh, J., Middleton, T., Rayburn, J., Sisk, M., Verdugo, D., Moniz, R., Gunter, J., Nielsen, J., Gasca, C., Rowland-Smith, A., Smith, M., Lovering, K., Campbell, K., Castenada, F., Gray, R., and Campbell, C., 2006, Geology of the Picacho State Recreation Area, SE California: Implications for the timing of formation of the Chocolate Mountains anticlinorium, *in* Girty, G.H., and Cooper, J., eds., Using Stratigraphy, Sedimentology, and Geochemistry to Unravel the Geologic History of the Southwestern Cordillera: A volume in honor of Patrick L. Abbott, Pacific Section SEPM Book 101, p. 255–274.

Glazner, A.F., 2008, Drawing the curtain on slab windows in southwestern North America: Geological Society of America Abstracts with Programs, v. 41, p. 8.

Glazner, A.F., and Bartley, J.M., 1984, Timing and tectonic setting of Tertiary low-angle normal faulting and associated magmatism in the southwestern United States: Tectonics, v. 3, p. 385–396.

Glazner, A.F., Walker, J.D., Bartley, J.M., and Fletcher, J.M., 2002, Cenozoic evolution of the Mojave block of southern California, *in* Glazner A.F., Walker, J.D., and Bartley, J.M, eds., Geologic evolution of the Mojave Desert and southwestern Basin and Range: Geological Society of America Memoir 195, p. 19–41.

Gordon, E., 2011, Metamorphic alteration of the Jurassic metavolcanic member, Winterhaven Formation, SE California: Implications for classification and interpreting tectonic setting [senior thesis]: San Diego State University, San Diego, California, 17 p.

Hamilton, W., 1987, Mesozoic geology and tectonics of the Big Maria Mountains region, southeastern California, *in* Dickinson, W. R., and Klute, M. A., eds., Mesozoic rocks of southern Arizona and adjacent areas: Arizona Geological Society Digest, v. 18, p. 33–47.

Hartung, E.A., 2005, The Little Picacho Wash formation and the timing of exposure of the Winterhaven Formation, Picacho State Recreation Area, SE California [senior thesis]: San Diego State University, San Diego, California, 31 p.

Haxel, G.B., 1977, The Orocopia Schist and the Chocolate Mountains thrust, Picacho-Peter Kane Mountain area, southeasternmost California [Ph.D. thesis]: Santa Barbara, California, University of California, 277 p.

Haxel, G.B., Tosdal, R.M., and Dillon, J.T., 1985, Tectonic setting and lithology of the Winterhaven Formation: A new Mesozoic stratigraphic unit in southeasternmost California and southwestern Arizona: U.S. Geological Survey Bulletin 1599, p. 19.

Haxel G.B., Jacobson, C.E., Richard, S.M., Tosdal, R.M., and Grubensky, M.J., 2002, The Orocopia Schist in southwest Arizona: Early Tertiary oceanic rocks trapped or transported far inland, *in* Barth, A., ed., Contributions to crustal evolution of the southwestern United States: Geological Society of America Special Paper 376, p. 99–128.

Holm, D.K., Fleck, R.J., and Lux, D.R., 1994, The Death Valley Turtlebacks Reinterpreted as Miocene-Pliocene folds of a major detachment surface: Journal of Geology, v. 102, p. 718–727.

Jacobson, C.E., 1990, The $^{40}Ar/^{39}Ar$ geochronology of the Pelona Schist and related rocks, southern California: Journal of Geophysical Research, v. 95, p. 509–528.

Jacobson, C.E., Oyarzabal, F.R., and Haxel, G.B., 1996, Subduction and exhumation of the Pelona-Orocopia-Rand schists, southern California: Geology, v. 24, p. 547–550.

Jacobson, C.E., Grove, M., Stamp., M.M., Vucic, A., Oyarzabal, F.R., Haxel., G.B., Tosdal, R.M., and Sherrod, D.R., 2002, Exhumation history of the Orocopia Schist and related rocks in the Gavilan Hills area of southeasternmost California, *in* Barth, A., ed., Contributions to crustal evolution of the southwestern United States: Geological Society of America Special Paper 365, p. 129–154.

Jacobson, C.E., Grove, M., Vucic, A., Pedrick, J.N., and Ebert, K.A., 2007, Exhumation of the Orocopia Schist and associated rocks of southeastern California: Relative roles of erosion, synsubduction, tectonic denudation, and middle Cenozoic extension, *in* Cloos, M., Carlson, W.D., Gilbert, M.C., Liou, J.G., and Sorensen, S.S., eds., Convergent margin terranes and associated regions — A tribute to W.G. Ernst,: Geological Society of America Special Paper 410, p.1–37.

Kay S., Coria, B.L., and Kay, R.W., 2009, Central Andean Galan ignimbrites; magma evolution from the mantle to eruption in a thickened crust: Geochimica et Cosmochimica, v. 73 (13S): A630.

Lease, R.O., McQuarrie, N., Oskin, M. and Leier, A., 2009, Quantifying dextral shear on the Bristol-Granite Mountains fault zone: successful geologic prediction from kinematic compatibility of the Eastern California Shear Zone: Journal of Geology, v. 117, p. 37–53.

Le Bas, M.J., Le Maitre, R.W., Streckeinsen, A. and Zanettin, B., 1986, A chemical classification of volcanic rocks based on the total alkali-silica diagram: Journal of Petrology, v. 27, p. 745–750.

Liu, M., Wang, H., and Li, Q., 2010, Inception of the eastern California shear zone and evolution of the Pacific-North American plate boundary: from kinematics to geodynamics: Journal of Geophysical Research, v. 115, p. B07401–B07413.

Losh, S., Purvance, D., Sherlock, R., and Jowett, E.C., 2005, Geologic and geochemical study of the Picacho gold mine, California: gold in a low-angle normal fault environment: Mineralium Deposita, v. 40, p. 137–155.

Lovering, K., 2005, Petrology, chemistry, and emplacement mechanisms of the Marcus Wash trachyte, Picacho State Recreation Area, SE California [senior thesis]: San Diego State University, San Diego, California, 18 p.

Marsh, J., 2003, Petrologic and geochemical evidence for an Early to Middle Jurassic extensional magmatic arc, southeastern

California [MS thesis]: San Diego State University, San Diego, California, 77 p.

McKenzie, D., and O'Nions, R.K., 1991, Partial melt distributions from inversion of rare earth elements: Journal of Petrology, v. 32, p. 1021–1091.

McQuarrie, N., and Wernicke, B.P., 2005, An animated tectonic reconstruction of southwestern North America since 36 Ma: Geosphere, v. 1, p. 147–172.

Miall, A.D., 1996, The geology of fluvial deposits: New York, New York, Springer, 582 p.

Muela, K.K., 2011, Timing and style of Miocene deformation, Indian Pass and Picacho State Recreation Area, SE California [MS thesis]: San Diego State University, San Diego, California, 44 p.

Needy, S.K., Wooden, J.L., Barth, A.P., and Jacobson, C.E., 2007, Geochronology of igneous rocks in the Chocolate Mountains region as a means to interpret the tectonic evolution of southeastern California: Geological Society of America Abstracts with Programs, v. 39, p. 407.

Nielson, J., 2004, Structural geology of Picacho State Recreation Areas (PSRA): implications for the transfer of Baja California to the Pacific plate [senior thesis]: San Diego State University, San Diego, California, 34 p.

Olson, H., 2010, The early Miocene Quechan volcanics, SE California, U.S.A.: slab-window or continental-margin arc volcanics? [MS thesis]: San Diego State University, San Diego, California, 56 p.

Oyarzabal, F.R., Jacobson, C.E., and Haxel, G.B., 1997, Extensional reactivation of the Chocolate Mountains subduction thrust in the Gavilan Hills of southeastern California: Tectonics, v. 16, p. 650–661.

Pearce, J.A., 1996, A user's guide to basalt discrimination diagrams *in* Wyman, D. A., ed., Trace Element Geochemistry of Volcanic Rocks: Applications for Massive Sulphide Exploration: Geological Association of Canada, Short Course Notes 12, p. 79–113.

Pelbath, T., 2011, Using texture and geochemical data to assess the stratigraphic relationships between the Marcus Wash and Rojo Grande units, Picacho State Recreation Area, SE California [senior thesis]: San Diego State University, San Diego, California, 25 p.

Plattner, C., 2009, Dynamic implications of Baja microplate kinematics on the North American-Pacific plate boundary, region [Ph.D. thesis], Ludwig-Maximilians-Universität München, 103 p.

Richard, S.M., 1989, The Chocolate Mountains anticlinorium in the Middle Mountains, SW Arizona: Geological Society of America Abstracts with Programs, v. 21, no. 6, p. A64.

Richard, S.M., 1993, Palinspastic reconstruction of southeastern California and southwestern Arizona for the middle Miocene: Tectonics, v. 12, p. 830–854.

Ricketts J.W., 2010, Miocene growth of a depression in the axis of the Chocolate Mountains anticlinorium recorded by the Bear Canyon conglomerate, SE California [MS thesis], San Diego State University, San Diego, California, 50 p.

Ricketts, J.W., Girty, G.H., Sainsbury, J.S., Muela, K.K., Sutton, L.A., Biggs, M.A., and Voyles, E.M., 2011, Episodic growth of the Chocolate Mountains anticlinorium recorded by the Neogene Bear Canyon conglomerate, SE California, U.S.A.: Journal of Sedimentary Research, v. 81, 859–873.

Robinson, K.L. and Frost, E.G., 1996, Orocopia Mountains detachment system: Progressive development of a tilted crustal slab and a half-graben sedimentary basin during regional extension, *in* Abbott, P.L. and Cooper, J.D., eds., Field Conference Guide 1996, Pacific Section American Association of Petroleum Geologists, GB 73, Pacific Section Society of Economic Paleontologists and Mineralogists, Book 80, p. 277–284.

Sainsbury, J.S., 2010, Timing of extension, NS shortening, and conjugate strike-slip faulting in the evolution of the Chocolate Mountains anticlinorium: Evidence from the Gavilan Hills, SE California [MS thesis]: San Diego State University, San Diego, California, 42 p.

Saleeby, J., 2003, Segmentation of the Laramide slab — evidence from the southern Sierra Nevada region: Geological Society of America Bulletin, v. 115, p. 655–668.

Saleeby, J., Farley, K.A., Kistler, R.W., and Fleck, R., 2007, Thermal evolution and exhumation of deep-level batholithic exposures, southernmost Sierra Nevada, California, *in* Cloos, M., Carlson, W.D., Gilbert, M.C., Liou, J.G., and Sorensen, S.S., eds., Convergent Margin Terranes and Associated Regions: A Tribute to W.G. Ernst: Geological Society of America Special Paper 419, p. 39–66.

Simpson, C., 1990, Microstructural evidence for northeastward movement on the Chocolate Mountains fault zone, southeastern California, Journal of Geophysical Research, v. 95, p. 529–537.

Smith, M., 2005, Assessing correlation of Marcus Wash trachyte across the Taylor Lake fault zone, Picacho State Recreation Area [senior thesis]: San Diego State University, San Diego, California, 21 p.

Spencer, J.E., 2010, Structural analysis of three extensional detachment faults with data from the 2000 space-shuttle radar topography mission: GSA Today, v. 20, p. 4–10.

Spencer, J.E., and Reynolds, S.J., 1990, Relationship between Mesozoic and Cenozoic tectonic features in west central Arizona and adjacent southeastern California: Journal of Geophysical Research, v. 95, p. 539–555.

Spencer, J.E., and Reynolds, S.J., 1991, Tectonics of mid-Tertiary extension along a transect through west central Arizona: Tectonics, v. 10, p. 1204–1221.

Spencer, J.E., Richard, S.M., Reynolds, S.J., Miller, R.J., Shafiquallah, M., Gilbert, W.G., and Grubensky, M.J., 1995, Spatial and temporal relationships between mid-Tertiary magmatism and extension in southwestern Arizona: Journal of Geophysical Research, v. 100, p. 10321–10351.

Stephenson, D., 2006, The structure, petrology, and chemistry of the Orocopia Schist, Picacho State Recreation Area, California [MS thesis]: San Diego State University, San Diego, California, 94 p.

Stephenson, D., and Girty, G.H., 2006, Constraining the source of the Orocopia Schist, SE California: A comparative study of the major and trace element chemistry of Cretaceous sandstones deposited in forearc settings, *in* Girty, G.H., and Cooper, J., eds., Using Stratigraphy, Sedimentology, and Geochemistry to Unravel the Geologic History of the Southwestern Cordillera: A volume in honor of Patrick L. Abbott, Pacific Section SEPM Book 101, p. 77–96.

Stock, J.M., and Hodges, K.V., 1989, Pre-Pliocene extension around the Gulf of California and the transfer of Baja California to the Pacific Plate: Tectonics 8, p. 99–115.

Su, B., 2008, Natural evidence for garnet-spinel transiton (GST) in the Earth's mantle: available from Nature Precedings, http://dx.doi.org/10.1038/npce.2008.1892.27

Sutton, L.A., 2010, The Copper Basin fault, a newly recognized reverse fault in the lower Colorado River region, SE California: Implications for Miocene-Pliocene N-S shortening within the Eastern California Shear Zone [M.S. thesis]: San Diego State University, San Diego, California, 42 p.

Taylor, S.R., and McLennan, S.M., 1985, The Continental Curst: its composition and evolution, Oxford, UK, Blackwell Scientific Publications, 312 p.

Twiss, R.J., and Moores, E.M., 2007, Structural geology: New York, New York, W.H. Freeman and Company, 736 p.

Wernicke, B., Axen, G.J., and Snow, J.K., 1988, Basin and Range extensional tectonics at the latitude of Las Vegas, Nevada: Geological Society of America Bulletin, v. 100, p. 1738–1757.

Winchester, J.A., and Floyd, P.A., 1977, Geochemical discrimination of different magma series and their differentiation products using immobile elements: Chemical Geology, v. 20, p. 325–343.

Winter J., 2010, An introduction to igneous and metamorphic petrology (2nd edition): Upper Saddle River, NJ, Prentice Hall, 702 p.

Wright, L.A., 1976, Late Cenozoic fault patterns and stress fields in the Great Basin and westward displacement of the Sierra Nevada block: Geology, v. 4, p. 489–494.

Yin, A., 2002, Passive-roof thrust model for the emplacement of the Pelona-Orocopia Schist in southern California, United States: Geology, v. 30, p. 183–186.

APPENDIX

Analytical Methods

Nineteen samples of metavolcanic rocks in the Winterhaven Formation and 79 samples from the various units of the Miocene volcanic and epiclastic sequence were collected for major, trace, and rare earth element analysis. Each sample was power-washed then rinsed with de-ionized water. Being careful not to contaminate samples, each specimen was wrapped in brown paper while they were broken down into pieces ≤0.5 cm². This procedure produced ~80 g of sample that was rendered into a powder in a tungsten carbide shatter box.

Each powdered sample was securely sealed in an air-tight container and transferred to the GeoAnalytical Lab, Washington State University (WSU), Pullman, Washington. Sample preparation at WSU included: mixing di-lithium tetraborate flux and powder in a 2:1 ratio, and then fusing the resulting mixture at 1000°C in a muffle oven. Once the fused bead cooled, it was reground, refused, and polished on diamond laps to provide a smooth flat surface for XRF analysis. R. M. Conrey and L. C. Wagoner then analyzed each sample utilizing a ThermoARL Advant'XP+ sequential X-ray fluorescence spectrometer.

Sample preparation for REE analysis included low-dilution fusion with di-lithium tetraborate followed by an open-vial mixed acid digestion. Dissolution with HF after the di-lithium tetraborate fusion quantitatively removes silica and more than 90% of the flux as gaseous fluorides, leaving clear, stable solutions for analysis on the inductively-coupled plasma-source mass spectrometer (ICP-MS). See http://www.sees.wsu.edu/Geolab/note/icpms for a comprehensive description of the methods used at WSU for trace and REE analysis. Final solutions are analyzed on an Agilent model 4500 ICP-MS.

The following 25 tables represent a summary of all XRF and ICP-MS analyses along with the latitudes and longitudes of all collected samples analyzed at WSU.

Table 1. *Locations of Winterhaven Formation metavolcanic member samples analyzed at WSU*

Sample	Latitude	Longitude
Winterhaven Formation–Metavolcanic member		
WHV-1	32° 38′ 51.90″	114° 36′ 40.23″
WHV-2	32° 58′ 48.55″	114° 36′ 39.70″
WHV-3	32° 58′ 47.75″	114° 36′ 40.23″
EG 1	32° 58′ 59.35′	114°38′ 38.27″
EG 4	32° 58′ 44.24″	114° 38′ 30.91″
EG 11	32° 59′ 1.36″	114° 38′ 31.19″
EG 12	32° 58′ 57.33″	114° 38′ 39.85″
EG 13	32° 59′ 3.67″	114° 38′ 37.62″
EG 14	32° 58′ 50.05″	114° 38′ 39.72
EG 15	32° 58′ 46.68″	114° 38′ 31.84″
EG 16	32° 58′ 51.45″	114° 38′ 17.83″
EG 17	32° 58′ 54.54″	114° 38′ 21.31″
EG 18	32° 59′ 0.83″	114° 38′ 27.87″
EG 19	32° 59′ 7.94″	114° 38′ 30.65″
EG 20	32° 59′ 4.75″	114° 38′ 34.87″
EG 21	32° 58′ 38.91″	114° 37″ 26.14″
EG 22	32° 59′ 3.60″	114° 37′ 39.96″
EG 23	32° 58′ 59.65″	114° 37′ 31.72″
EG 24	32° 59′ 3.97″	114° 37′ 34.86″
EG 25	32° 59′ 7.14″	114° 37′ 41.36″
EG 26	32° 58′ 49.61″	114° 37′ 18.66″
EG 27	32° 58′ 53.29″	114° 36′ 45.41″

Table 2. *Winterhaven Metavolcanic Member—Results from XRF analysis of major elements*[1]

Sample	SiO_2	TiO_2	Al_2O_3	FeO_2	MnO	MgO	CaO	Na_2O	K_2O	P_2O_5	LOI	Sum
WHV-1	45.9	1.07	13.3	9.52	0.15	12.8	5.86	1.24	0.84	0.34	9.44	100.4
WHV-2	47.5	2.75	14.5	13.9	0.19	4.43	4.98	5.36	0.95	0.66	5.08	100.2
WHV-3	42.3	1.82	16.6	10.1	0.20	4.20	9.32	2.76	3.32	0.33	9.57	100.5
EG1	44.8	2.02	17.4	11.7	0.17	6.64	5.68	2.56	2.21	0.49	7.15	100.8
EG 4	48.0	2.46	14.9	13.2	0.16	3.43	5.56	3.71	2.30	0.62	6.42	100.8
EG 11	54.5	2.26	14.0	9.81	0.18	0.32	6.21	6.23	1.28	1.28	4.21	100.3
EG 12	49.1	2.37	14.0	9.56	0.30	0.59	8.70	5.22	2.17	1.60	6.21	99.9
EG 13	50.8	2.62	16.4	11.0	0.15	1.24	4.35	3.46	4.41	1.64	3.67	99.8
EG 14	51.9	1.72	14.9	10.3	0.13	7.00	3.48	2.93	1.59	0.44	6.30	100.7
EG 15	55.4	1.47	14.3	6.66	0.27	0.92	7.14	3.32	3.47	0.75	6.50	100.2
EG 16	45.0	2.04	17.2	11.9	0.15	6.61	5.32	3.88	1.05	0.50	6.51	100.2
EG 17	50.8	1.30	11.6	5.40	0.33	1.08	12.7	3.10	2.26	0.77	10.6	100.0
EG 18	51.6	2.22	16.8	9.47	0.15	0.81	5.28	4.39	3.69	1.32	4.37	100.0
EG 19	43.4	2.72	16.2	9.99	0.36	3.97	6.38	1.78	4.95	1.78	8.19	99.7
EG 20	52.9	2.06	14.1	8.67	0.24	1.18	6.56	2.96	3.86	1.30	5.61	99.4
EG 21	67.4	0.60	15.2	4.57	0.04	1.42	1.80	2.33	3.38	0.12	3.51	100.4
EG 22	44.1	1.95	13.2	11.11	0.25	3.42	11.6	3.77	0.72	0.47	9.85	100.5
EG 23	51.0	1.98	12.7	9.85	0.28	2.87	9.00	4.93	0.32	0.68	7.02	100.7
EG 24	53.8	1.96	14.1	12.1	0.15	2.72	4.79	5.91	0.37	0.67	4.20	100.7
EG 25	43.5	1.93	13.6	11.0	0.21	5.55	9.82	4.13	0.52	0.47	10.0	100.8
EG 26	56.4	0.87	15.3	5.23	0.10	4.22	6.35	3.68	1.70	0.23	6.50	100.6
EG 27	55.3	0.80	17.8	7.02	0.09	3.49	3.94	3.69	2.93	0.37	5.56	101.0

[1] Concentrations in weight percent (wt.%) [2] Total iron as FeO

Table 3. *Winterhaven Metavolcanics—Results from XRF analysis trace elements*[1]

Sample	Ni	Cr	Sc	V	Ba	Rb	Sr	Zr	Y	Nb	Ga	Cu	Zn	Pb	La	Ce	Th	Nd	U	Cs
WHV-1	312.8	582.3	21.0	175.2	135.7	26.1	252.8	145.4	13.2	11.8	18.3	172.5	90.4	3.5	22.7	45.1	2.0	23.4	1.9	5.4
WHV-2	17.6	0.0	31.7	254.2	291.0	19.7	427.3	317.4	41.0	21.5	18.7	5.4	132.7	25.8	35.2	72.4	0.2	39.9	0.4	3.9
WHV-3	63.7	38.4	29.6	239.2	745.7	90.0	613.3	178.6	27.9	9.5	19.5	3.2	120.4	24.8	12.2	29.9	0.6	20.8	0.9	4.6
EG1	77.1	95.2	29.0	281.1	435.5	45.4	364.2	196.0	29.2	16.7	20.4	74.9	95.1	11.1	24.9	48.1	1.0	25.2	1.0	1.8
EG 4	12.3	0.0	32.4	324.0	936.2	59.1	233.5	295.4	41.6	19.1	20.7	11.6	116.8	15.4	28.6	68.1	2.3	34.9	2.3	5.0
EG 11	5.8	0.0	26.0	127.6	437.4	30.1	323.0	411.9	57.6	23.6	14.9	19.1	57.2	27.3	50.8	125.6	1.5	67.7	2.4	3.2
EG 12	5.2	0.0	22.9	103.5	921.7	43.5	297.9	454.1	71.7	29.3	17.7	20.5	60.8	39.7	60.6	148.4	2.4	76.2	1.9	1.8
EG 13	4.2	0.0	22.9	119.9	1157.7	120.0	179.9	514.7	63.7	32.1	22.6	0.2	136.0	25.9	60.1	155.1	3.1	81.3	3.6	9.3
EG 14	71.5	78.6	22.8	179.9	457.5	32.0	192.5	169.6	20.1	14.9	16.0	0.0	94.8	10.8	16.3	36.8	1.3	21.4	1.3	6.3
EG 15	7.2	1.5	17.3	79.7	1138.7	62.3	344.8	475.5	60.3	29.1	19.6	1.6	93.2	19.7	49.0	124.0	2.7	63.0	0.8	6.3
EG 16	79.6	92.4	30.3	261.7	344.9	20.4	593.6	196.0	29.5	17.4	19.0	3.2	99.3	28.2	22.7	53.5	1.6	27.5	1.2	3.1
EG 17	3.9	1.3	13.4	54.1	929.3	46.5	335.2	438.2	57.3	23.1	15.0	2.7	109.2	19.4	40.0	100.3	2.3	51.1	1.1	5.0
EG 18	4.3	0.0	22.8	70.2	1151.2	75.1	283.7	630.3	74.4	35.1	23.6	1.8	98.6	21.0	66.7	164.7	4.9	88.1	1.5	6.5
EG 19	6.1	0.0	25.5	137.5	2012.2	106.7	254.0	531.7	80.1	33.3	28.3	26.5	266.6	19.1	75.6	175.0	2.9	96.8	2.2	9.2
EG 20	5.2	0.0	19.6	100.6	1003.9	101.8	201.9	465.1	53.0	26.4	20.2	1.8	126.1	32.0	64.5	150.4	3.0	73.0	2.5	6.2
EG 21	18.6	46.7	11.2	109.8	840.7	107.1	159.3	215.4	23.0	11.1	16.0	9.5	73.9	11.1	19.4	41.3	8.5	20.5	1.3	7.3
EG 22	21.3	17.3	26.3	277.1	132.4	19.5	709.2	202.0	37.4	14.1	18.5	33.3	77.7	9.4	27.5	62.0	2.6	33.3	2.4	4.2
EG 23	3.2	0.0	24.5	150.3	147.8	6.8	370.9	283.9	50.8	20.6	14.7	63.5	96.3	11.6	37.1	83.6	1.5	44.6	0.0	2.7
EG 24	2.9	0.0	22.8	189.6	125.1	9.7	262.7	282.0	41.2	18.8	16.6	5.0	75.7	9.1	39.7	85.2	2.4	43.7	1.3	0.7
EG 25	50.6	24.8	25.8	246.8	522.9	18.0	403.6	199.0	36.1	12.8	15.8	21.4	109.8	4.6	19.2	50.7	0.8	28.3	1.2	1.1
EG 26	110.2	171.8	17.6	131.5	685.2	40.1	902.0	143.5	19.1	9.8	17.7	40.2	80.6	16.2	22.4	47.8	7.0	24.8	2.6	1.5
EG 27	13.6	1.9	11.8	153.5	747.3	63.6	429.6	148.4	14.6	6.9	22.6	33.8	84.4	7.9	23.9	49.2	2.3	25.4	1.2	4.6

[1] Concentrations in parts per million (ppm)

Table 4. *Winterhaven Metavolcanic Member—ICP-MS analysis of rare earth elements (REE)*[1]

Sample	La	Ce	Pr	Nd	Sm	Eu	Gd	Tb	Dy	Ho	Er	Tm	Yb	Lu
WHV-1	22.57	48.42	6.21	24.60	4.74	1.37	3.67	0.54	3.07	0.57	1.50	0.22	1.35	0.23
WHV-2	35.16	76.98	10.21	42.26	9.28	2.62	8.62	1.37	8.26	1.65	4.38	0.62	3.85	0.62
WHV-3	13.20	32.72	4.63	20.43	5.04	1.71	5.39	0.91	5.62	1.13	2.97	0.42	2.58	0.39
EG-1	23.33	50.50	6.69	28.05	6.31	1.93	6.08	0.97	5.90	1.18	3.07	0.44	2.73	0.42
EG-4	32.40	69.58	8.92	36.66	7.88	2.27	7.89	1.28	7.86	1.60	4.23	0.61	3.73	0.57
EG-11	49.37	134.72	15.74	66.68	14.62	4.35	13.54	2.02	11.52	2.23	5.67	0.76	4.56	0.69
EG-12	62.89	164.20	19.34	81.15	17.54	4.45	16.05	2.44	14.37	2.85	7.46	1.04	6.23	0.98
EG-13	67.46	169.01	20.48	84.57	17.55	4.55	15.21	2.28	13.23	2.58	6.76	0.93	5.70	0.89
EG-14	18.36	39.08	5.05	20.82	4.62	1.42	4.34	0.67	4.10	0.84	2.25	0.33	2.08	0.35
EG-15	54.82	133.90	16.34	66.55	14.09	3.73	12.54	1.96	11.72	2.34	6.30	0.91	5.71	0.91
EG-16	53.19	129.55	15.76	64.12	13.78	3.66	12.30	1.94	11.56	2.35	6.29	0.90	5.59	0.89
EG-17	44.57	108.36	12.88	51.43	10.56	2.70	10.01	1.67	10.27	2.19	6.10	0.87	5.53	0.88
EG-18	66.88	171.54	21.90	90.05	18.79	4.60	16.79	2.58	15.46	3.11	8.20	1.20	7.28	1.15
EG-19	74.77	183.55	23.27	97.03	20.39	5.28	18.65	2.81	16.27	3.21	8.29	1.16	6.90	1.07
EG-20	64.32	152.64	18.22	73.38	14.81	3.86	12.71	1.89	10.87	2.16	5.49	0.77	4.63	0.72
EG-21	22.49	42.36	5.19	19.94	4.09	0.99	3.83	0.64	4.01	0.85	2.32	0.35	2.24	0.36
EG-22	27.51	59.17	7.70	32.09	7.37	2.32	7.36	1.16	7.03	1.43	3.69	0.53	3.13	0.47
EG-23	38.90	84.05	10.81	45.48	9.99	2.95	10.05	1.57	9.50	1.91	5.01	0.71	4.19	0.63
EG-24	40.54	83.95	10.72	43.76	9.39	2.66	8.81	1.33	7.98	1.61	4.20	0.60	3.69	0.59
EG-25	23.18	51.44	6.88	29.33	6.80	2.06	6.82	1.11	6.76	1.38	3.58	0.51	3.06	0.48
EG-26	23.46	47.66	5.99	24.21	5.02	1.37	4.28	0.64	3.70	0.74	1.95	0.29	1.76	0.27
EG-27	24.27	50.85	6.43	25.48	4.84	1.33	3.87	0.53	2.99	0.57	1.46	0.21	1.28	0.21

[1] Concentrations in parts per million (ppm)

Table 5. *Winterhaven Metavolcanic Member—ICP-MS analysis of selected trace elements*[1]

Sample	Ba	Th	Nb	Y	Hf	Ta	U	Pb	Rb	Cs	Sr	Sc	Zr
WHV-1	136	3.45	11.22	13.62	3.56	0.72	0.79	4.56	24.6	6.34	242	22.2	144
WHV-2	288	2.16	20.39	41.24	7.13	1.50	0.88	27.31	19.0	3.50	423	31.7	327
WHV-3	745	1.06	8.83	27.85	3.92	0.62	0.87	24.18	84.4	3.88	594	28.6	173
EG-1	430	2.25	16.35	29.45	4.34	1.06	0.67	11.84	44.1	2.82	349	29.4	192
EG-4	919	3.13	18.45	40.59	6.55	1.23	0.81	15.61	54.0	3.60	217	30.6	298
EG-11	440	2.76	23.04	57.60	9.08	1.37	1.37	27.36	28.0	3.00	304	27.2	424
EG-12	940	2.65	28.87	74.57	9.98	1.61	1.49	41.59	43.3	3.78	294	24.2	483
EG-13	1164	4.54	31.21	65.92	11.35	1.74	1.92	27.23	118.2	7.80	176	24.7	540
EG-14	459	1.89	14.31	21.03	3.77	1.47	0.74	10.82	31.4	3.65	188	23.8	169
EG-15	1156	4.11	28.84	61.93	10.40	1.68	1.89	20.33	62.7	7.24	341	19.8	506
EG-16	1135	4.08	27.59	60.84	10.23	1.63	1.87	19.90	61.8	7.23	348	20.0	501
EG-17	922	3.41	21.93	55.66	9.32	1.23	1.46	18.27	43.4	4.02	296	14.0	441
EG-18	1153	5.19	33.61	79.60	13.85	1.96	1.61	21.38	75.6	6.70	293	26.5	678
EG-19	2052	3.04	31.32	81.64	11.25	1.79	1.07	20.71	100.3	6.50	248	28.1	545
EG-20	1004	3.94	21.05	55.53	9.65	1.17	1.45	32.85	101.3	6.02	207	22.0	490
EG-21	842	8.64	9.59	23.04	5.51	0.78	1.75	11.36	105.6	6.00	161	11.8	214
EG-22	126	2.32	13.44	37.38	4.51	0.85	0.64	10.61	17.7	3.33	730	27.9	200
EG-23	144	3.14	18.30	48.74	6.29	1.20	0.93	11.15	6.5	1.13	350	24.4	284
EG-24	122	3.13	17.41	40.69	6.31	1.18	0.96	8.82	9.2	0.85	260	23.3	291
EG-25	513	1.12	12.23	35.38	4.41	0.78	0.63	5.74	17.1	3.15	409	27.4	196
EG-26	676	7.52	8.47	18.45	3.65	0.58	2.74	15.47	36.9	2.53	891	17.6	140
EG-27	736	2.45	6.31	14.99	3.65	0.40	0.63	7.77	61.4	3.15	438	12.8	147

[1] Concentrations in parts per million (ppm)

Table 6. *Locations of Lower and Middle Quechan Volcanics Analyzed at WSU*

Sample	Latitude	Longitude
Quechan Volcanics		
Lower		
30: 3 m flow	N32°58′30.65″	W114°35′42.75″
31: 5.25 m flow	N32°58′23.35′	W114°35′48.90″
32: 6.7 m flow	above 31	
33: massive flow	N32°58′18.36″	W114°35′51.56″
34: 2.75 m flow	N32°58′17.79″	W114°35′53.30″
35: 12 m flow	N32°58′16.78″	W114°35′55.57″
36: 2 m flow	N32°58′18.66″	W114°35′58.91″
37: 2 m flow	above 36	
Middle		
38a-lower part: ~13–14 m flow	N32°58′20.81″	W114°35′29.63″
38b-upper part: ~13–14 m flow		
39-clast in 7–8 m flow breccia	above 38	
40: 16–17 m flow	above 39	
41: 17–18 m flow	above 40	
42: 16–18 m flow	N32°58′15.88″	W114°36′30.72″
43: >2 m flow	N32°58′14.23″	W114°36′51.83″
44: 20 m flow	N32°59′10.92″	W114°36′59.08″
45: 20 m flow	N32°59′13.85″	W114°36′57.47″
46: 20 m flow	N32°59′15.02″	W114°36′55.10″
47: >10 m flow	N32°59′16.05″	W114°36′54.66″
21: massive flow	N32°59′18.35″	W114°37′51.89″
22: clast in flow breccia	7′ flow breccia above 21	
23: massive flow	2′ massive flow above 22	
24: flow breccia	5′ flow breccia above 23	
25: matrix in flow breccia	N32°59′29.48″	W114°37′54.08″
26: vesiculated flow	>10′ flow–4th flow above 25	
20: laminated flow?	N32°59′18.35′	W114°37′45.44″

Table 7. *Lower and Middle Quechan Volcanics—Results from XRF analysis of major elements*[1]

Sample	SiO_2	TiO_2	Al_2O_3	FeO_2	MnO	MgO	CaO	Na_2O	K_2O	P_2O_5	LOI	Sum
Lower Member												
30	44.5	1.62	14.1	9.42	0.11	5.25	8.00	2.83	2.23	0.40	10.4	98.8
31	34.9	1.15	10.5	6.89	0.24	10.0	12.6	2.06	1.47	0.30	18.9	99.1
32	50.4	1.73	17.0	6.43	0.06	6.42	4.59	3.11	2.00	0.40	7.32	99.5
33	45.7	1.54	14.1	8.78	0.14	7.54	8.23	2.75	1.42	0.36	8.55	99.2
34	50.4	1.56	15.2	9.00	0.10	5.82	8.83	3.14	1.48	0.37	3.42	99.4
35	56.9	1.34	15.9	6.51	0.08	4.89	6.22	3.46	1.89	0.25	2.24	99.7
36	49.9	1.51	15.5	8.54	0.07	6.61	8.22	2.90	1.27	0.31	4.54	99.4
Middle Member												
37	57.6	1.21	15.3	5.79	0.12	3.87	5.97	3.59	2.56	0.32	2.77	99.2
38a	60.0	1.11	14.7	4.67	0.09	2.19	5.18	2.78	3.33	0.30	4.70	99.0
38b	56.5	1.26	17.0	5.02	0.10	2.21	4.31	3.16	3.85	0.35	4.82	98.6
39	58.4	1.12	13.4	5.16	0.08	4.76	6.18	2.80	2.11	0.33	5.23	99.7
40	61.8	1.27	16.2	5.26	0.09	1.32	4.33	3.68	3.15	0.35	2.16	99.6
41	66.9	0.88	13.5	4.72	0.03	1.37	3.72	2.57	2.50	0.24	2.83	99.3
42	66.7	0.87	14.5	4.23	0.07	0.84	3.76	3.13	3.43	0.24	1.90	99.6
43	58.2	1.32	17.3	6.26	0.06	1.73	5.22	3.92	2.95	0.36	2.43	99.8
44	58.6	1.23	16.0	5.97	0.11	2.78	5.39	3.63	2.64	0.29	2.54	99.1
45	64.2	0.94	15.4	4.72	0.05	1.12	3.63	3.83	3.14	0.25	1.62	98.9
46	57.6	1.36	17.2	6.27	0.10	1.92	5.31	3.98	3.03	0.36	2.43	99.6
47	57.9	1.40	17.1	6.57	0.10	2.28	4.83	4.16	2.97	0.36	2.33	100.1
20	61.3	0.69	16.5	4.19	0.08	2.42	3.86	3.80	3.05	0.19	3.93	100.0

[1] Concentration in weight percent (wt.%) [2] Total iron as FeO

Table 8. *Lower and Middle Quechan Volcanics—Results from XRF analysis trace elements*[1]

Sample	Ni	Cr	Sc	V	Ba	Rb	Sr	Zr	Y	Nb	Ga	Cu	Zn	Pb	La	Ce	Th	Nd	U	Cs
Lower Member																				
30	271.6	382.0	21.5	174.1	641.7	61.2	641.9	212.1	22.9	31.0	17.8	14.1	80.8	6.7	39.5	74.3	7.6	32.7	2.9	3.2
31	122.5	248.4	16.7	94.6	452.2	39.3	477.7	143.5	19.4	23.5	13.2	36.6	59.3	6.8	29.9	58.2	3.9	26.2	1.3	0
32	130.6	278.2	24.4	233.9	625.9	40.4	560.0	209.4	25.4	29.9	19.1	86.8	97.4	8.3	39	77.5	6.4	33.7	3.1	5.4
33	103.5	253.2	23.3	141.2	500.4	39.7	533.2	171.1	23.0	27.2	16.9	22.8	76.9	5.6	31.1	69.0	4.0	31.3	1.6	0
34	111.3	280.0	23.7	185.2	533.6	28.4	572.5	180.8	23.2	27.4	18.3	28.2	82.8	8.3	34.1	73.6	6.3	34.6	2.7	3
35	54.7	153.0	19.0	142.1	594.1	60.1	465.5	174.4	24.6	16.9	18.6	23.7	81.2	9.9	27.9	50.8	6.0	24.5	0.6	2.2
36	127.6	350.8	24.9	201.8	497.7	20.0	501.4	177.1	23.6	24.6	19.0	25.4	78.7	7.0	26.5	55.3	4.7	26.6	1.1	5.5
Middle Member																				
37	69.5	217.2	17.9	126.4	708.5	96.9	535.2	199.5	21.6	19.4	20.5	19.6	73.1	11.3	29.8	62.8	8.1	29.8	2.6	6.7
38a	7.2	14.9	9.5	61.4	814.9	204.1	448.4	204.1	23.2	22.0	17.8	11.8	65.6	11.8	30.8	73.9	8.2	30.7	3.5	37.4
38b	6.0	10.8	11.9	79.9	1034.7	182.9	471.0	242.0	26.2	25.7	20.0	12.7	73.2	12.8	35.6	79.9	9.9	34.6	3.6	26.9
39	84.9	301.9	14.5	111.5	622.5	145.3	512.9	175.5	17.8	23.8	16.1	26.2	64.1	8.6	33.5	64.9	6.9	27.8	3.1	19.1
40	6.1	24.3	12.4	74.6	841.9	130.2	477.6	233.3	24.0	24.9	20.0	11.9	70.6	12.4	36.9	74.8	9.0	34.7	2.6	11.7
41	15.0	13.0	10.0	39.7	687.9	127.4	424.0	200.4	23.2	18.7	15.8	13.3	83.7	10.6	31.6	67.5	10.3	31.2	1.1	8.7
42	7.2	13.4	9.7	52.0	776.3	103.3	527.2	204.1	21.7	20.2	17.0	8.1	63.4	11.8	29.1	62.9	8.3	29.3	3.4	5.3
43	8.1	17.8	12.1	83.6	860.6	104.0	547.4	248.7	33.2	26.7	20.3	58.2	66.1	21.1	44.8	81.1	10.4	36.8	3.0	3.0
44	11.7	32.4	15.3	109.9	711.1	95.8	450.4	224.8	25.1	25.0	20.8	12.8	73.9	14.1	31.5	72.8	10.2	33.6	2.8	3.7
45	6.2	12.0	9.2	42.4	783.1	74.3	361.9	223.7	25.0	21.8	17.6	15.0	64.1	13.3	36.1	64.3	9.2	28.1	2.1	0.8
46	10.2	20.4	13.5	85.7	823.2	116.2	526.9	247.5	26.8	26.3	21.0	12.0	77.0	15.1	40.8	77.0	10.4	33.2	3.0	4.5
47	10.5	23.2	14.6	95.6	852.3	108.1	485.4	249.4	27.4	27.2	21.0	13.5	77.9	14.8	38.2	83.2	10.2	37.0	3.8	5.7
20	46.5	65.5	9.7	76.3	836.8	99.2	489.0	181.8	17.8	12.9	20.6	19.3	80.1	15	24.9	55.5	9.3	23.4	2.3	3.7

[1] Concentrations in parts per million (ppm)

Table 9. *Lower and Middle Quechan Volcanics—ICP-MS analysis of rare earth elements (REE)*[1]

Sample	La	Ce	Pr	Nd	Sm	Eu	Gd	Tb	Dy	Ho	Er	Tm	Yb	Lu
Lower Quechan														
30	37.69	73.78	8.85	34.21	6.67	2.01	5.67	0.86	4.76	0.92	2.35	0.32	1.92	0.30
31	28.12	53.71	6.61	25.35	4.97	1.50	4.38	0.66	3.80	0.73	1.90	0.26	1.61	0.24
32	37.55	74.29	8.90	34.52	6.85	2.11	5.80	0.88	5.12	1.00	2.60	0.37	2.26	0.35
33	29.70	59.08	7.53	29.75	6.17	1.87	5.46	0.83	4.72	0.89	2.27	0.31	1.89	0.28
34	32.96	64.58	8.13	31.95	6.56	1.96	5.68	0.86	4.88	0.94	2.40	0.33	2.00	0.31
35	25.31	49.11	6.24	24.45	5.27	1.61	5.03	0.81	4.80	0.93	2.48	0.35	2.18	0.33
36	27.90	53.85	6.85	27.25	5.69	1.81	5.21	0.81	4.68	0.91	2.38	0.33	2.05	0.32
Middle Quechan														
37	31.93	61.94	7.87	30.47	6.12	1.76	5.20	0.78	4.44	0.83	2.18	0.31	1.87	0.29
38a	34.85	66.75	8.27	31.38	6.04	1.57	5.17	0.79	4.52	0.87	2.32	0.32	2.01	0.31
38b	39.75	80.89	9.49	35.96	6.97	1.85	5.97	0.91	5.18	1.01	2.66	0.38	2.36	0.36
39	32.89	62.12	7.72	29.27	5.59	1.58	4.62	0.68	3.73	0.70	1.77	0.24	1.50	0.22
40	38.97	77.14	9.14	34.83	6.72	1.75	5.69	0.86	4.86	0.94	2.50	0.35	2.20	0.34
41	33.42	63.30	7.82	29.42	5.88	1.44	5.11	0.79	4.51	0.88	2.32	0.33	2.00	0.31
42	32.67	61.74	7.62	28.57	5.63	1.41	4.76	0.74	4.29	0.85	2.22	0.31	1.96	0.31
43	42.75	85.52	10.00	37.69	7.32	1.89	6.46	1.00	5.87	1.18	3.10	0.44	2.74	0.43
44	36.10	68.98	8.46	31.78	6.35	1.63	5.63	0.86	4.96	0.97	2.53	0.36	2.22	0.34
45	35.74	68.48	8.35	31.51	6.13	1.58	5.37	0.84	4.86	0.95	2.53	0.35	2.26	0.34
46	41.29	83.07	9.73	36.89	7.14	1.87	6.07	0.94	5.34	1.03	2.70	0.38	2.41	0.36
47	41.47	83.57	9.85	37.42	7.30	1.90	6.42	0.95	5.51	1.07	2.85	0.40	2.49	0.38
20	28.77	52.89	6.45	23.94	4.78	1.33	4.12	0.62	3.48	0.65	1.73	0.25	1.48	0.22
37	31.93	61.94	7.87	30.47	6.12	1.76	5.20	0.78	4.44	0.83	2.18	0.31	1.87	0.29
38a	34.85	66.75	8.27	31.38	6.04	1.57	5.17	0.79	4.52	0.87	2.32	0.32	2.01	0.31
38b	39.75	80.89	9.49	35.96	6.97	1.85	5.97	0.91	5.18	1.01	2.66	0.38	2.36	0.36
39	32.89	62.12	7.72	29.27	5.59	1.58	4.62	0.68	3.73	0.70	1.77	0.24	1.50	0.22
40	38.97	77.14	9.14	34.83	6.72	1.75	5.69	0.86	4.86	0.94	2.50	0.35	2.20	0.34
41	33.42	63.30	7.82	29.42	5.88	1.44	5.11	0.79	4.51	0.88	2.32	0.33	2.00	0.31

[1] Concentrations in parts per million (ppm)

Table 10. *Lower and Middle Quechan Volcanics—ICP-MS analysis of selected trace elements*[1]

Sample	Ba	Th	Nb	Y	Hf	Ta	U	Pb	Rb	Cs	Sr	Sc	Zr
Lower Quechan													
30	639	8.36	28.89	22.84	5.02	1.82	2.42	7.09	57.4	2.67	666	20.5	205
31	446	5.23	22.81	19.28	3.28	1.34	1.48	7.13	36.9	0.81	493	16.9	139
32	618	7.75	28.67	25.15	4.91	1.73	2.64	7.83	37.5	3.29	572	24.7	202
33	484	4.93	24.55	22.28	3.86	1.52	1.03	5.39	37.0	0.80	546	23.0	161
34	520	6.22	26.01	23.46	4.28	1.60	1.77	6.96	26.1	2.46	595	22.6	175
35	587	6.97	16.34	23.98	4.42	1.13	2.13	10.01	56.9	2.70	463	18.9	171
36	496	6.31	22.59	22.68	4.26	1.41	1.48	6.72	18.8	3.17	500	23.1	170
Middle Quechan													
37	702	7.67	18.87	21.45	4.91	1.20	2.30	10.81	94.3	5.30	559	19.0	199
38a	810	8.63	21.74	23.01	5.00	1.38	2.88	11.00	198.9	30.93	447	11.9	204
38b	1037	10.14	25.26	26.65	5.84	1.62	2.99	13.19	176.5	23.07	483	12.6	252
39	619	6.77	22.89	17.87	4.18	1.45	3.08	8.78	137.5	20.88	525	16.8	172
40	837	9.75	24.53	24.19	5.67	1.58	3.22	12.74	124.0	7.91	472	13.1	234
41	688	9.33	18.32	22.84	4.99	1.28	1.73	10.82	124.2	9.88	422	10.0	200
42	771	8.94	19.51	21.46	5.08	1.28	3.03	12.00	100.2	5.08	543	9.7	203
43	847	10.35	25.72	32.69	6.01	1.67	2.64	19.51	97.5	3.34	563	13.4	251
44	705	10.74	24.69	25.16	5.84	1.69	2.78	13.06	90.0	2.83	446	15.5	221
45	789	10.18	21.37	24.76	5.58	1.43	2.47	12.95	70.9	1.04	360	10.0	224
46	816	10.32	25.84	26.43	6.02	1.66	3.66	14.10	109.0	4.00	539	13.6	249
47	840	10.43	26.09	28.19	6.14	1.73	2.64	14.23	100.7	3.37	476	14.0	246
20	836	9.32	12.22	17.58	4.53	0.84	2.58	14.50	95.1	2.98	485	10.4	178

[1] Concentrations in parts per million (ppm)

Table 11. *Locations of upper Quechan volcanics analyzed at WSU*

Sample	Latitude	Longitude
Upper		
1:clast in lahar	N32°58′42.11″	W114°39′1.62″
2: clast in flow breccia	N32°58′43.8″	W114°39′0.4″
3: clast in flow breccia	N32°58′45.96″	W114°38′56.76″
4b: massive flow	N32°58′47.28″	W114°38′55.69″
5a: clast in flow breccia	N32°58′42″	W114°38′37.57″
5b: intrusive plug?	30 ft plug above flow at 5a	
6: massive flow	N32°58′31.69″	W114°38′41.27″
7:matrix flow breccia	12 ft of flow breccia above 6	
8: clast flow breccia	N32°58′31.84″	W114°38′43.93″
9: massive flow	60 ft flow above 8	
11: massive flow	78 ft flow above 45 ft flow above 9	
12: clast in flow breccia	Flow above 11	
13: visculated massive flow	Flow above 13	
14a: clast	N32°59′17.82″	W114°37′48.99″
14b: flow breccia	Flow breccia containing 14a (clast)	
15: flow breccia	5′ flow breccia above 14	
16: massive flow	7′ flow above 15	
17: clast in breccia	8′ lahar flow breccia above 16	
18: clast in breccia	10′ flow breccia above 17	
19: massive flow	10′ massive flow above 18	
21: massive flow	N32°59′18.35″	W114°37′51.89″
22: clast in flow breccia	7′ flow breccia above 21	
23: massive flow	2′ massive flow above 22	
24: flow breccia	5′ flow breccia above 23	
25: matrix in flow breccia	N32°59′29.48″	W114°37′54.08″
26: vesiculated flow	>10′ flow–4th flow above 25	

Table 12. *Upper Quechan Volcanics—Results from XRF analysis of major elements*[1]

Sample	SiO_2	TiO_2	Al_2O_3	FeO[2]	MnO	MgO	CaO	Na_2O	K_2O	P_2O_5	LOI	Sum
1	51.5	1.19	13.2	7.38	0.09	4.08	10.10	2.52	2.42	0.42	7.68	100.6
2	43.8	0.97	11.4	5.97	0.22	3.39	14.72	1.99	3.39	0.37	13.1	99.3
3	50.1	1.33	14.9	8.19	0.10	5.24	7.82	2.88	2.67	0.36	5.78	99.4
4b	52.1	1.38	15.5	7.30	0.15	5.76	6.79	2.94	2.76	0.39	3.96	99.0
5a	47.2	1.63	14.7	9.12	0.08	7.05	6.30	3.00	2.13	0.38	7.30	98.9
5b	51.0	1.95	15.7	8.65	0.09	5.57	6.60	3.26	2.03	0.41	4.24	99.5
6	50.9	1.40	15.5	7.72	0.13	6.09	7.22	3.00	2.65	0.37	3.81	98.8
7	50.4	1.33	15.0	8.18	0.10	6.39	5.69	2.76	3.03	0.36	5.64	98.9
8	50.9	1.34	15.2	8.29	0.12	6.55	5.63	2.57	2.23	0.32	6.19	99.4
9	52.4	1.42	15.8	7.43	0.07	5.39	6.43	2.99	2.73	0.38	4.44	99.4
11	52.5	1.37	15.4	7.39	0.08	5.37	7.12	3.00	2.69	0.38	3.88	99.1
12	51.6	1.37	15.4	7.35	0.07	5.65	6.20	2.98	2.88	0.39	5.15	99.0
13	50.5	1.32	15.0	7.17	0.06	6.95	6.50	2.64	2.56	0.44	5.94	99.1
14a	48.0	1.79	14.7	9.05	0.13	7.04	6.73	2.99	2.01	0.37	6.04	98.9
14b	47.8	1.70	14.0	8.58	0.13	6.58	6.99	2.73	1.76	0.35	8.14	98.7
15	48.5	1.76	14.1	9.22	0.10	7.69	5.60	2.67	1.87	0.37	7.08	99.0
16	51.1	1.88	15.2	8.64	0.11	5.45	7.43	3.11	1.88	0.40	3.95	99.1
17	49.8	1.69	13.6	8.96	0.23	7.77	5.52	2.62	2.04	0.35	6.88	99.4
18	48.8	1.66	13.6	8.70	0.16	6.95	7.41	2.70	2.03	0.32	7.03	99.3
19	50.5	1.67	14.8	8.74	0.17	5.26	7.94	3.04	2.06	0.42	5.10	99.7
21	51.0	1.74	15.0	8.22	0.19	5.17	8.14	3.13	2.04	0.40	4.78	99.8
22	49.5	1.68	14.5	8.75	0.11	5.66	7.55	2.88	2.20	0.39	6.12	99.3
23	50.8	1.59	15.4	8.21	0.11	5.47	8.18	3.16	2.08	0.39	4.89	100.4
24	49.5	1.45	14.3	9.15	0.09	7.87	4.69	2.94	2.15	0.34	6.77	99.2
25	47.4	1.51	13.6	8.73	0.12	5.01	6.78	2.46	2.46	0.37	11.3	99.8
26	48.7	1.80	16.4	7.35	0.04	7.14	5.14	3.19	2.21	0.41	6.64	99.0

[1] Concentration in weight percent (wt.%) [2] Total iron as FeO

Table 13. *Upper Quechan Volcanics—Results from XRF analysis trace elements*[1]

Sample	Ni	Cr	Sc	V	Ba	Rb	Sr	Zr	Y	Nb	Ga	Cu	Zn	Pb	La	Ce	Th	Nd	U	Cs
1	125.2	241.1	18.5	94.6	741.6	57.1	565.2	198.4	18.3	18.0	16.0	29.7	67.6	12.8	35.0	74.0	9.9	36.1	1.2	2.6
2	131.8	206.2	13.8	78.3	1432.1	81.0	562.6	224.8	19.4	14.3	15.5	31.4	57.2	14.9	44.9	85.9	12.3	40.4	4.3	0
3	146.5	291.2	20.2	146.8	775.6	80.3	566.8	210.7	22.8	19.6	18.2	26.3	77.2	8.9	37.9	72.4	10.6	35.2	2.5	1.6
4b	143.6	287.6	22.6	192.2	806.6	86.6	573.2	220.1	21.3	21.5	18.6	20.3	78.4	9.8	37.1	78.0	12	35.7	3.5	3.6
5a	183.0	282.3	20.9	163.4	566.5	63.9	561.4	203.9	20.8	26.0	16.5	20.4	84.3	7.0	33.1	67.5	6.6	27.9	1.6	2.4
5b	123.2	267.7	24.4	343.1	643.1	61.2	607.1	207.6	27.4	27.9	19.8	33.6	86.5	9.7	36.4	74.8	8.6	35.1	2.0	7.0
6	140.5	289.1	22.9	247.9	754.0	80.6	587.8	217.5	22.7	21.9	18.4	42.6	72.9	9.5	41.5	86.5	10.8	38.8	2.4	1.5
7	133.8	262.5	22.0	105.8	799.1	95.9	585.7	208.5	23.0	20.7	19.1	36.7	67.4	9.7	37.4	77.8	10.7	35.6	2.3	3.4
8	132.0	239.8	21.9	60.5	705.6	44.7	589.9	202.8	22.3	20.0	19.3	34.4	71.9	10.0	35.3	71.9	10.3	34.6	3.7	3.8
9	137.4	287.6	23.4	178.5	777.2	86.3	578.5	219.9	22.7	22.2	19.8	27.9	77.4	10.2	38.4	77.2	10.6	36.2	3.7	3.6
11	153.3	288.0	23.1	169.9	772.0	81.4	582.0	215.7	23.1	20.9	19.2	37.2	81.8	10.0	38.8	80.8	10.7	36.0	2.6	0.9
12	144.0	279.9	20.8	180.6	789.9	91.4	612.7	220.9	21.7	21.5	18.1	26.7	82.2	10.0	40.6	81.0	11.0	38.9	4.0	1.0
13	139.3	268.3	20.4	163.5	857.4	66.7	695.5	226.5	27.3	19.7	17.1	41.6	81.0	9.8	47.2	92.7	10.5	43.9	1.9	1.3
14a	128.8	240.2	21.8	175.6	597.4	62.5	528.1	193.6	25.1	26.0	18.1	16.4	78.3	5.5	31.3	64.8	6.5	30.0	1.8	3.3
14b	132.7	245.0	20.5	181.9	523.1	52.5	484.4	184.7	24.2	24.5	17.9	29.5	81.7	6.7	31.2	60.1	6.1	29.2	1.3	3.0
15	131.0	237.8	20.6	151.1	546.3	56.8	497.6	185.3	26.3	24.2	17.0	23.7	85.6	9.1	35.1	67.1	7.5	32.7	3.2	3.1
16	131.9	255.9	23.6	188.5	607.2	54.8	526.4	197.6	27.1	26.2	20.2	38.0	81.1	8.9	35.7	69.0	8.7	35.6	0.3	1.6
17	127.6	229.7	20.6	111.7	742.6	64.1	538.3	178.7	22.4	23.6	18.0	35.5	83.6	9.1	31.2	62.6	7.6	30.4	1.3	0
18	126.5	237.2	20.4	138.4	579.0	61.1	533.2	179.5	23.2	23.7	16.8	34.3	80.0	9.0	34.7	67.3	7.4	32.0	2.0	0
19	159.2	293.5	22.2	207.7	764.0	63.5	567.8	196.1	26.5	25.1	17.5	33.9	85.0	11.3	44.7	85.2	12.5	40.0	2.8	3.5
21	147.7	285.5	23.7	175.2	732.8	60.8	603.0	198.3	26.5	24.4	18.4	35.8	83.9	10.7	44.2	85.3	11.1	39.1	3.7	3.4
22	143.3	277.2	22.2	178.6	709.6	64.7	614.8	191.2	25.4	24.5	17.6	16.9	75.8	9.3	44.5	77.3	11.1	35.8	4.0	5.8
23	156.3	287.3	22.8	179.2	751.2	61.9	625.3	198.7	26.0	24.9	19.4	43.0	79.0	10.2	42.7	82.3	11.7	36.2	3.7	0
24	149.4	269.4	21.1	136.9	710.2	66.1	518.5	181.7	19.6	23.2	17.4	17.6	81.2	9.2	36.4	76.8	10.5	32.6	2.9	2.0
25	160.4	302.9	19.4	152.8	675.7	67.3	638.0	196.3	22.9	25.7	16.5	48.8	70.6	8.9	32.0	69.8	7.0	33.0	2.5	2.4
26	192.2	284.6	22.5	175.4	614.3	65.4	586.9	221.9	20.1	28.6	18.5	54.7	95.8	8.3	36.9	69.7	6.1	32.2	1.8	2.9

[1] Concentrations in parts per million (ppm)

Table 14. *Upper Quechan Volcanics—ICP-MS analysis of rare earth elements (REE)*[1]

Sample	La	Ce	Pr	Nd	Sm	Eu	Gd	Tb	Dy	Ho	Er	Tm	Yb	Lu
1	33.86	72.29	8.57	34.18	6.64	1.86	5.08	0.70	3.70	0.69	1.73	0.23	1.47	0.22
2	45.36	86.20	10.35	39.57	7.34	1.85	5.44	0.73	3.72	0.69	1.74	0.24	1.45	0.22
3	40.36	78.16	9.45	36.58	7.14	2.02	5.72	0.83	4.62	0.87	2.26	0.31	1.93	0.29
4b	39.62	77.09	9.40	36.32	7.03	2.05	5.68	0.81	4.46	0.84	2.11	0.29	1.72	0.25
5a	33.32	63.58	7.93	30.87	6.11	1.89	5.16	0.79	4.52	0.87	2.30	0.33	2.03	0.30
5b	35.72	71.75	8.65	33.69	6.87	2.08	5.97	0.93	5.24	1.02	2.62	0.37	2.22	0.34
6	40.41	79.88	9.69	37.34	7.28	2.11	5.98	0.87	4.76	0.90	2.26	0.31	1.88	0.27
7	36.63	74.37	8.94	35.10	6.95	2.02	5.69	0.83	4.60	0.88	2.29	0.32	1.93	0.31
8	35.40	67.99	8.21	31.79	6.30	1.85	5.25	0.80	4.51	0.87	2.27	0.31	1.93	0.30
9	39.25	77.19	9.35	36.36	7.16	2.06	5.82	0.86	4.69	0.91	2.32	0.32	1.94	0.30
11	39.63	78.52	9.50	36.93	7.15	2.01	5.79	0.85	4.67	0.88	2.26	0.31	1.93	0.29
12	43.06	85.47	10.30	40.02	7.75	2.18	6.03	0.86	4.71	0.89	2.24	0.31	1.90	0.29
13	46.88	92.80	11.36	44.29	8.44	2.43	6.75	0.96	5.31	1.01	2.58	0.35	2.12	0.32
14a	31.49	60.51	7.63	30.26	6.19	1.93	5.62	0.87	4.98	0.99	2.55	0.36	2.20	0.34
14b	30.90	59.13	7.50	29.49	6.08	1.86	5.43	0.84	4.87	0.95	2.46	0.35	2.12	0.32
15	34.98	65.42	8.34	32.57	6.70	2.02	6.06	0.92	5.31	1.05	2.67	0.38	2.32	0.35
16	36.02	68.71	8.67	34.25	6.96	2.13	6.23	0.96	5.51	1.07	2.79	0.39	2.37	0.36
17	31.49	61.27	7.56	29.39	5.97	1.79	5.20	0.81	4.65	0.90	2.34	0.33	2.04	0.31
18	31.36	61.05	7.64	29.87	6.01	1.83	5.35	0.82	4.78	0.93	2.43	0.33	2.12	0.31
19	43.58	82.45	10.06	38.52	7.53	2.17	6.29	0.92	5.26	1.00	2.61	0.37	2.22	0.34
21	42.06	80.66	9.80	38.16	7.57	2.20	6.38	0.95	5.36	1.03	2.68	0.37	2.27	0.34
22	39.37	77.64	9.50	36.89	7.35	2.16	6.18	0.93	5.29	1.02	2.66	0.37	2.25	0.35
23	41.50	78.85	9.55	36.51	7.24	2.12	6.12	0.90	5.20	1.00	2.58	0.36	2.18	0.34
24	36.93	73.56	8.50	32.34	6.26	1.88	5.22	0.76	4.26	0.80	2.07	0.28	1.75	0.26
25	33.34	64.86	8.15	31.81	6.35	1.90	5.42	0.83	4.64	0.88	2.30	0.32	1.91	0.29
26	34.85	67.00	8.48	32.84	6.56	2.05	5.58	0.82	4.52	0.85	2.10	0.28	1.74	0.26

[1] Concentrations in parts per million (ppm)

Table 15. *Upper Quechan Volcanics—ICP-MS analysis of selected trace elements*[1]

Sample	Ba	Th	Nb	Y	Hf	Ta	U	Pb	Rb	Cs	Sr	Sc	Zr
1	711	10.06	15.53	17.19	4.55	0.98	1.82	13.40	54.0	1.00	573	19.3	185
2	1417	12.07	12.13	18.41	5.27	0.78	3.17	15.08	74.1	0.66	571	15.2	226
3	772	11.41	18.67	21.89	5.10	1.19	2.68	9.22	76.0	1.78	589	22.0	205
4b	788	11.87	19.51	20.56	5.34	1.24	2.85	10.26	81.7	2.36	589	22.2	211
5a	560	7.67	25.13	21.06	4.82	1.57	2.12	7.72	60.4	1.89	585	20.5	198
5b	619	9.39	25.79	26.02	4.83	1.63	2.48	8.69	57.8	2.28	626	25.6	200
6	758	11.62	20.08	22.64	5.27	1.28	2.95	10.34	77.4	1.82	613	23.5	211
7	790	11.16	19.10	22.26	5.09	1.22	2.66	10.14	90.6	2.37	613	21.6	202
8	679	10.87	18.11	21.74	4.89	1.17	2.61	9.78	41.0	1.54	605	22.1	194
9	764	11.94	20.23	22.05	5.32	1.30	2.87	10.09	82.1	3.10	601	23.0	212
11	754	11.33	19.27	22.24	5.11	1.21	2.77	9.58	77.6	2.03	602	23.0	208
12	790	12.15	19.60	21.91	5.33	1.24	3.00	10.29	87.7	2.28	648	23.0	219
13	852	11.97	18.39	26.05	5.42	1.16	2.95	10.44	63.6	2.79	727	21.1	225
14a	588	8.22	24.70	24.79	4.71	1.59	2.04	6.69	60.0	1.58	559	22.3	188
14b	515	7.74	23.62	24.05	4.43	1.50	2.04	7.27	51.0	1.30	512	21.9	180
15	537	8.79	23.27	26.07	4.48	1.51	2.03	8.85	54.2	1.12	516	21.1	180
16	605	9.51	25.23	26.93	4.81	1.62	2.39	8.60	52.5	1.26	551	22.3	194
17	732	8.49	22.25	22.22	4.29	1.46	1.64	9.26	60.7	1.19	552	21.7	174
18	570	8.85	22.22	23.53	4.32	1.45	2.06	9.13	57.8	1.18	546	21.8	174
19	756	12.76	23.34	26.00	4.72	1.53	2.96	10.46	59.2	1.92	582	23.6	191
21	731	12.09	23.93	26.25	4.76	1.59	2.85	10.14	57.1	1.21	618	23.2	192
22	703	11.67	23.16	26.11	4.71	1.53	2.79	8.71	61.5	1.24	628	22.5	186
23	736	12.17	22.65	25.35	4.76	1.51	2.86	10.11	57.3	1.42	632	23.1	188
24	694	11.40	20.94	19.72	4.42	1.40	2.56	9.38	61.3	1.54	531	21.6	174
25	661	7.61	24.05	22.58	4.68	1.51	2.00	8.84	62.5	1.12	649	19.9	189
26	609	8.12	27.36	20.46	5.27	1.73	2.13	7.77	61.4	1.25	599	22.4	215

[1] Concentrations in parts per million (ppm)

Table 16. *Sample locations for Rojo Grande unit analyzed at WSU*

Sample	Latitude	Longitude
Rojo Grande Unit		
RE-3	N33° 1′ 1.22″	W114° 36′ 33.10″
RE-4	N33° 1′ 3.22″	W114° 36′ 28.05″
RE-5	N33° 1′ 3.55″	W114° 36′ 24.39″
RE-7	N33° 1′ 7.0″	W114° 36′ 36.78″
RE-12	N33° 0′ 49.05″	W114° 36′ 47.79″
RE-13	N33° 1′ 8.75″	W114° 36′ 44.12″
RE-14	N33° 1′ 11.92″	W114° 36′ 44.08″
RE-15	N33° 1′ 1.32″	W114° 36′ 45.60″
RE-16	N33° 1′ 8.82″	W114° 36′ 13.02″
RE-17	N33° 1′ 2.87″	W114° 36′ 13.58″
RE-18	N33° 1′ 9.07″	W114° 36′ 43.68″
RE-19	N33° 1′ 8.21″	W114° 36′ 14.40″
RE-20	N33° 0′ 57.29″	W114° 36′ 8.50″
RE-21	N33° 0′ 55.73″	W114° 36′ 17.04″
RE-22	N33° 0′ 58.24″	W114° 36′ 24.74″
RE-23	N33° 1′ 0.95″	W114° 36′ 29.60″

Table 17. *Rojo Grande Unit—Results from XRF analysis of major elements*[1]

Sample	SiO_2	TiO_2	Al_2O_3	FeO_2	MnO	MgO	CaO	Na_2O	K_2O	P_2O_5	LOI	Sum
RE-2	74.4	0.28	12.3	2.32	0.01	0.35	0.50	1.80	5.30	0.05	2.00	99.3
RE-3	78.0	0.06	8.92	0.87	0.01	0.05	2.07	0.41	6.40	0.03	2.57	99.4
RE-4	77.7	0.06	9.72	0.64	0.00	0.03	1.58	0.38	7.26	0.02	2.20	99.6
RE-5	75.1	0.06	9.49	1.44	0.02	0.04	2.69	0.37	7.22	0.05	2.83	99.3
RE-7	67.5	0.36	15.6	2.75	0.01	0.47	1.15	2.45	6.54	0.06	2.64	99.5
RE-12	72.9	0.32	13.2	2.29	0.02	0.23	0.34	0.94	7.49	0.07	2.03	99.8
RE-13	67.2	0.41	15.5	2.98	0.01	0.38	0.54	0.98	8.49	0.08	2.39	99.0
RE-14	67.0	0.40	15.3	3.06	0.01	0.63	0.65	2.40	6.08	0.08	3.88	99.5
RE-15	67.0	0.45	15.6	3.05	0.03	0.44	0.57	1.73	7.93	0.08	2.15	99.0
RE-16	64.5	0.42	16.2	3.03	0.05	0.43	1.73	3.51	5.77	0.08	3.47	99.2
RE-17	70.7	0.38	13.5	2.66	0.01	0.41	0.38	1.96	4.74	0.07	3.15	98.0
RE-18	65.7	0.40	16.0	2.91	0.01	0.38	0.61	1.15	9.22	0.08	2.43	98.9
RE-19	71.9	0.31	12.2	2.06	0.04	0.28	1.89	2.50	4.21	0.06	3.41	98.8
RE-20	67.0	0.39	15.4	2.91	0.01	0.34	1.08	2.43	6.77	0.07	2.78	99.3
RE-21	71.3	0.33	13.0	2.43	0.02	0.37	0.57	1.55	6.85	0.06	2.12	98.6
RE-22	67.3	0.38	15.2	3.10	0.01	0.37	0.32	1.17	8.88	0.07	1.66	98.4
RE-23	72.2	0.32	13.5	2.55	0.01	0.25	0.37	1.65	6.41	0.06	1.68	99.0

[1] Concentration in weight percent (wt.%) [2] Total iron as FeO

Table 18. *Rojo Grande Unit—Results from XRF analysis trace elements*[1]

Sample	Ni	Cr	Sc	V	Ba	Rb	Sr	Zr	Y	Nb	Ga	Cu	Zn	Pb	La	Ce	Th	Nd	U	Cs
RE-2	4.3	4.2	4.6	12.6	2166.2	207.5	116.2	395.1	19.6	20.6	16.5	6.2	44.4	18.3	34.5	68.7	14.6	25.5	4.7	11.4
RE-3	3.8	4.8	1.8	7.7	446.7	243.0	104.1	140.0	26.1	27.2	11.0	2.9	6.4	14.9	26.4	56.7	17.3	21.2	4.9	4.1
RE-4	3.8	2.6	0.9	8.6	400.2	309.9	55.1	142.3	25.4	28.8	13.8	1.3	2.5	10.5	24.9	52.7	18.1	20.3	5.7	3.2
RE-5	5.4	5.3	1.8	13.9	615.6	300.6	123.3	146.6	31.4	29.8	12.0	3.2	8.2	26.9	30.8	66.3	19.4	23.6	5.0	5.5
RE-7	5.1	2.9	5.3	11.4	2459.4	298.6	191.0	511.2	26.1	26.0	19.3	3.4	55.1	25.3	41.8	85.6	18.6	34.2	5.9	15.8
RE-12	5.5	5.6	4.6	14.7	1039.2	334.6	75.3	299.6	24.6	23.1	16.1	4.8	58.8	13.9	43.7	82.7	18.8	31.2	8.1	16.4
RE-13	4.3	4.1	5.7	20.9	2861.7	415.3	95.5	541.5	27.9	25.0	20.4	5.9	48.2	21.9	40.1	84.2	16.8	33.3	4.7	13.8
RE-14	3.9	2.9	6.1	23.8	2682.4	244.7	193.7	522.6	24.0	24.4	19.7	5.5	44.1	25.8	32.3	66.1	15.3	27.0	4.6	9.6
RE-15	7.1	2.6	5.1	18.1	3096.8	409.4	166.8	554.9	25.7	25.5	20.5	5.5	135.0	23.1	43.0	80.0	16.4	33.9	6.2	17.3
RE-16	4.4	5.9	5.7	20.8	2691.6	201.4	225.7	574.5	27.0	26.6	20.9	4.3	72.3	30.5	47.7	87.7	16.9	33.3	7.8	17.0
RE-17	4.9	3.4	5.8	12.7	2252.4	189.3	182.5	479.2	22.2	22.4	17.3	5.5	48.3	22.6	37.1	69.0	13.4	29.1	5.2	11.0
RE-18	5.3	4.3	5.2	17.1	2965.6	433.3	103.0	538.9	25.6	26.6	20.6	6.1	69.3	22.4	43.1	83.2	17.7	31.1	7.5	14.0
RE-19	4.0	3.3	3.9	11.0	2221.8	142.5	210.2	394.2	20.1	18.9	14.7	5.1	44.4	19.4	32.2	60.7	12.5	22.3	4.7	11.4
RE-20	5.2	3.2	4.9	13.1	2722.0	284.2	191.1	524.8	27.6	26.2	19.1	4.8	49.7	23.5	41.4	82.2	16.6	29.9	5.4	16.8
RE-21	6.6	4.4	5.3	8.8	2275.4	280.3	115.9	428.4	21.9	22.3	17.4	6.0	55.5	21.8	34.8	71.9	15.4	28.9	6.4	16.5
RE-22	8.6	3.3	5.8	17.3	2868.0	425.7	97.4	528.8	25.0	25.3	20.0	4.8	261.0	22.9	45.7	81.0	17.1	33.5	4.9	9.2
RE-23	4.7	3.2	4.4	13.6	2332.6	301.7	93.0	449.6	25.6	22.7	18.4	5.3	68.8	24.1	41.0	78.8	15.7	29.7	3.8	17.9

[1] Concentrations in parts per million (ppm)

Table 19. *Rojo Grande Unit—ICP-MS analysis of rare earth elements (REE)*[1]

Sample	La	Ce	Pr	Nd	Sm	Eu	Gd	Tb	Dy	Ho	Er	Tm	Yb	Lu
RE-2	38.23	72.87	8.30	28.78	5.38	1.18	4.10	0.63	3.66	0.71	1.98	0.30	1.95	0.30
RE-3	28.80	62.21	6.93	23.47	4.84	0.07	3.99	0.71	4.36	0.89	2.64	0.42	2.77	0.43
RE-4	27.40	58.03	6.83	23.36	4.89	0.05	4.05	0.73	4.53	0.95	2.79	0.44	2.89	0.44
RE-5	30.09	64.60	7.06	23.61	5.45	0.13	4.60	0.84	5.30	1.09	3.24	0.49	3.23	0.49
RE-7	46.72	89.62	9.96	35.34	6.47	1.50	5.37	0.83	4.96	0.98	2.68	0.39	2.55	0.40
RE-12	38.75	76.57	8.80	31.51	5.98	0.98	4.87	0.79	4.67	0.94	2.57	0.38	2.43	0.38
RE-13	42.65	84.82	9.64	34.85	6.60	1.79	5.68	0.89	5.23	1.04	2.83	0.42	2.69	0.42
RE-14	36.55	71.72	8.21	29.80	5.77	1.83	4.80	0.78	4.46	0.89	2.48	0.37	2.39	0.38
RE-15	37.93	76.99	8.81	31.76	6.12	1.98	5.08	0.82	4.72	0.96	2.58	0.38	2.46	0.40
RE-16	49.45	92.56	10.43	37.15	6.91	1.95	5.55	0.89	5.20	1.02	2.82	0.42	2.66	0.44
RE-17	38.03	73.17	8.18	29.14	5.55	1.66	4.61	0.72	4.19	0.83	2.30	0.33	2.20	0.35
RE-18	36.95	75.37	8.75	31.79	6.07	1.57	5.05	0.80	4.62	0.90	2.48	0.37	2.47	0.39
RE-19	35.23	68.11	7.39	26.20	4.96	1.47	4.07	0.66	3.76	0.75	2.04	0.30	1.95	0.31
RE-20	44.59	86.55	9.63	33.73	6.32	1.78	5.37	0.87	5.20	1.06	2.89	0.43	2.78	0.42
RE-21	39.58	77.92	8.47	29.90	5.63	1.32	4.63	0.73	4.31	0.85	2.35	0.34	2.24	0.35
RE-22	47.15	89.42	10.02	34.65	6.43	1.74	5.21	0.83	4.88	0.97	2.65	0.39	2.54	0.41
RE-23	43.60	85.00	9.21	32.31	6.16	1.52	5.06	0.80	4.58	0.94	2.60	0.38	2.39	0.38

[1]Concentrations in parts per million (ppm)

Table 20. *Rojo Grande Unit—ICP-MS analysis of selected trace elements*[1]

Sample	Ba	Th	Nb	Y	Hf	Ta	U	Pb	Rb	Cs	Sr	Sc	Zr
RE-2	2324	15.24	21.07	18.75	9.27	1.74	4.49	18.75	212.9	8.39	114	5.7	412
RE-3	463	17.69	27.98	24.62	5.19	1.96	4.65	14.12	252.1	3.37	102	1.5	143
RE-4	415	18.76	30.60	25.90	5.43	2.53	4.88	11.05	325.8	3.63	56	1.4	148
RE-5	640	20.18	31.01	30.91	5.49	2.18	5.38	27.02	311.2	3.64	122	1.5	149
RE-7	2633	18.14	25.87	25.47	11.68	1.78	5.24	25.39	287.5	13.36	179	7.1	518
RE-12	1060	18.69	23.48	23.51	7.92	1.92	5.63	14.45	325.0	12.87	71	5.5	321
RE-13	3091	16.80	25.46	27.68	12.44	1.80	5.15	21.71	406.3	13.61	92	7.7	573
RE-14	2887	15.81	24.32	22.86	11.63	1.66	5.22	24.85	236.9	7.81	181	7.2	528
RE-15	3349	15.75	25.63	24.32	12.67	1.89	5.14	22.92	394.8	15.73	158	7.9	590
RE-16	2910	18.26	26.90	27.27	12.97	1.89	7.17	29.98	200.2	13.84	219	7.9	588
RE-17	2463	14.29	22.24	22.11	10.62	1.55	5.36	22.01	190.3	10.87	177	6.8	493
RE-18	3212	17.22	25.95	22.03	12.37	1.85	5.98	21.09	409.1	12.05	96	7.4	554
RE-19	2418	13.05	19.53	19.92	9.30	1.63	4.39	19.50	146.6	10.20	209	6.2	425
RE-20	2949	17.68	25.48	27.76	12.00	1.98	5.76	23.36	278.2	13.56	182	7.7	540
RE-21	2469	15.97	22.22	22.11	10.30	1.90	4.72	21.24	280.0	12.00	113	6.7	450
RE-22	3116	17.96	25.68	25.70	12.23	2.07	5.22	23.18	426.1	12.21	96	7.9	549
RE-23	2532	16.40	22.77	24.79	10.67	1.99	5.20	22.56	299.0	14.57	90	6.8	471

[1] Concentrations in parts per million (ppm)

Table 21. *Sample locations for ignimbrite in lower part of White Wash unit analyzed at WSU*

Sample	Latitude	Longitude
White Wash Unit–Tan to White Ignimbrite		
RE-3	N33° 1′ 1.22″	W114° 36′ 33.10″
RE-4	N33° 1′ 3.22″	W114° 36′ 28.05″
RE-5	N33° 1′ 3.55″	W114° 36′ 24.39″
RE-6	N33° 1′ 11.52″	W114° 36′ 32.80″
RE-8	N33° 1′ 7.73″	W114° 36′ 42.60″
RE-9	N33° 1′ 1.01″	W114° 36′ 24.07″
RE-11	N33° 1′ 7.88″	W114° 36′ 27.49″
TT-1	N33° 1′ 4.70″	W114° 36′ 38.24″
TT-2	N33° 1′ 3.28″	W114° 36′ 36.57″
TT-3	N33° 1′ 0.99″	W114° 36′ 34.36″
TT-4	N33° 1′ 3.40″	W114° 36′ 34.18″
TT-5	N33° 1′ 4.98″	W114° 36′ 33.30″
TT-6	N33° 1′ 4.08″	W114° 36′ 29.88″
TT-7	N33° 1′ 4.82″	W114° 36′ 26.06″
TT-8	N33° 1′ 7.81″	W114° 36′ 29.83″
TT-9	N33° 1′ 0.61″	W114° 36′ 22.18″

Table 22. *White Wash Unit Ignimbrite—Results from XRF analysis of major elements*[1]

Sample	SiO_2	TiO_2	Al_2O_3	FeO_2	MnO	MgO	CaO	Na_2O	K_2O	P_2O_5	LOI	Sum
RE-3	78.0	0.06	8.92	0.87	0.01	0.05	2.07	0.41	6.40	0.03	2.57	99.4
RE-4	77.7	0.06	9.72	0.64	0.00	0.03	1.58	0.38	7.26	0.02	2.20	99.6
RE-5	75.1	0.06	9.49	1.44	0.02	0.04	2.69	0.37	7.22	0.05	2.83	99.3
RE-6	80.7	0.06	9.10	0.68	0.00	0.04	0.54	0.27	6.87	0.02	1.51	99.8
RE-8	78.6	0.07	10.1	0.64	0.00	0.12	0.76	0.24	7.29	0.01	1.61	99.4
RE-9	78.8	0.06	9.41	0.58	0.00	0.03	1.26	0.37	7.27	0.01	1.74	99.6
RE-11	77.3	0.07	10.1	1.04	0.00	0.06	1.12	0.30	7.79	0.02	1.77	99.5
TT-1	77.8	0.06	10.0	0.31	0.00	0.03	1.12	0.29	7.65	0.02	1.78	99.1
TT-2	79.4	0.06	8.96	0.79	0.00	0.04	0.78	0.43	6.65	0.02	1.69	98.8
TT-3	78.2	0.07	9.72	0.70	0.00	0.01	1.25	0.45	7.43	0.02	1.81	99.7
TT-4	74.7	0.06	9.60	0.75	0.00	0.13	2.53	0.42	7.23	0.02	2.98	98.5
TT-5	79.6	0.06	9.07	0.80	0.00	0.03	0.70	0.23	7.22	0.01	1.55	99.3
TT-6	77.4	0.06	8.91	0.55	0.00	0.04	2.10	0.21	7.00	0.01	2.68	98.9
TT-7	75.9	0.06	9.44	0.50	0.01	0.03	2.73	0.21	7.40	0.02	3.10	99.4
TT-8	78.2	0.06	9.91	0.60	0.00	0.02	0.80	0.17	7.54	0.02	2.39	99.7
TT-9	78.4	0.06	8.81	0.63	0.00	0.07	1.68	0.24	6.82	0.02	1.60	98.4

[1] Concentration in weight percent (wt.%) [2] Total iron as FeO

Table 23. *White Wash Unit Ignimbrite—Results from XRF analysis trace elements*[1]

Sample	Ni	Cr	Sc	V	Ba	Rb	Sr	Zr	Y	Nb	Ga	Cu	Zn	Pb	La	Ce	Th	Nd	U	Cs
RE-3	3.8	4.8	1.8	7.7	446.7	243.0	104.1	140.0	26.1	27.2	11.0	2.9	6.4	14.9	26.4	56.7	17.3	21.2	4.9	4.1
RE-4	3.8	2.6	0.9	8.6	400.2	309.9	55.1	142.3	25.4	28.8	13.8	1.3	2.5	10.5	24.9	52.7	18.1	20.3	5.7	3.2
RE-5	5.4	5.3	1.8	13.9	615.6	300.6	123.3	146.6	31.4	29.8	12.0	3.2	8.2	26.9	30.8	66.3	19.4	23.6	5.0	5.5
RE-6	4.0	4.4	1.4	21.3	230.7	289.5	74.4	140.5	30.6	26.6	10.1	1.4	6.5	10.5	31.9	59.0	17.3	23.2	5.2	3.6
RE-8	3.9	2.7	1.6	6.1	415.2	298.8	71.3	152.8	29.9	30.2	14.2	1.6	15.8	11.8	25.4	51.1	20.0	20.6	5.9	10.6
RE-9	3.1	3.6	0.7	4.4	575.5	290.7	49.8	141.5	25.5	28.3	12.1	1.1	5.3	15.9	25.5	52.3	18.6	21.1	3.7	2.4
RE-11	5.0	5.4	1.6	2.4	307.0	317.6	87.5	156.8	26.0	31.4	12.8	3.1	9.4	24.6	30.1	62.7	20.0	23.4	4.3	7.0
TT-1	3.9	3.4	1.9	2.8	273.7	322.7	39.5	150.5	29.4	30.9	14.0	1.1	9.1	13.0	25.2	57.2	19.5	23.2	4.4	2.4
TT-2	3.6	3.9	1.4	5.3	343.4	268.0	41.7	141.8	32.3	28.1	11.9	1.5	8.2	17.5	29.5	57.3	16.8	23.9	4.1	5.2
TT-3	3.4	3.8	1.4	10.3	371.6	291.2	97.8	149.9	27.3	29.1	11.4	1.1	5.7	24.6	27.1	66.6	18.3	23.7	5.3	4.9
TT-4	4.1	4.0	1.7	8.4	338.5	281.9	109.9	146.7	33.6	31.3	12.1	3.7	6.4	22.1	27.1	63.6	18.8	24.9	4.2	4.4
TT-5	4.4	4.5	2.3	6.8	460.7	290.1	73.2	144.8	31.2	29.3	9.3	1.6	10.4	15.7	25.0	57.3	17.8	22.4	5.8	4.0
TT-6	3.7	3.9	1.6	6.8	299.0	279.1	64.6	133.7	24.5	27.1	9.5	1.5	6.8	11.1	25.2	58.7	16.9	22.4	4.4	3.9
TT-7	4.6	3.9	1.5	8.3	550.6	311.8	88.7	142.5	24.9	28.9	11.2	2.9	4.9	15.5	22.9	43.5	18.2	16.4	4.2	1.1
TT-8	4.0	3.7	1.7	6.7	211.4	313.1	77.8	148.6	22.7	29.6	12.8	1.6	3.3	12.9	27.0	47.1	18.5	15.4	5.0	4.8
TT-9	4.6	3.0	1.5	6.4	301.2	291.6	65.0	134.1	22.2	27.3	9.7	2.3	4.0	10.6	23.5	58.3	17.0	21.7	3.8	4.8

[1] Concentrations in parts per million (ppm)

Table 24. *White Wash Unit Ignimbrite—ICP-MS analysis of rare earth elements (REE)*[1]

Sample	La	Ce	Pr	Nd	Sm	Eu	Gd	Tb	Dy	Ho	Er	Tm	Yb	Lu
RE-3	28.80	62.21	6.93	23.47	4.84	0.07	3.99	0.71	4.36	0.89	2.64	0.42	2.77	0.43
RE-4	27.40	58.03	6.83	23.36	4.89	0.05	4.05	0.73	4.53	0.95	2.79	0.44	2.89	0.44
RE-5	30.09	64.60	7.06	23.61	5.45	0.13	4.60	0.84	5.30	1.09	3.24	0.49	3.23	0.49
RE-6	28.89	60.82	6.94	23.64	5.13	0.06	4.43	0.82	5.26	1.10	3.16	0.48	3.23	0.50
RE-8	26.89	59.52	6.64	22.72	5.02	0.08	4.29	0.80	5.15	1.06	3.10	0.50	3.36	0.51
RE-9							no ICP-MS							
RE-11	31.19	65.66	7.73	26.19	4.79	0.06	3.59	0.62	3.99	0.90	2.73	0.46	3.16	0.51
TT-1	29.70	60.26	7.24	24.55	5.45	0.06	4.49	0.81	5.10	1.06	3.12	0.49	3.18	0.50
TT-2	30.87	66.04	7.66	25.95	5.39	0.06	4.61	0.85	5.31	1.10	3.11	0.46	3.06	0.48
TT-3	31.47	67.03	7.60	25.05	5.02	0.06	4.14	0.76	4.81	1.02	3.02	0.47	3.09	0.48
TT-4	29.50	68.44	7.58	25.88	5.94	0.07	4.90	0.90	5.77	1.20	3.57	0.56	3.65	0.56
TT-5	28.95	61.65	7.30	25.20	5.46	0.06	4.83	0.87	5.47	1.12	3.19	0.49	3.26	0.51
TT-6	28.38	61.33	6.96	22.18	4.68	0.05	3.83	0.67	4.23	0.90	2.67	0.43	2.92	0.47
TT-7	21.81	45.71	5.28	17.06	3.58	0.04	2.82	0.57	4.01	0.90	2.74	0.45	3.05	0.47
TT-8	26.57	57.24	6.58	22.01	4.76	0.06	3.83	0.64	3.92	0.81	2.43	0.40	2.66	0.42
TT-9	25.75	50.93	5.41	16.90	3.74	0.04	3.09	0.58	3.90	0.84	2.50	0.42	2.91	0.46

[1] Concentrations in parts per million (ppm)

Table 25. *White Wash Unit Ignimbrite—ICP-MS analysis of selected trace elements*[1]

Sample	Ba	Th	Nb	Y	Hf	Ta	U	Pb	Rb	Cs	Sr	Sc	Zr
RE-3	463	17.69	27.98	24.62	5.19	1.96	4.65	14.12	252.1	3.37	102	1.5	143
RE-4	415	18.76	30.60	25.90	5.43	2.53	4.88	11.05	325.8	3.63	56	1.4	148
RE-5	640	20.18	31.01	30.91	5.49	2.18	5.38	27.02	311.2	3.64	122	1.5	149
RE-6	234	17.87	27.40	30.04	5.26	2.56	4.67	10.05	292.2	3.87	74	1.6	143
RE-8	431	20.07	30.66	28.96	5.75	2.63	5.45	10.65	301.5	6.02	69	1.6	156
RE-9							no ICP-MS						
RE-11	315	20.51	30.45	24.34	5.93	2.11	4.43	23.71	320.2	3.49	85	1.7	159
TT-1	279	20.17	32.55	28.73	5.70	3.21	5.35	11.98	332.9	3.99	40	1.7	154
TT-2	358	18.03	29.02	30.81	5.17	3.21	4.51	17.17	280.3	3.79	42	1.7	146
TT-3	381	19.57	29.95	27.05	5.60	2.93	4.53	24.20	293.0	3.61	95	1.7	154
TT-4	355	20.55	33.20	33.61	5.72	2.77	4.71	22.33	293.0	3.00	110	1.7	154
TT-5	474	18.93	30.86	31.69	5.35	2.99	4.72	15.02	307.3	3.52	75	1.6	147
TT-6	309	18.15	29.32	24.84	5.10	2.79	4.64	11.45	296.9	2.87	67	1.6	140
TT-7	574	18.45	30.91	24.41	5.34	3.39	4.58	14.43	316.7	3.43	87	1.7	144
TT-8	309	17.47	28.81	21.86	5.05	3.77	4.49	10.11	301.8	4.06	65	1.5	138
TT-9	222	19.10	32.13	22.89	5.64	3.12	4.50	12.77	330.2	3.59	80	1.9	156

[1] Concentrations in parts per million (ppm)

Constraining the Source of the Orocopia Schist, SE California:
A Comparative Study of the major and trace element Chemistry of Cretaceous Sandstones deposited in Forearc Settings

DOUGLAS B. STEPHENSON AND GARY H. GIRTY
Department of Geological Sciences
San Diego State University
San Diego, California 92182

ABSTRACT

The Orocopia Schist and its equivalents are primarily quartzofeldspathic Late Cretaceous–Early Tertiary metagraywackes of oceanic origins. Current-day exposures of the schist occur along the San Andreas fault system and within the core of the Chocolate Mountains anticlinorium. The marine origin of the Orocopia Schist has led to numerous models for its emplacement, most involving subduction of sediments of either the Franciscan Complex or Great Valley sequence along the low-angle trajectory of a thickened segment of the Farallon plate. In order to evaluate the source, compositions, and protolith of the Orocopia Schist we collected chemical data from the literature and analyzed 29 samples from the Orocopia schist and 6 from the Point Loma Formation of the Rosario Group.

On a wide variety of major and trace element tectonic discrimination diagrams most samples analyzed from the Orocopia Schist consistently plot in the active continental margin field or spread from this field into the adjacent continental margin island arc field. These results indicate that the protolith of the Orocopia Schist was likely deposited in an active continental margin setting lying adjacent to a continental margin island arc.

The average composition of sandstone samples analyzed from the Franciscan Complex, the Great Valley sequence, and the Point Loma Formation clearly show a compositional change in a north to south direction, with samples from the Great Valley sequence in the Sacramento Valley forming one end member to the north and samples from the Point Loma Formation of the Rosario Group another to the south. Within this spectrum of compositions, on the pre-Neogene reconstruction of J. Saleeby, samples from the San Joaquin Valley lie just to the north and samples from the Point Loma Formation lie just to the south of the breached portion of the western Cordilleran margin lying above the low-angle subducting segment of the Farallon plate.

Most of the data from the Point Loma Formation fall within the compositional field defined by samples analyzed from the Orocopia Schist on plots of TiO_2 *versus* $Fe_2O_3^* + MgO$ *and La versus Th while data from the San Joaquin Valley do so less frequently. This observation suggests that the source of the protolith of the Orocopia Schist probably was located somewhere within the breached portion of the Cordilleran margin, and probably somewhere along its southern half.*

As discussed in the literature, during Late Cretaceous to Eocene time the forearc region of the breached segment of the Cordilleran margin likely consisted of a complex network of active normal faults. Across such a terrane sediments were

Adapted with permission from: Stephenson, D., and Girty, G., 2006, Constraining the source of the Orocopia Schist, SE California: A comparative study of the major and trace element chemistry of Cretaceous sandstones deposited in forearc settings, in Girty, G. H., and Cooper, J. D., eds., *Using Stratigraphy, Sedimentology, and Geochemistry to Unravel the Geologic History of the Southwestern Cordillera: A Volume in Honor of Patrick L. Abbott*: Pacific Section SEPM (Society for Sedimentary Geology), Book 101, p. 77–96.

fed westward into the trench. In this complex setting, resulting sediment would have been derived from both continental island arc and active margin sources such as fault-controlled uplifted volcanic and plutonic ridges. Given the documented history and likely latest Cretaceous to Paleogene age of the protolith of the Orocopia Schist, some of the sediment that made its way to the trench was carried to great depths along the low-angle but increasingly inclined trajectory of the subducting Farallon plate. Hence, the pre-Neogene reconstruction of J. Saleeby offers a plausible tectonic framework in which to view the otherwise problematic origins of the Orocopia Schist.

INTRODUCTION

THE OROCOPIA SCHIST, AND ITS LATERAL equivalents to the west and northwest, the Pelona and Rand schists, are a primarily quartzofeldspathic, homogenous, flaggy, meta-graywacke of Late Cretaceous to Early Tertiary age (Haxel et al., 2002; Jacobson et al., 2002). In southeastern California, the Orocopia and portions of the Pelona Schist (from here on referred to as the Orocopia Schist, unless otherwise noted) are exposed along the early to middle Miocene Chocolate Mountains anticlinorium (Fig. 1) (Grove et al., 2003; Haxel et al., 2002; Jacobson et al., 2002; Yin, 2002; Girty et al., 2005; this volume). Outcrops of the Pelona and Rand schists (as well as the schist of Sierra de Salinas and Portal Ridge) occur further to the north, along the Vincent thrust, San Andreas fault, and Garlock fault (Fig. 1A).

Approximately 10% of the Orocopia and related schists are made up of metamorphosed MORB-like basalts, metachert (quartzite), and marble (Haxel et al., 2002; Jacobson et al., 2002). As discussed in Haxel et al. (2002), the non-graywacke portion of the schist implies a eugeoclinal origin, an interpretation that is further supported by flysch-like layering, and a negative Ce anomaly that is characteristic of marine waters and sediments lying in intraplate to mid-ocean ridge oceanic settings (cf. Girty et al., 1996). Metamorphic mineral assemblages imply recrystallization depths as great as 25–35 km, with temperatures between 400–600°C very shortly after initial deposition of the protolith of the Orocopia Schist (Jacobson et al., 2002; Grove et al., 2003).

Current models for emplacement of the Orocopia Schist beneath the southwestern North American margin involve subduction of marine sediments along the low-angle trajectory of the Farallon plate (Jacobson et al., 1996; 2002; in press;

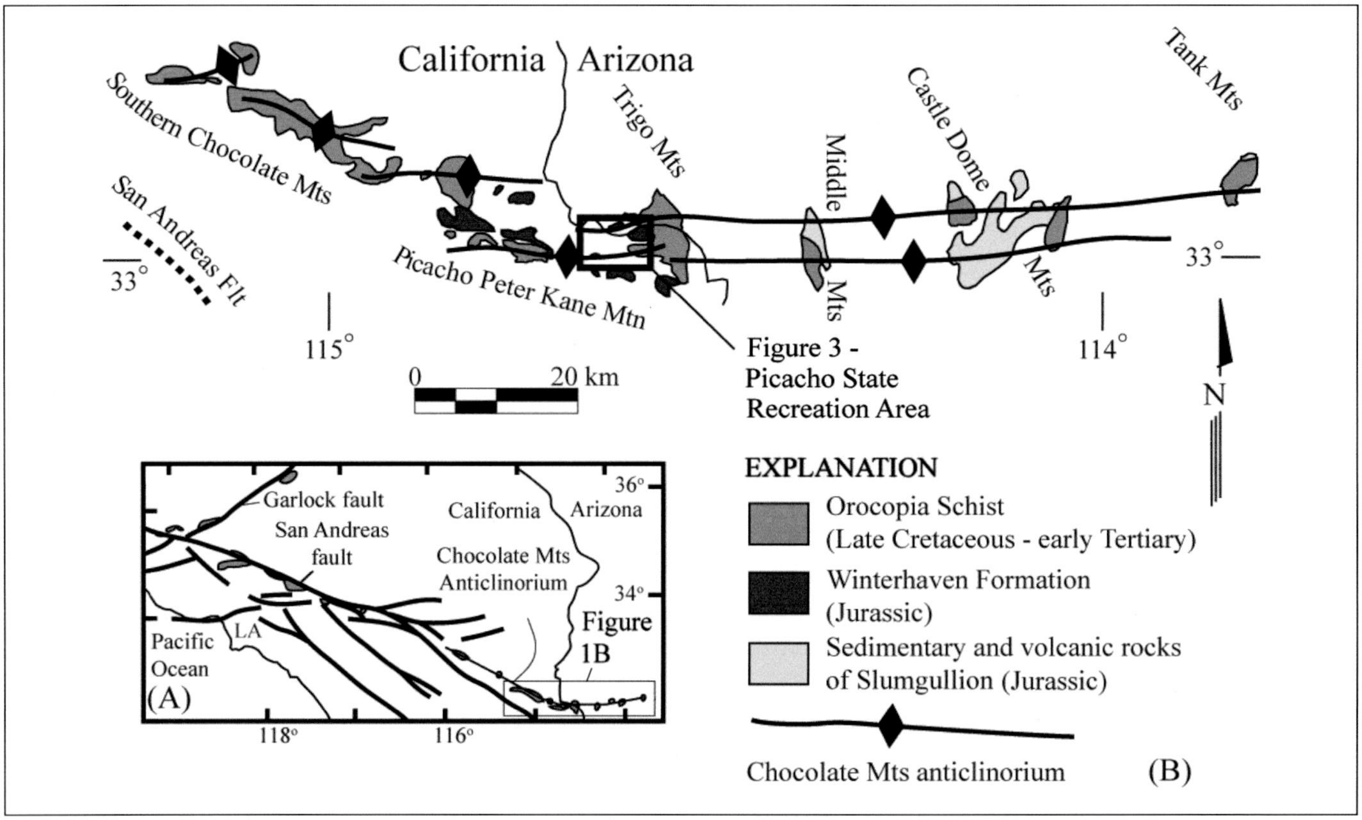

Figure 1. *(A) Location of (B) with respect to other prominent structural features in southern California. (B) Generalized map showing distribution of Orocopia Schist and equivalents in southeastern California and southwestern Arizona. Modified from Jacobson et al. (2002).*

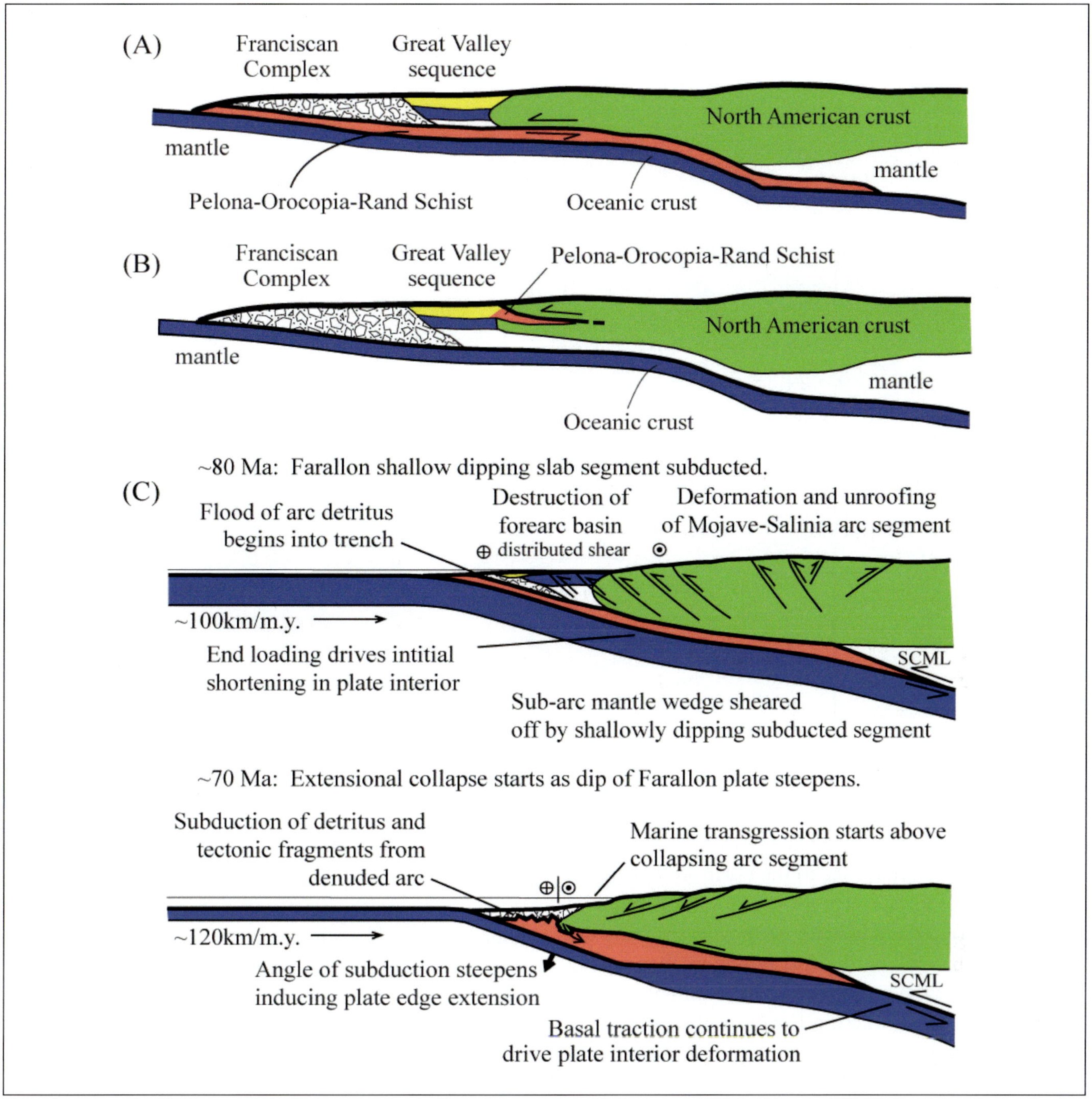

Figure 2. *Generalized tectonic models depicting the Orocopia Schist as derived from (A) trench sediments of the Franciscan Complex, (B) forearc basin sediments of the Great Valley sequence, and (C) trench sediments derived from an extensional borderland terrane and continental margin volcanic arc. Models (A) and (B) are from Haxel et al. (2002) while model (C) is from Saleeby (2003). SCML in (C) is sub-continental mantle lithosphere.*

Haxel et al., 2002; Grove et al., 2003; Saleeby, 2003). However, the exact nature and depositional setting of the protolith of the Orocopia Schist has yet to be resolved. For example, Haxel et al. (2002) proposed two tectonic models to explain the origin of the Orocopia. In one model, the protolith of the Orocopia Schist is trench fill of the Franciscan Complex (Fig. 2A), while in the other the protolith is represented by forearc basin deposits of the Great Valley sequence (Fig. 2B).

In a third model, subduction of a buoyant over thickened segment of the Farallon plate results in uplift and erosion of the forearc area and sediment transportation westward into the trench (Fig. 2C) (Saleeby, 2003). In this model, the Late Cretaceous to Paleogene forearc region undergoes extensional collapse in the wake of the low-angle subduction of the over thickened segment of the Farallon plate, and, as a result, does not represent a

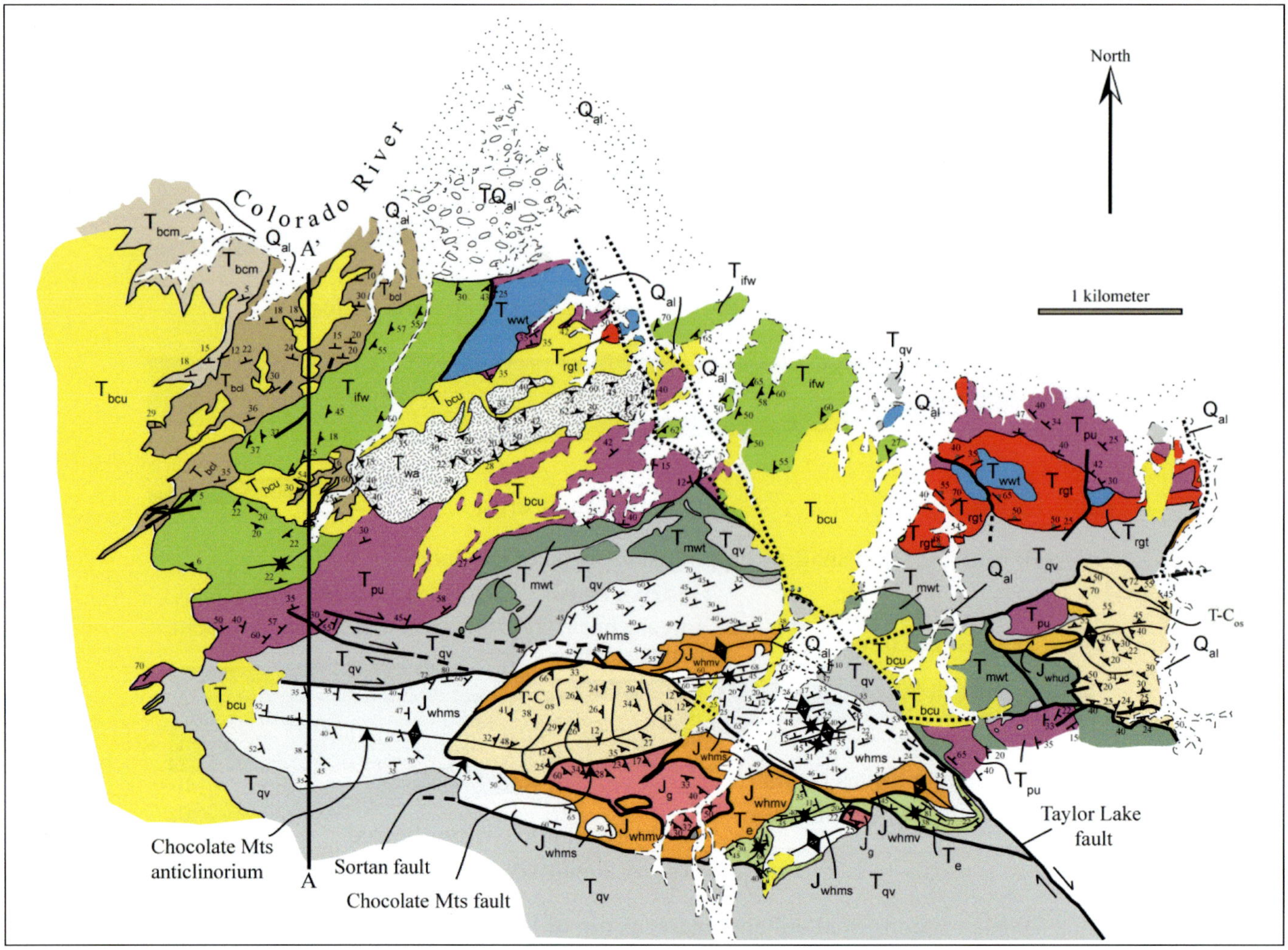

forearc basin like that in which temporally equivalent strata of the Great Valley sequence to the north were deposited. A more appropriate term for such a setting might be extensional borderland terrane. Sediments reaching the trench through such a borderland terrane are then carried downward along the low-angle but increasingly inclined trajectory of the subducting Farallon plate (Fig. 2C).

In this paper we compare the major and trace element chemistry of the Orocopia Schist to its most plausible protoliths, sandstones of the Franciscan Complex, Great Valley sequence, and Rosario Group. Due to the likelihood of metasomatism attending the amphibolite grade metamorphism of the Orocopia Schist, we here focus most of our discussion on elements that were most likely immobile under such conditions, and are routinely reported in the literature; i.e., Ti, Fe, Mg, La, and Th. These elements represent the major components of the tectonic provenance-discrimination models of Bhatia (1983) and Bhatia and Crook (1986). A thorough discussion of the full array of analyzed and collected data for this study can be found in Stephenson (2006).

PICACHO STATE RECREATION AREA

Lying along the northern flank of the Chocolate Mountains anticlinorium in southeastern most California is Picacho State Recreation Area (Fig. 1). Immediately to the south of Picacho, the Orocopia Schist lies in the core of the Chocolate Mountains anticlinorium, and can be subdivided into western and eastern belts, with each belt lying on opposite sides of the dextral-sense Taylor Lake fault (Fig. 3) (Girty et al., 2005). S1, the primary foliation in the schist, dips moderately to the west in the western belt, but to the east in the eastern belt. This variation in dip is most likely the result of rotations associated with movement along the Taylor Lake fault. In the field, S1 is a penetrative continuous foliation that is locally folded by rootless tight to isoclinal folds and in some areas is overprinted by a secondary spaced cleavage, S2. In both eastern and western belts, S1 defines several map-scale E–W folds. In the western belt, the largest of these folds is the Chocolate Mountains anticlinorium (Fig. 3).

FIGURE 3 LEGEND

	Q_{al} Holocene alluvium
	TQ_{al} old Colorado River gravels and sands
Bear Canyon conglomerate	T_{bcu} upper member
	T_{bcm} middle member
	T_{bcl} lower member
Oligocene volcanic sequence	T_{ifw} ignimbrite of Ferguson Wash (27–25 Ma)
	T_{wa} Walker andesite
	T_{wwt} White Wash trachyte
	T_{pu} pyroclastic sequence
	T_{rgt} Rojo Grande trachyte
	T_{mwt} Marcus Wash trachyte
	T_{qv} Quechan volcanics (33–27 Ma)
	T_e Eocene (?) sedimentary rocks
Basement complex	Orocopia Schist–latest Cretaceous–earliest Tertiary
	$T\text{-}C_{os}$
	Winterhaven Fm.—Jurassic
	J_{whms} metasedimetary
	J_{whmv} metavolcanic
	J_{whud} undifferentiated
	mylonitic orthogneiss
	J_g

Figure 3. *Geologic map (opposite) showing the western and eastern belts of the Orocopia Schist cropping out on opposite sides of the Taylor Lake fault. See Figure 4 for cross section along A-A'.*

In both eastern and western belts, the Orocopia Schist is dominated by a polygenetic metamorphic mineral assemblage consisting of quartz + albite + K-feldspar + biotite + white mica + chlorite ± epidote ± calcite ± graphite ± garnet. The prograde assemblage was most likely quartz + albite + K-feldspar + biotite ± garnet while textural evidence suggests that white mica + chlorite ± epidote ± calcite probably represent a retrograde event. S1 is defined by the alignment of lensoid shaped masses of recrystallized quartz and quartzofeldspathic aggregates and the alignment of the {001} cleavage plane of biotite. White mica and chlorite are a replacement of biotite. Porphyroblastic albite is riddled with inclusions, some of which contains an earlier foliation oriented at oblique angles to S1.

ANALYTICAL METHODS

Twenty-nine representative samples of the Orocopia Schist were collected from the western and eastern belts and were prepared for whole rock trace and major element geochemistry following conventional XRF techniques (see Girty et al., this volume for a detailed description). Of the 29 collected samples, 13 specimens were from the eastern belt and 16 were from the western belt. Sample locations are provided in Stephenson (2006). Resulting data are provided in an appendix as Table 1, and were analyzed using the tectonic provenance-discrimination models of Bhatia (1983) and Bhatia and Crook (1986).

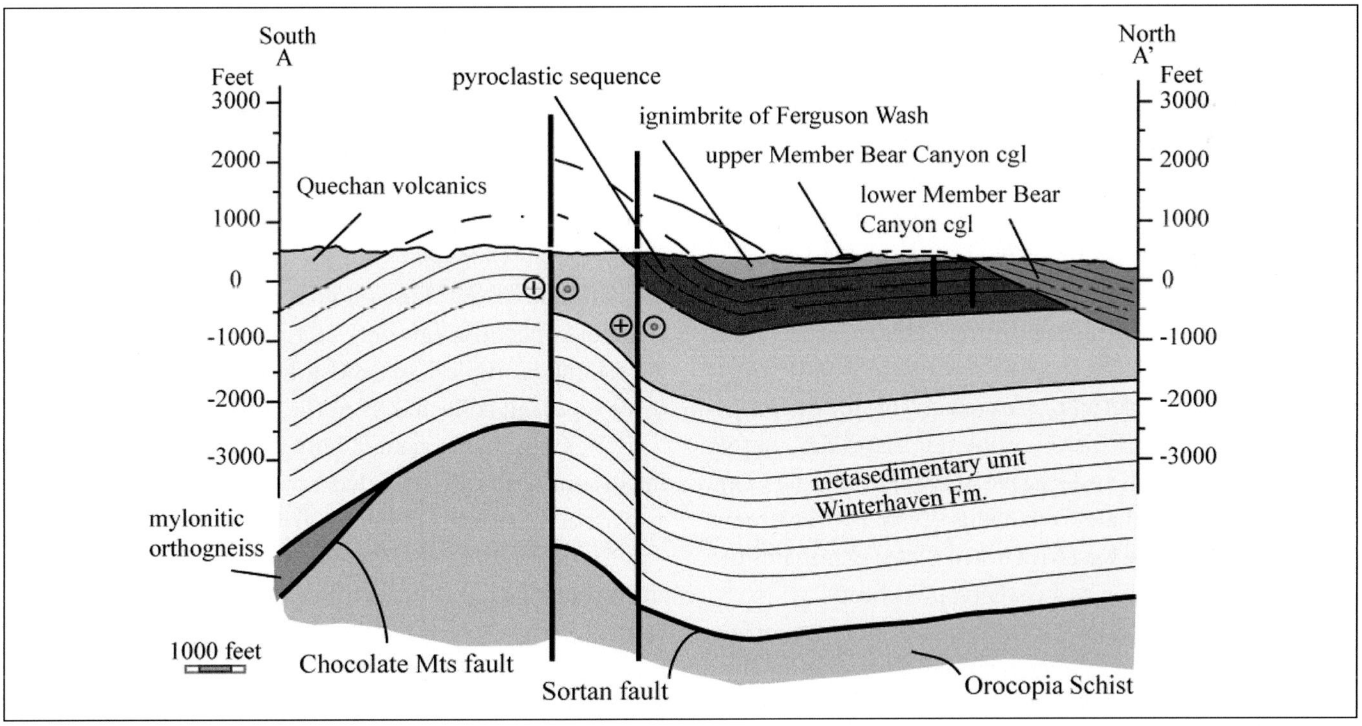

Figure 4. *Cross-section A-A'. See Figure 3 for location.*

By analyzing variations of the primarily immobile elements in Australian turbidities derived from four end-member tectonic settings, Bhatia (1983) and Bhatia and Crook (1986) developed a series of discrimination diagrams that allow turbidities of unknown tectonic affinity to be classified in terms of their likely paleo-tectonic setting. The four tectonic settings recognized in these discrimination diagrams are: Ocean Island Arc (OIA), Continental Island Arc (CIA), Active Continental Margin (AM), and Passive Margin (PM) (Bhatia and Taylor, 1981; Bhatia, 1983; 1985; Bhatia and Crook, 1986). The dominant type of depositional basin, type of crust adjacent to basin, provenance type, and modern analog for each tectonic setting are listed in the appendix as Table 2.

CHEMISTRY OF OROCOPIA SCHIST

Major element concentrations in the Orocopia Schist at Picacho are very similar to those of other correlative schist bodies exposed throughout the Chocolate Mountains anticlinorium (Haxel et al., 1987). At Picacho, and at other localities throughout the Chocolate Mountains, the schists are quartzofeldspathic and were likely derived from sandstones (Haxel et al., 1987). Al_2O_3 concentrations are in the 12–15% range, with some rocks in the western belt approaching concentrations of 17% (Table 1). Such compositions are commonly referred to as semi-pelitic. Average Al_2O_3 concentrations in the western belt are slightly higher than those derived from samples in the eastern belt, at the expense of SiO_2 concentrations, which are slightly lower in the west (Table 1). At Picacho, Na_2O is the only other major element that shows any distinct variation from one belt to the next. The average difference between analyzed samples from the eastern and western belts for both Al_2O_3 and SiO_2 is minor, only ~1.5%, while Na_2O is only ~1.25% higher in the western belt.

The chemical classification scheme of Herron (1988) indicates that the protolith of the Orocopia Schist is primarily graywacke (Fig. 5F) while its quartzofeldspathic nature suggests a continental source. This interpretation implies that three of the tectonic settings of Bhatia (1983) and Bhatia and Crook (1986) are conceivable for the protolith of the Orocopia Schist: Continental Island Arc, Active Continental Margin, and Passive Margin (Table 2). Considering the known Late Cretaceous–Tertiary convergent margin setting of the western United States eliminates the Passive Margin setting as a realistic possibility.

Analyzed samples of the Orocopia Schist plot mostly in the Active Continental Margin field but scatter into the Continental Island Arc and to a lesser degree into the Oceanic Island Arc and Passive Margin fields on the (1) TiO_2 versus $Fe_2O_3^* + MgO$ (Fig. 5A), (2) Al_2O_3/SiO_2 versus $Fe_2O_3^* + MgO$ (Fig. 5B), (3) $Al_2O_3/(CaO + Na_2O)$ versus $Fe_2O_3^* + MgO$ (Fig. 5C), (4) K_2O/Na_2O versus $Fe_2O_3^* + MgO$ (Fig. 5D), and (5) the CaO–K_2O–Na_2O (Fig. 5E) discrimination diagrams of Bhatia (1983) and Bhatia and Crook (1986). As would be expected for major element data derived from amphibolite grade rocks considerable scatter of data is evident on discrimination plots involving the mobile components K_2O, Na_2O, and CaO (Figs.5C, 5D, and 5E). Nevertheless, the average of the 29 analyzed samples consistently plots in the Active Continental Margin field regardless of which discrimination plot is considered (Fig. 5).

On plots utilizing the more immobile trace elements La, Th, Sc, Zr, and V most samples plot in the Continental Island Arc field or spread from this field into the Active Continental Margin field (Fig. 6). The spread of some data just outside the Continental Island Arc field in Figures 6C and 6E may simply reflect the approximate nature of the boundary for this field.

The results shown in Figures 5 and 6 are consistent with the idea that the protolith of the Orocopia Schist originated within a complex forearc active borderland-like setting like that proposed by Saleeby (2003) for the forearc region of the breached segment of the Cordilleran margin, but given the known Late Cretaceous to Paleogene trench-parallel dispersion of some accreted terranes (e.g., Saleeby, 2003, and references therein), can we rule out a setting elsewhere along the Cordilleran margin?

COMPARATIVE CHEMISTRIES

Chemical data derived from the Franciscan Complex and Great Valley sequence were extracted from Blake Jr. and Jones (1981) and Linn et al. (1992), respectively. Surprisingly, to our knowledge little other geochemical data exist in the literature for these two well-known units. The Franciscan Complex data of Blake Jr. and Jones (1981) consists of average major element concentrations derived from 23 meta-graywackes, 66 melange samples, 49 sandstones, and an average of all three types, all collected from the San Francisco Bay area.

The Great Valley data of Linn et al. (1992) are separated into a southern suite collected from the San Joaquin Valley, and a northern suite collected from the Sacramento Valley. There are distinct chemical differences between each suite, a result that reflects differing provenance inputs: a central Sierra Nevada source for sandstones of the San Joaquin Valley, and a northern Sierra Nevada and Klamath Mountains source for sandstones in the Sacramento Valley (Linn et al., 1991; 1992).

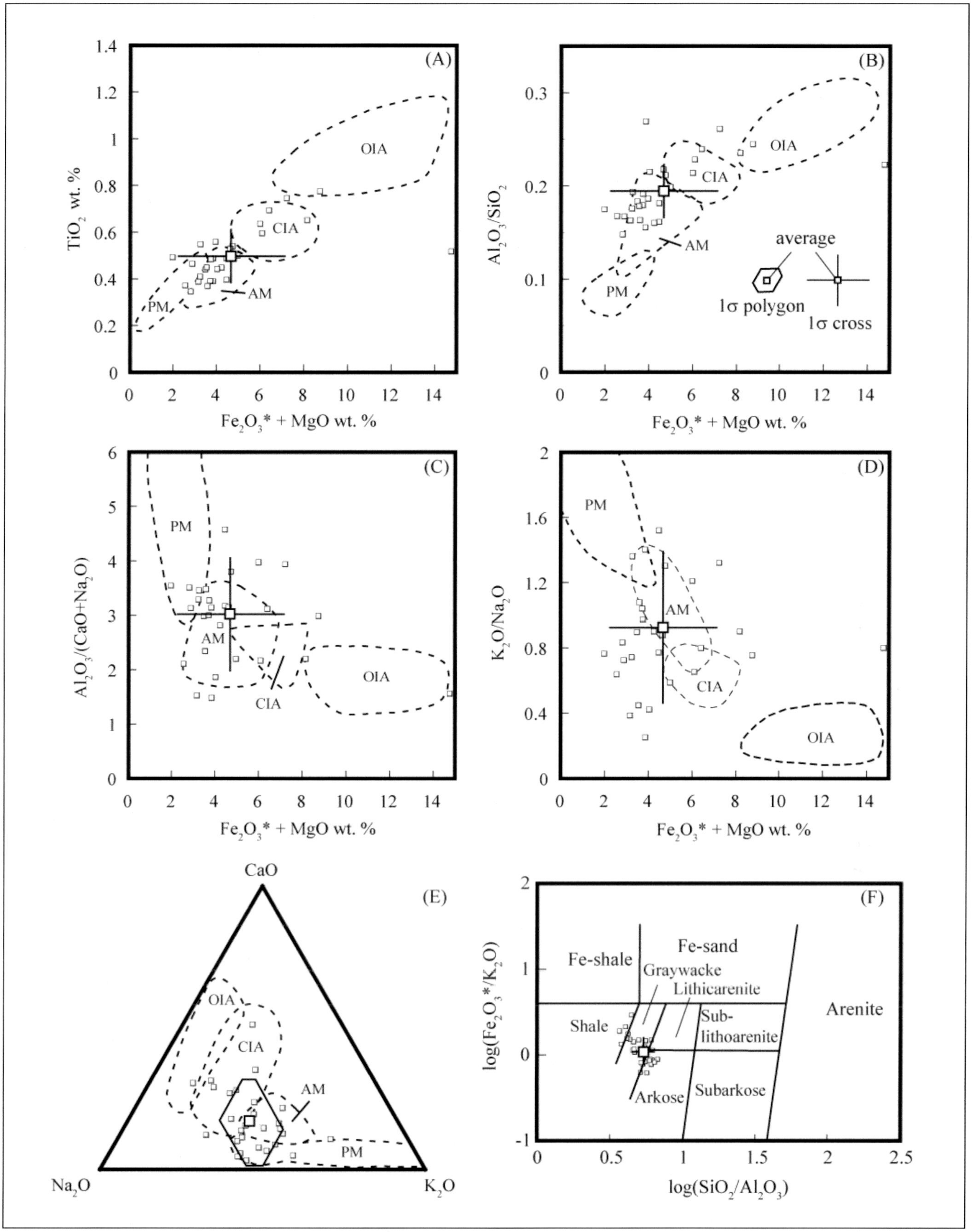

Figure 5. *Major element data from Orocopia Schist plotted on provenance-discrimination diagrams of Bhatia (1983) and Bhatia and Crook (1986), and on sandstone classification diagram of Herron (1988).* Fe_2O_3* *represents total iron. OIA = oceanic island arc. CIA = continental island arc. AM = active continental margin. PM = passive margin.*

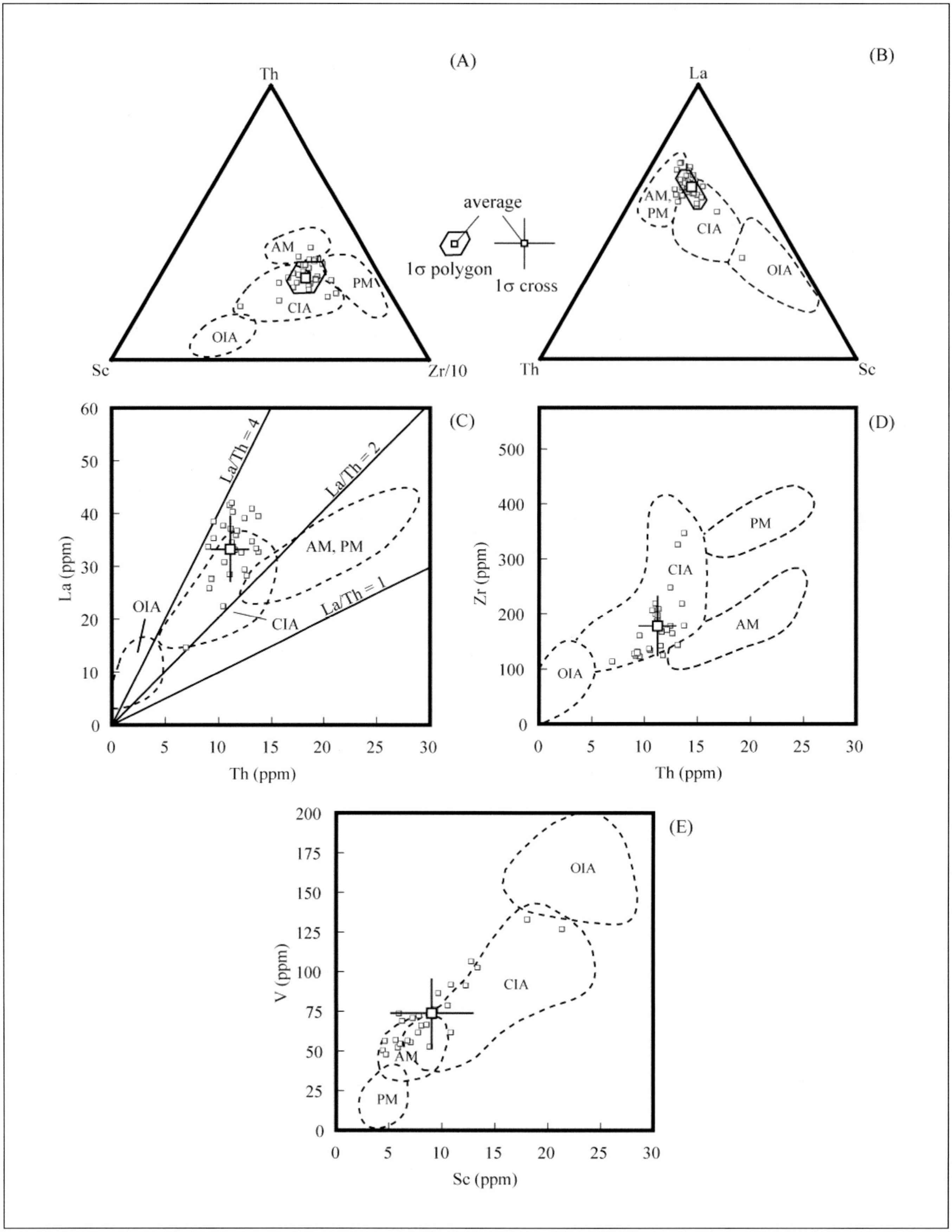

Figure 6. *Trace element data derived from Orocopia Schist plotted on tectonic-provenance discrimination diagrams of Bhatia (1983) and Bhatia and Crook (1986). OIA = oceanic island arc. CIA = continental island arc. AM = active continental margin. PM = passive margin.*

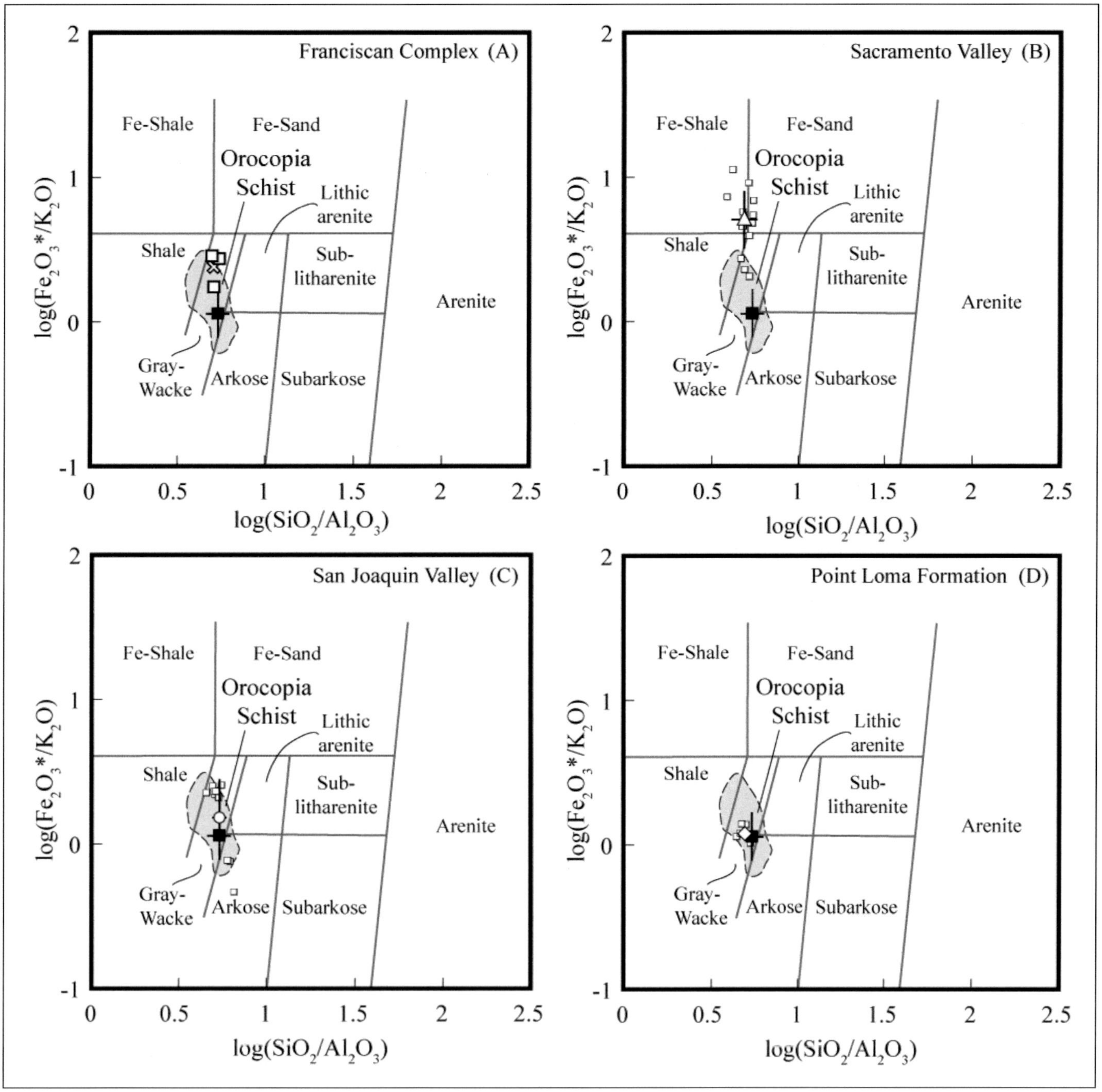

Figure 7. *Analyses reported in the literature or derived from this study plotted on the chemical classification diagram of Herron (1988). Crosses are 1σ about means. Black square is average Orocopia. White circle, diamond, triangle, square, and cross are averages of data plotted in individual diagram or cited in literature.*

On the sandstone discrimination diagram of Herron (1988) data from the Franciscan Complex and San Joaquin Valley plot in the graywacke field and fall within or adjacent to the compositional field defined by samples from the Orocopia Schist (Figs. 7A, 7C). In contrast, data from the Sacramento Valley plot in the fields of iron-rich shales and sands (Fig. 7B).

If ~300 km of displacement is removed from the San Andreas fault, then the Rosario Group currently exposed along the coast of southern California (Kennedy and Peterson, 1975) represents the southern continuation of the Great Valley depositional system (Saleeby, 2003). Lying along the coast of San Diego, the Upper Cretaceous Point Loma Formation is a major component of the Rosario Group (Nilsen and Abbott, 1981), and as a result 6 arenite samples were collected for major and trace element analysis. Resulting data are provided in Table 1 of the appendix. Of the 6 samples analyzed from the Point Loma Formation,

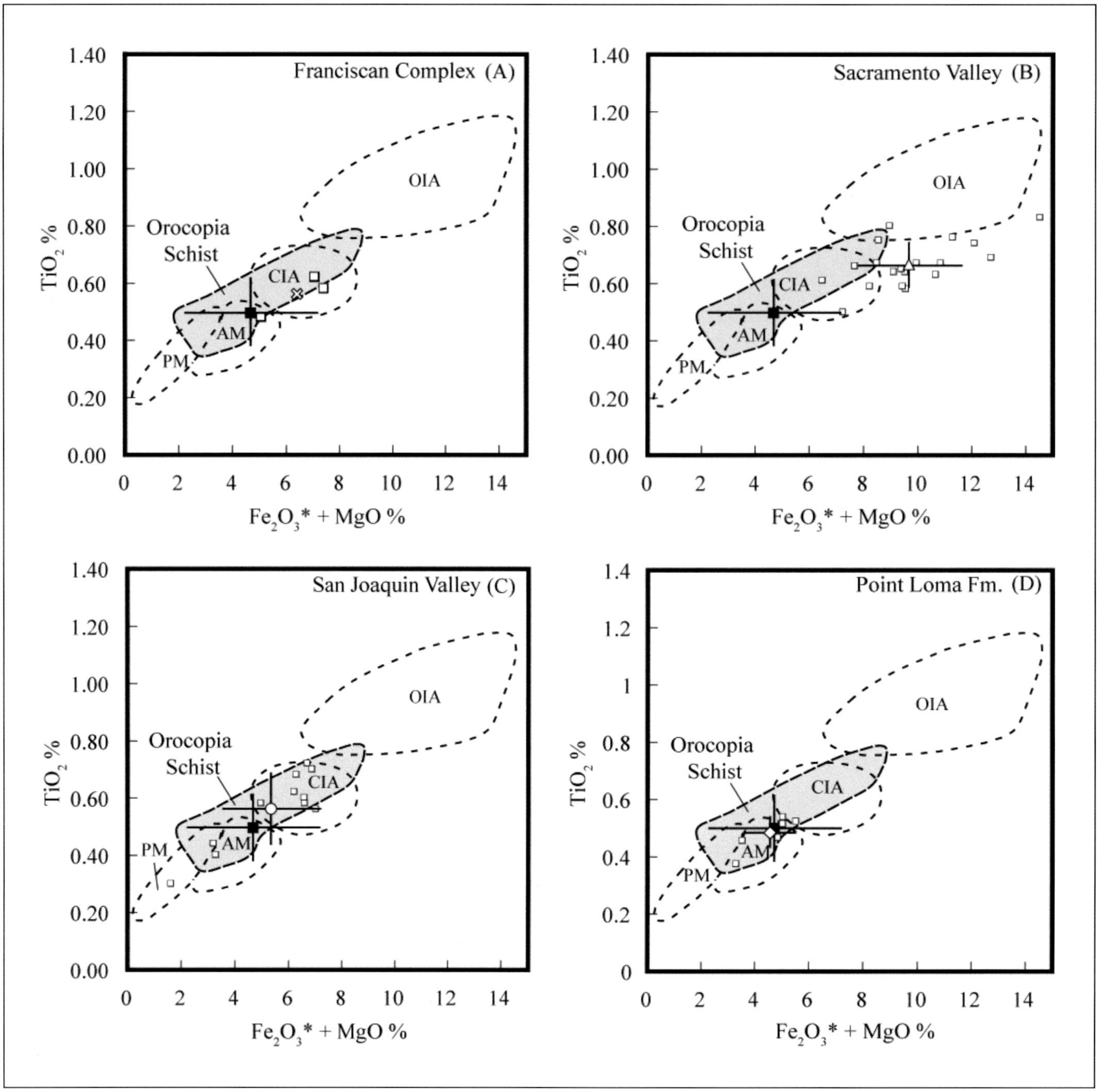

Figure 8. *Major element data plotted on the TiO_2 versus Fe_2O_3* + MgO tectonic discrimination diagram of Bhatia (1983). Crosses are 1σ values. Black square is average Orocopia. White circle, diamond, triangle, square, and "x" are averages of data plotted in individual diagrams or averages cited in literature. Fe_2O_3* represents total iron as Fe_2O_3. See Figure 5A for names of fields.*

three are from the shelf facies and three are from the outer submarine fan facies of Nilsen and Abbott (1981). Sample locations are provided by Stephenson (2006).

On the sandstone discrimination diagram of Herron (1988), the six samples analyzed from shelf and submarine fan deposits are graywackes (Fig. 7D).

Franciscan Complex

The various averages of the metagraywacke, mélange, and sandstone samples analyzed from the Franciscan Complex by Blake Jr. and Jones (1981) plot within the Continental Island Arc and Active Margin fields on the TiO_2 versus Fe_2O_3* + MgO discrimination diagram (Fig. 8A). This result is not surprising as it is well known that the Franciscan Complex includes trench sediments that were in part derived from the Cretaceous magmatic arc lying to the east, and in part from a complex borderland terrane produced by the accretion and trench-parallel dispersion of seamounts, oceanic plateaus, and other far-traveled oceanic terranes (e.g., Saleeby, 2003).

The various averages of samples analyzed from the Franciscan Complex consistently plot within or very near the boundary of the compositional field defined by samples analyzed from the Orocopia Schist (Fig. 8A). However, the averages from Franciscan metagraywacke and mélange samples display slightly more enriched $Fe_2O_3^* + MgO$ and TiO_2 compositions than do the average Franciscan sandstone analyzed by Blake Jr. and Jones (1981), and all three averages are higher in $Fe_2O_3^* + MgO$, and TiO_2 than is the average derived from the 29 samples analyzed from the Orocopia Schist.

Great Valley Sequence

The average of TiO_2 and $Fe_2O_3^* + MgO$ data along with 14 out of 19 samples from the Sacramento Valley plot outside of the various tectonic provenance fields outlined in Figure 8B. In contrast, the average of 8 out of 11 specimens analyzed from the San Joaquin Valley plot within the Continental Island Arc field while 3 spread into the Active and Passive margin fields (Fig. 8C). On bivariate plots of La versus Th available data from the Sacramento Valley cluster within the Oceanic Island Arc field (Fig. 9A) while data from the San Joaquin Valley spread from this field into the Continental Island Arc field (Fig. 9B). These disparate results reflect higher $Fe_2O_3^* + MgO$ and TiO_2 concentrations, and lower amounts of La and Th in samples from the Sacramento Valley than in those specimens analyzed from the San Joaquin Valley. Hence, they are interpreted to be a direct reflection of the sources of the analyzed sediment. For example, sandstones from the Sacramento Valley are composed of detrital material derived mostly from Paleozoic and Mesozoic ensimatic island arc and trench assemblages along with Cretaceous continental margin magmatic arc material that evolved within such already accreted oceanic materials. In contrast, the source of San Joaquin Valley samples was predominantly the Cretaceous continental ensialic margin island arc lying to the east along with Phanerozoic crustal materials that it had interacted with during its evolution.

The majority of samples from the San Joaquin Valley plot in or very near the compositional field defined by samples analyzed from the Orocopia Schist on the TiO_2 versus $Fe_2O_3^* + MgO$ (10 out of 11) (Fig. 8C). In contrast, on the La versus Th tectonic discrimination diagram only 3 samples from the San Joaquin Valley fall within the compositional field of the Orocopia Schist (Figs. 9B).

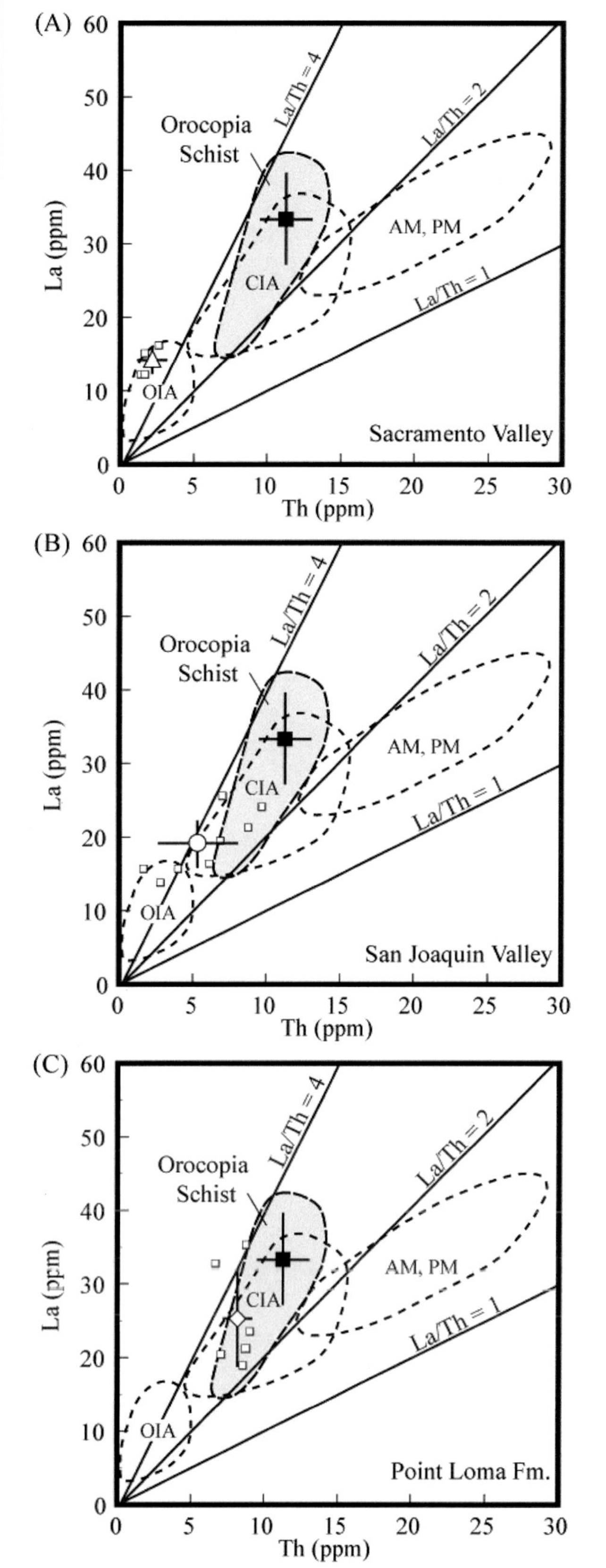

Figure 9. *La-Th data plotted on the tectonic discrimination diagram of Bhatia and Crook (1986). Crosses are 1σ; black square is average Orocopia; white diamond, circle, and triangle are averages of data plotted in each diagram. See Figure 6C for names of fields.*

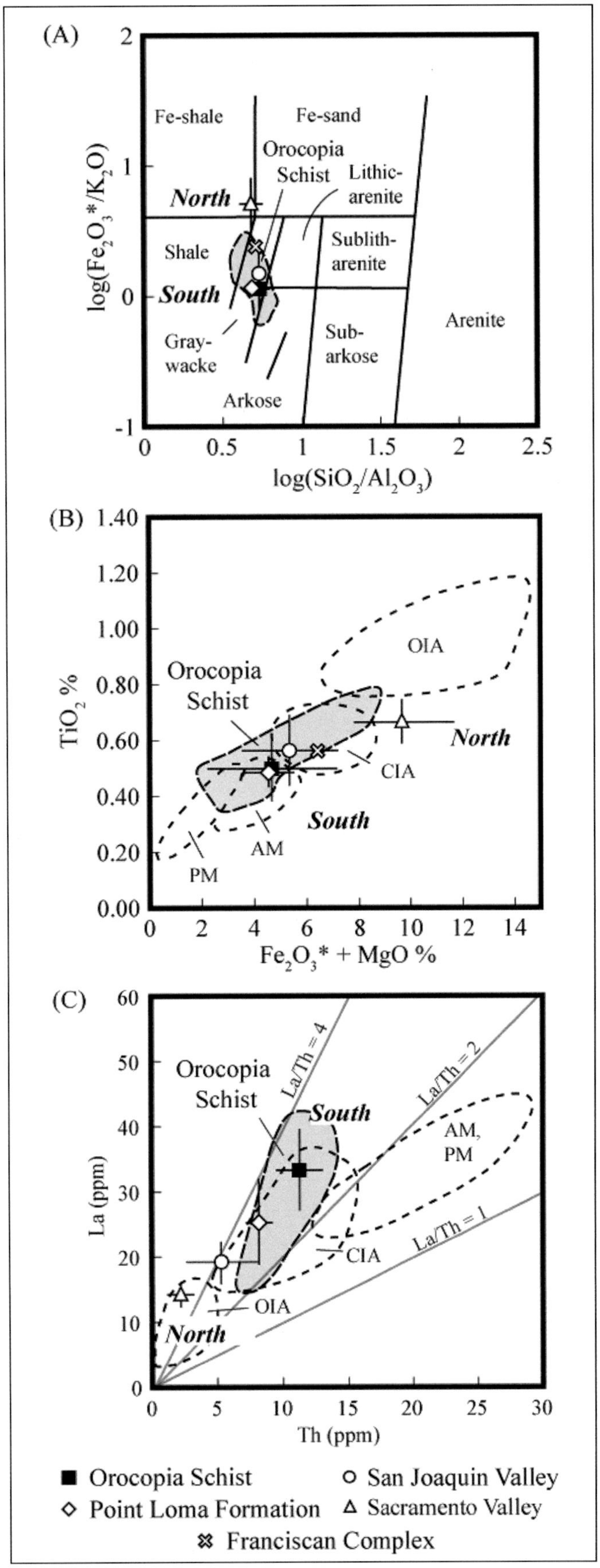

Figure 10. *North to south variation displayed by data discussed with in the text. Crosses are 1σ about means. See Figures 5A and 6C for names of fields shown in B and C.*

Most data from the Sacramento Valley data set do not plot in the compositional field defined by the analyzed specimens of the Orocopia Schist (14 out of 19 for TiO_2 versus $Fe_2O_3^*$ + MgO, and all available La versus Th data) (Figs. 8B, 9A). Hence, forearc basin sediments in the San Joaquin Valley are plausible protoliths of the Orocopia Schist while the compositions of sedimentary material analyzed from the Sacramento Valley are too unlike the composition of the Orocopia Schist for them to be a likely protolith.

Rosario Group–Point Loma Formation

On the TiO_2 versus $Fe_2O_3^*$ + MgO discrimination diagram, data from the Point Loma Formation spread about the boundary, separating the Active Margin and Continental Island Arc fields (Fig. 8D). In contrast, on the La versus Th diagram all but one sample plots within or along the boundary of the Continental Island Arc field (Fig. 9C). The lone anomalous sample does not fall in any of the discrimination fields outlined on the La-Th discrimination diagram.

All but the one anomalous sample noted above from the Point Loma Formation plots on the TiO_2 versus $Fe_2O_3^*$ + MgO and La versus Th diagrams within the compositional field defined by the Orocopia Schist (Figs. 8D, 9C). In addition, when the averages and 1σ values for data from the Great Valley sequence and Rosario Group are compared to the same

DISCUSSION AND CONCLUSIONS

The averages of sandstone samples analyzed from the Franciscan Complex, the Sacramento Valley, the San Joaquin Valley, and the Point Loma Formation clearly show a compositional change in a north to south direction, with samples from the Sacramento Valley forming one end member to the north and the Point Loma Formation another to the south (Fig. 10). Within this spectrum of compositions on the pre-Neogene reconstruction of Saleeby (2003), samples from the San Joaquin Valley lie just to the north and samples from the Point Loma Formation lie just to the south of the breached portion of the western Cordilleran margin (Fig. 11).

Given that compositional data described in previous sections of this paper suggest that the protolith of the Orocopia Schist was compositionally most like graywackes in the Point Loma Formation and to a lesser degree like those in the San Joaquin Valley, the above observation implies that the source of the protolith of the Orocopia

Schist might have lain within the breached portion of the Cordilleran margin and perhaps somewhere along its southern half.

As noted by Saleeby (2003), during Late Cretaceous to Eocene time within the breached portion of the Cordilleran margin a normal forearc basin did not exist (see Figure 2C). Instead the forearc region likely consisted of a complex network of active normal faults across which sediment was transported westward into the trench. Such a setting would have yielded sediment with compositions that would plot in both continental island arc and active margin fields on the discrimination diagrams of Bhatia (1983) and Bhatia and Crook (1986). Given the documented history and likely Late Cretaceous to Paleogene age of the protolith of the Orocopia Schist, some of this sediment must have made its way quickly to the trench where it was carried to great depths along the low-angle but increasingly inclined trajectory of the subducting Farallon plate. Subsequently, the sediment, now converted to schist, was returned to the surface of the Earth by Oligocene extension followed by erosional denudation associated with ~NS shortening sometime during the early to middle Miocene (Girty et al., 2005; 2006; Jacobson et al., in press). In short, though much work remains to be done, the model proposed by Saleeby (2003) provides a plausible framework in which to view the otherwise enigmatic origin of the Orocopia Schist.

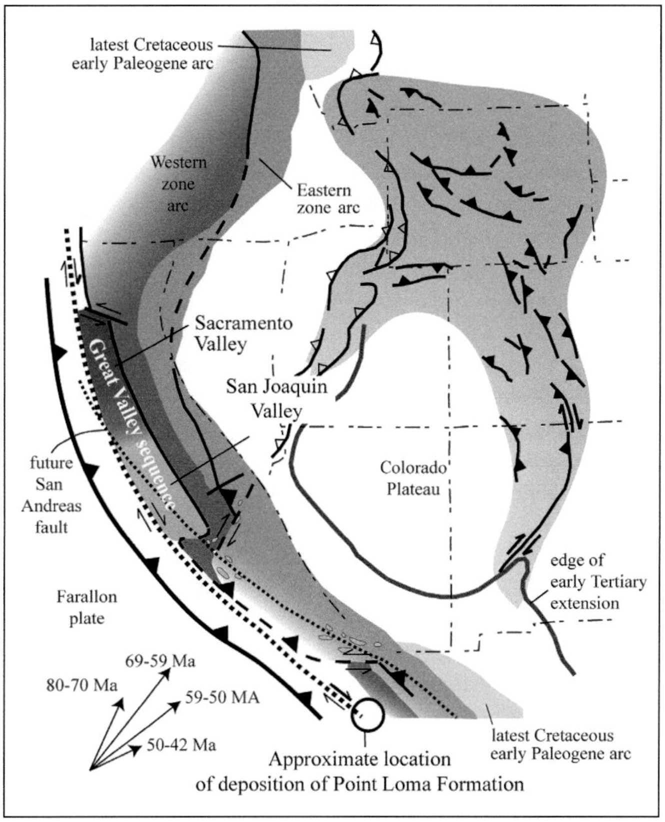

Figure 11. *Pre-Neogene reconstruction by Saleeby (2003) showing relative position of depositional settings of samples discussed in the text.*

ACKNOWLEDGEMENTS

Pat Abbott has always had an intense interest in using and developing a variety of techniques to get at the tectonic settings and sources of clastic sedimentary rocks. We hope that this small contribution is viewed as continuation of those efforts at Pat's home institute San Diego State University.

Carl Jacobson and Marty Groove graciously shared their immense wealth of knowledge of the Orocopia Schist and its correlatives while George Jefferson provided the initial impetus for our work at Picacho. John Cooper provided a timely and constructive critique of the manuscript.

REFERENCES CITED

Bhatia, M.R., and Taylor, S.R., 1981, Trace-element geochemistry and sedimentary provinces: A study from the Tasman geosyncline, Australia: Chemical Geology, v. 33, p. 115–125.

Bhatia, M.R., 1983, Plate tectonics and geochemical composition of sandstones: The Journal of Geology, v. 91, no. 6, p 611–627.

Bhatia, M.R., 1985, Rare Earth element geochemistry of Australian Paleozoic graywackes and mudrocks: Provenance and tectonic control, v. 45, p. 97–113.

Bhatia, M.R. and Crook, K.A.W., 1986, Trace element characteristics of graywackes and tectonic setting discrimination of sedimentary basins: Contributions to Mineralogy and Petrology, v. 92, p. 181–193.

Blake, M.C., Jr. and Jones, D.L., 1981, The *Franciscan assemblage* and *related rocks in northern California: a reinterpretation in Ernst, W.G., ed.,* The geotectonic development of *California, Rubey v. 1, p. 307–328.*

Girty, G.H., Ridge, D.L., Knaack, C., Johnson, D., and Al-Riyami, R.K, 1996, Provenance and depositional setting of Paleozoic chert and argillite, Sierra Nevada, California: Journal of Sedimentary Research, v. 66, p. 107–118.

Girty, G.H., Canfield, A., Stephenson, D., Middleton, T., Moniz, R., Rayburn, J., Sisk, M., and Verdugo, D., 2005, The Bear Canyon conglomerate and the timing of formation of the Chocolate Mountains anticlinorium, SE California: Geological Society of America Abstracts with Programs, v. 37, p. 481.

Girty, G.H., Stephenson, D., Canfield, A., Marsh, J., Middleton,

T., Rayburn, J., Sisk, M., Verdugo, D., Moniz, R., Gunter, J., Nielsen, J., Gasca, C., Rowland-Smith, A., Smith, M., Lovering, K., Campbell, K., Castaneda, F., Gray, R., Campbell, C., 2006, Geology of Picacho State Recreation Area, SE California: Implications for the timing of formation of the Chocolate Mountains anticlinorium, *in* Girty, G.H., and Cooper, J., eds., Using Stratigraphy, Sedimentology, and Geochemistry to Unravel the Geologic History of the Southwestern Cordillera : A Volume in Honor of Patrick L. Abbott, p. 255–274.

Grove, M., Jacobson, C.E., Barth, A.P., and Vucic, A., 2003, Temporal and spatial trends of Late Cretaceous-early Tertiary underplating of Pelona and related schist beneath southern California and southwestern Arizona: Geological Society of America Special Paper 374, p. 381–406.

Haxel, G.B., Budahn, J.R., Fries, T.L., King, B.W., White, L.D., and Aruscabage, P.J., 1987, Geochemistry of the Orocopia Schist, Southeastern California: Summary, *in* Dickinson, W.R., and Klute, M.A., ed., Mesozoic Rocks of Southern Arizona and Adjacent Areas: Arizona Geological Society Digest, v. 18, p. 49–64.

Haxel, G.B, Jacobson, C.E., Richard, S.M., Tosdal, R.M., and Grubensky, M.J., 2002, The Orocopia Schist in southwest Arizona: Early Tertiary oceanic rocks trapped or transported far inland, *in* Barth, A., eds., Contributions to crustal evolution of the southwestern United States: Geological Society of America Special Paper 365, p. 99–128.

Herron, M.M., 1988, Geochemical classification of terrigenous sands and shales from core or log data: Journal of Sedimentary Petrology, v. 58, p. 820–829.

Jacobson, C.E., Oyarzabal, F.R., Haxel, G.B., 1996, Subduction and exhumation of the Pelona-Orocopia-Rand schists, southern California: Geology, v. 24, p. 547–550.

Jacobson, C.E., Grove, M., Stamp, M.M., Vucic, A., Oyarzabal, F.R., Haxel, G.B., Tosdal, R.M., and Sherrod, D.R., 2002, Exhumation history of the Orocopia Schist and related rocks in the Gavilan Hills area of southeasternmost California, *in* Barth, A., ed., Contributions to crustal evolution of the southwestern United States: Geological Society of America Special Paper 365, p. 129–154.

Jacobson, C.E., Grove, M., Vucic, A., and Pedrick, J.N., *in press*, Exhumation of the Orocopia Schist and associated rocks of southeastern California: Relative roles of erosion, synsubduction tectonic denudation, and middle Cenozoic extension *in* Cloos, M., Carlson, W.D., Gilbert, M.C., Liou, J.G., and Sorensen, S.S., eds., Convergent margin terranes and associated regions — A tribute to W.G. Ernst: Geological Society of America Special Paper.

Kennedy, M.P., and Peterson, P.L., 1975, Geology of the San Diego metropolitan area, California: California Division of Mines and Geology Bulletin 200, 56 p.

Linn, A.M., DePaolo, D.J., and Ingersoll, R.V., 1991, Nd-Sr isotopic provenance analysis of Upper Cretaceous Great Valley fore-arc sandstones: Geology, v. 19, p. 803–806

Linn, A.M., DePaolo, D.J., and Ingersoll, R.V., 1992, Nd-Sr isotopic, geochemical, and petrographic stratigraphy and paleotectonic analysis: Mesozoic Great Valley forearc sedimentary rocks of California: Geological Society of America Bulletin, v. 104, p. 1264–1279.

Nilsen, T.H., and Abbott, P.L., 1981, Paleogeography and sedimentology of Upper Cretaceous turbidites, San Diego, California: American Association of Petroleum Geologists, v, 65, p. 1256–1284.

Saleeby, J., 2003, Segmentation of the Laramide Slab — evidence from the southern Sierra Nevada region: Geological Society of America Bulletin, v. 115, p. 655–668.

Stephenson, D.B., 2006, The structure, petrology, and chemistry of the Orocopia Schist, Picacho State Recreation Area, SE California [unpublished MS thesis]: San Diego, San Diego State University, 97 p.

Yin, A., 2002, Passive-roof thrust model for the emplacement of the Pelona-Orocopia Schist in southern California, United States: Geology, v. 30, p. 183–186.

APPENDIX

Table 1. *Major and trace element analyses, Orocopia Schist and Point Loma Formation*

Sample	DS-01	DS-02	DS-03	DS-04	DS-05	DS-06	DS-07	DS-08	DS-09	DS-10	DS-11	DS-12	DS-13
wt.%													
SiO_2	74.6	75.0	77.8	72.6	74.6	75.0	75.3	74.3	68.9	73.0	62.0	53.8	71.8
Al_2O_3	12.0	12.2	11.5	13.5	11.5	12.5	13.1	13.0	14.6	11.7	14.6	12.0	11.6
Fe_2O_3[1]	3.08	2.54	2.26	2.99	2.53	2.15	1.57	2.76	3.68	2.64	5.15	6.20	2.68
CaO	0.29	0.43	0.17	0.62	1.46	0.63	0.34	0.29	1.17	1.53	3.36	5.03	2.95
MgO	1.41	1.08	0.58	1.00	1.34	0.76	0.43	0.51	1.09	1.63	3.04	8.59	0.51
K_2O	3.55	3.32	2.58	3.87	3.12	2.43	2.57	2.72	3.47	2.36	2.96	2.15	1.81
Na_2O	2.34	3.09	3.11	1.39	2.23	3.37	3.38	3.68	2.67	2.63	3.30	2.70	4.74
MnO	0.04	0.03	0.03	0.03	0.04	0.03	0.01	0.03	0.03	0.03	0.05	0.12	0.04
TiO_2	0.39	0.37	0.34	0.56	0.39	0.46	0.49	0.41	0.54	0.45	0.65	0.52	0.39
P_2O_5	0.08	0.07	0.06	0.10	0.08	0.07	0.09	0.07	0.13	0.10	0.17	0.12	0.08
LOI[2]	1.71	1.36	1.33	2.75	2.69	2.18	1.87	1.95	3.48	3.94	4.18	8.87	2.87
Total	99.5	99.5	99.8	99.4	100.0	99.5	99.2	99.7	99.7	100.0	99.5	100.0	99.5
ppm													
Sc	7.9	5.9	4.5	9.4	10.9	7.3	9.4	7.1	9.7	9.3	12.8	18.1	6.1
V	71.9	51.9	50.1	74.8	61.4	70.4	76	55.1	86.1	73.3	106.2	132.5	54.2
Cr	83.6	94.3	123.4	81.6	127.8	111.9	92.3	126.8	141.5	226.5	255.1	847.1	179.2
Co	47.4	52.8	56.3	45.3	44.2	47.2	52.4	52.7	40.4	48.1	34.4	53.8	34.8
Ni	21.3	24.3	17.3	19	41.2	15.8	13.9	15	26.5	43.5	62.6	531	22.7
Cu	15.2	11	10.8	16.4	16.7	25	16.6	11	21.2	19.1	48.2	11.3	11.2
Zn	36.4	29.7	27	37.1	35.7	43.8	23.5	26.3	47.3	34.7	61.8	80.6	31.2
Rb	118.2	115.8	72.2	124.2	109.5	84.6	94.1	90.5	128.6	88.2	124.4	118.5	58.6
Sr	132.7	115.1	99.8	58.2	174.1	60.5	79.1	130.5	81.3	90.8	264.3	133.9	301.9
Y	17.1	17.4	17.7	26.9	18.7	21.6	21.2	20.5	21.4	19.9	28	18.6	19.7
Zr	122.6	129.5	121.8	324.8	126.8	185.3	198.3	159.8	166.6	128.7	191.8	112.6	141.2
Nb	10.5	11.4	11.6	16.4	11	14.1	14.8	13	13.1	11.6	13.7	7.3	11.4
Ba	1020.4	819.9	679.7	1164.8	937.2	718.5	785.1	681	1003.2	804.5	906.8	284.9	887.1
La	25.7	27.6	35.2	40.8	33.6	37	41.5	38.4	35.8	27.5	34.2	14.5	33
Ce	50.5	54.9	56.6	76.4	56	66.9	62	58.4	61.4	51.6	49.3	22	49.7
Nd	20.3	21	21.9	30.7	21.2	26.4	26.5	24.8	26.9	22.1	26.1	11.5	20.7
Sm	3.6	4.1	2.7	6.8	4.2	3	6.4	2.9	2.1	4.7	3.3	5.4	4.8
Yb	7.2	8.1	8.2	8.5	5.9	6.2	8.4	7.9	5.6	5.8	5.2	2.3	6
Hf	3.1	3.3	2.5	8.5	3.3	4.5	5.1	3.8	5.8	3.4	7.1	4.9	4.3
Pb	12.3	10.2	12.8	11.4	9.6	13.9	8.1	12.6	10.2	8.9	13.3	3.7	12.9
Th	9.2	9.4	9.6	13.2	9.1	11.2	11.1	9.6	11.7	9.4	11.3	7	11.6
U	0.7	0.9	1.2	1.5	0.3	1.7	0.7	1.5	1.6	2	2	1	2.4

[1] Total Iron as Fe_2O_3 [2] Loss on ignition

Table 1 *(continued). Major and trace element analyses, Orocopia Schist and Point Loma Formation*

Sample	DS-14	DS-15	DS-16	DS-17	DS-18	DS-19	DS-20	DS-21	DS-22	DS-23	DS-24	DS-25	DS-26
wt.%													
SiO_2	73.9	68.5	72.9	62.3	73.6	66.5	73.0	73.4	70.1	64.1	69.3	71.5	72.2
Al_2O_3	12.3	14.7	13.3	16.7	13.3	15.1	13.0	13.0	13.9	16.7	15.0	13.8	13.8
Fe_2O_3[1]	2.36	2.95	2.75	3.42	3.26	4.19	2.77	3.00	3.46	5.35	3.43	2.53	2.83
CaO	1.52	2.89	1.39	4.00	0.76	2.72	1.11	0.91	2.32	1.17	1.31	1.00	0.58
MgO	0.22	1.11	0.76	0.46	1.23	1.93	0.98	0.57	1.54	1.90	1.24	0.76	0.94
K_2O	2.77	2.11	2.75	1.83	2.65	2.79	3.37	2.08	2.35	4.06	3.05	4.09	3.53
Na_2O	4.36	5.05	3.08	7.40	3.45	4.30	3.25	4.68	4.03	3.08	3.49	3.01	3.64
MnO	0.02	0.05	0.02	0.06	0.04	0.06	0.03	0.03	0.05	0.06	0.05	0.03	0.03
TiO_2	0.37	0.44	0.44	0.49	0.49	0.59	0.39	0.45	0.50	0.74	0.53	0.55	0.48
P_2O_5	0.07	0.12	0.08	0.15	0.08	0.17	0.08	0.08	0.12	0.20	0.15	0.13	0.09
LOI[2]	1.71	1.98	2.67	4.015	1.6	1.91	1.98	1.39	1.6	2.94	2.34	2.49	1.78
Total	99.6	99.8	100.1	100.8	100.4	100.3	100.0	99.6	100.0	100.3	99.9	99.9	99.8
ppm													
Sc	4.8	5.7	6.8	8.9	8.6	6	4.7	7.8	6.3	13.4	9.2	10.6	8.1
V	47.5	56.6	56.2	52.4	66.2	73.3	56.1	61.3	68.6	102.2	71.6	78.3	65.7
Cr	206.8	139.7	113.7	63.1	95.2	107.6	115.2	83.7	133.1	64.5	63	77.9	52
Co	47	44.8	44.2	26.4	38.3	36	37.8	41.5	39.7	30.5	33.4	40.4	40.6
Ni	9.5	5.1	22.9	14.4	10.7	16.5	17.6	7.4	10	21.5	13	14.6	11.2
Cu	11.2	9	12.4	19.9	18.1	17.9	12.1	16	14.6	38.7	17.2	17.9	14.7
Zn	21.7	51.1	41.4	33	33.8	73.3	35.5	37.4	54.8	95.7	64.1	34.5	42.7
Rb	71.1	84.8	93.6	66.1	97.9	100.2	117.1	77.2	88.4	206.1	129.7	126.8	118
Sr	170.1	492.8	213.6	321.7	214.5	343	174.8	224.6	288.3	245.1	252.8	171.7	153.8
Y	18.6	18.2	19	20.3	21.1	20.5	18.6	20.2	19.6	28.3	20.5	25.8	22.9
Zr	123.8	177.2	133.2	169.9	177.6	205.4	135.8	142.2	163.5	217.3	178.9	207.7	195.8
Nb	12	11.9	13.5	12.4	14.3	12.5	11.2	13.6	11.1	15.9	12.7	14.3	13.7
Ba	1023.9	884.7	806	540.5	740.9	851.6	806.3	746.2	763.1	1083.7	855.5	1235	948.1
La	36.6	29.3	30.7	32.5	32.6	22.3	37.6	34.6	28.1	33.3	34.4	40.2	41.9
Ce	52.9	51.7	56.4	64.4	67.5	43.9	51.1	63.1	46	64.7	60.4	64.6	65.1
Nd	19.6	21.1	23.4	26.1	26.1	16.7	21.5	21.8	16.9	27.3	23.9	26.7	27.5
Sm	3.2	4.3	2.8	6.1	3.6	3	4.3	4.4	4.1	1.9	4.2	3.1	4
Yb	7.5	6.1	7.4	4.4	5.8	5.6	6.2	7	6.7	6.6	5.6	5.9	6.5
Hf	3.5	4.9	3.9	5.6	5.5	6.2	4.3	3.4	4.8	7.3	5.7	5.2	3.8
Pb	11.3	19.6	15.3	12.1	13.8	13.2	11.9	9.3	13.1	18.6	21.4	17.8	13.8
Th	11.8	12.5	10.6	12.2	13.8	10.8	10.5	13.2	12.7	13.6	11.3	11.4	11.3
U	0.7	3	1.5	1.4	1.3	5.1	1.9	1.9	2.9	3.3	2.7	2.9	1.4

[1] Total Iron as Fe_2O_3 [2] Loss on ignition

Table 1 *(continued). Major and trace element analyses, Orocopia Schist and Point Loma Formation*

Sample	DS-27	DS-28	DS-39	PL-01	PL-02	PL-03	PL-07	PL-09	PL-10	PL-11
wt.%										
SiO_2	68.7	64.3	66.4	70.9	67.3	74.3	56.8	67.6	70.0	69.2
Al_2O_3	14.7	15.7	15.9	13.4	13.4	13.9	15.0	15.2	15.0	14.6
Fe_2O_3[1]	4.27	6.57	4.71	2.57	3.86	2.79	5.88	3.36	3.23	3.75
CaO	0.63	1.13	1.25	4.31	4.55	1.43	8.5	2.55	2.85	2.72
MgO	1.76	2.20	1.73	0.75	1.19	0.76	1.91	1.69	1.64	1.79
K_2O	3.70	3.12	3.07	2.62	2.93	2.59	1.87	3.1	2.81	2.82
Na_2O	3.07	4.15	3.86	2.83	3.23	3.19	3.21	4.71	4.03	3.89
MnO	0.06	0.10	0.07	0.06	0.09	0.03	0.06	0.05	0.05	0.05
TiO_2	0.63	0.77	0.69	0.37	0.54	0.46	0.71	0.51	0.47	0.52
P_2O_5	0.14	0.21	0.18	0.07	0.06	0.09	0.10	0.16	0.11	0.12
LOI[2]	2.24	1.83	1.97	3.65	5.25	2.31	7.95	3.83	2.57	3.16
Total	99.9	100.1	99.8	101.5	102.4	101.9	102.0	102.7	102.8	102.6
ppm										
Sc	10.9	21.4	12.3	17.3	12.1	14	22.8	9.4	10.2	10.8
V	91.4	126.5	90.9	86	93.9	92.2	117.1	73.6	72.4	84.1
Cr	70.3	137.5	53.5	80.5	59	78.1	70.9	98.1	95.3	113.7
Co	37.5	45.5	31.2	34.5	29.2	30	23.8	15.7	24.9	24.2
Ni	17.8	60.2	11.3	6.3	5.5	4.3	20.6	6.7	6.4	9.7
Cu	25.9	42.2	33.4	6.7	9.7	9.5	24.5	14.9	12.5	17.4
Zn	64	97.6	70.1	59	67	56.1	80.2	68.5	64.4	73.2
Rb	154.8	135.6	123.2	85.8	111.2	84.4	85.4	97.8	91.8	96
Sr	115.1	207.6	243.2	200.2	179.7	188.8	412.7	263.5	269.5	259.3
Y	21.3	29.5	30.4	23.1	17.6	21.3	24.1	19.1	16.5	18.2
Zr	217.9	246.6	345.8	95.3	82.6	123.5	198.2	158.6	134.4	138.6
Nb	13.9	13.6	16.6	5.6	7.8	7.6	5.1	8.3	7.5	7.9
Ba	1121.8	643.9	1047.7	698.8	684.6	703.1	443.4	739.4	734.2	786.9
La	28.4	39	39.4	32.7	20.3	35.2	13	18.8	21.1	23.4
Ce	63	72.7	77.3	45.9	31.7	55.6	25.6	32.8	30	39.8
Nd	28.8	31.9	33.1	21.3	15.9	26	16.1	14.9	13	18.8
Sm	3.9	7.3	6.4	2.7	5.1	6.2	2.7	4.1	3.9	6.5
Yb	6.4	4.5	5.6	5.4	5.3	5.1	4.6	4.7	4.7	3.2
Hf	4.9	5.2	7.6	1.5	2.1	3.5	4.8	3.9	3.2	3.8
Pb	13.1	13.6	15.9	10.6	8.4	8.5	5.2	7.7	8.7	7.9
Th	11.1	12.5	13.8	6.7	7.1	8.8	4.7	8.6	8.8	9.1
U	2.3	2.5	3	1.3	1.5	1	1.6	2.1	0.4	0.8

[1] Total Iron as Fe_2O_3 [2] Loss on ignition

Table 2. *Provenance-discrimination types discussed in text*

Tectonic setting	Dominant Depositional basin(s)	Nature of adjacent crust	Provenance type	Modern analog
Oceanic island arc	Fore-arc, back-arc	Oceanic island arc or island arc formed on thin continental crust	Undissected magmatic arc	Western Aleutian Islands; Marianas; Lesser Antilles
Continental island arc	Inter-arc, fore-arc, back-arc	Island arc formed on well-developed continental crust or on this continental margin	Dissected magmatic arc-recycled orogen	Cascades of northwestern U.S.A; Sea of Japan; Puerto Rico
Active continental margin	Retro-arc foreland; marginal basins, oblique-slip basins	Thick continental margin; crystalline basement	Uplifted basement	Northern Chile, Peru; Southern California borderland
Passive margin	Major cratonic depositional basins on trailing edges; rifted continental margin (miogeosynclines and abyssal plains)	Normal continental crust; extended continental crust	Recycled and collision orogens; craton interior	Bengal-Nicobar Fan and Nicobar Basin; Atlantic Ocean

Potholes, California. Based on the number of buildings, this image of Potholes may likely be a work camp for the construction of the Laguna Dam, built from 1905 through 1908. Photo courtesy of the Arizona Historical Society, Yuma

Dry Diggings

A History of the Potholes Mining District

STEPHAN VAN WORMER

svanwor@cox.net

LOCATED IN SOUTHEASTERN IMPERIAL County, California, the Potholes Mining District is situated at the southwestern end of the Chocolate Mountains, and along the northwestern bank of the Colorado River, at the location of present day Laguna Dam. The All American Canal now defines the southwest boundary of the area. A little over a mile square, the district's mineral resources consist largely of placer gold in Quaternary gravels. Some gold bearing quartz veins occur on the northern edge of the district at the Trio and Three C's mines (Morton 1977:58). The sparsely vegetated terrain consists of sand and gravel filled washes, dark brown desert pavement terraces, and small steep granitic hills. Abandoned shafts, old dry washer tailing piles, and "rat hole" tunnels dug into banks and terraces represent the work of a class of miners that remains relatively undocumented by western historians: the independent Mexican prospector and self employed small mine owner.

European settlement first occurred in the Lower Colorado River Valley in the late 18th century.

In 1780, under the direction of Father Garcés, two missions were erected near present-day Yuma, Arizona on the California side of the Colorado River: Purísima de la Concepción was built on the site where present-day Fort Yuma now stands, and San Pedro y San Pablo de Bicuñer was about 15 miles upriver near present-day Laguna Dam and about 0.5 miles southwest of the placer deposits at Potholes (Trafzer 1980:19). Various geological reports have claimed that the earliest mining by Europeans in the region occurred at this time, especially at present-day Potholes where Spanish presidial soldiers discovered and worked the placer deposits (Morton 1977:58).

Although unconfirmed by reliable sources this belief is probably correct. Mexican Colonial soldiers who had been raised in the desert regions of northern Mexico were familiar with dry placer deposits and the methods required to work them (Rodriguez and Ortega-Soto 1985; Officer 1991:4–5). It seems unlikely that these settlers would have occupied the location and not have engaged in mining. Their efforts, however, would have been limited and brief, for the Spanish occupied the area for less than two years. In July 1781, the local Quechan Indians revolted and destroyed the two missions and the presidio at Yuma. The Spanish never returned to establish permanent settlements in the region (Trafzer 1980:20).

Mining returned to the area following the discovery of gold at Sutter's Mill in northern California in January 1848. Within a year, 80,000 people had traveled to California from around the world (Greeley 1987:14). Many gold rush immigrants from the U.S. and Mexico used the Gila River Overland Trail, prompting establishment of Fort Yuma where the Gila River joined the Colorado River (Trafzer 1980:52–53).

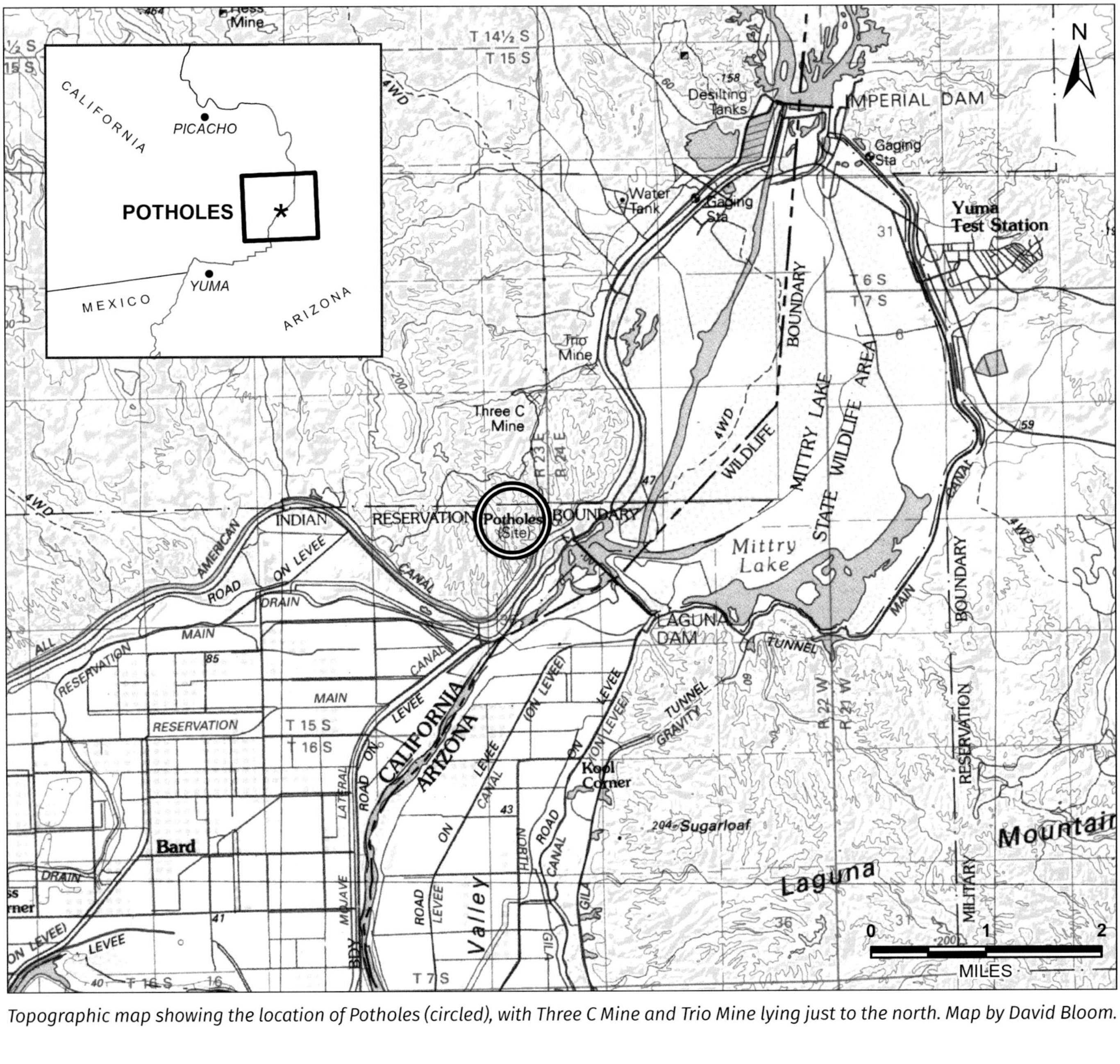

Topographic map showing the location of Potholes (circled), with Three C Mine and Trio Mine lying just to the north. Map by David Bloom.

Mexicans were among the first to learn of the strike, receiving the news from passing ships. Between six and ten thousand Sonorans followed the Gila River Route to California during 1849 and 1850. In contrast to Anglo-American "Forty-Niners", the Sonorans often traveled and worked as family units. They also were more familiar with prospecting and mining techniques, so that they found gold where Americans could not. Anglo-American antagonism against Mexican miners, along with deep-seated anti-Mexican nativism, led to abuses and discrimination, culminating in the Foreign Miners Tax of April 1850. By 1854 most Hispanics had left the northern California gold fields (Park 1961:101; Roske 1961:189–99; Kenny 1967:582; Peterson 1977; Officer 1987:245, 249; Robrock 1991:21).

Some Mexican miners, after being forced out of northern California, began prospecting along the lower Colorado River in the mid-1850s and discovered most of the major placer deposits in the area, including Potholes (Van Wormer & Newland 1996; Park 1961:101) In 1858, miners found placer deposits on the Gila River, 12 miles east of Yuma. A year later prospectors found another rich placer field at Chimney Peak, 14 miles north of Yuma. In June 1860, the *Alta Californian* reported that prospectors on the Gila and Colorado rivers included 350 Mexicans and only five Americans. They recovered gold from the arid sands using dry washing techniques developed in Sonora. Several large Hispanic strikes that have been recorded (many were not) were those made by Jose María Mendivil at Picacho in 1861, where Mexican families had already been prospecting, and by Jose María Remodino at La Paz, 60 miles north of Yuma in 1862 (Renner 1983:120; Love 1974:29–33, 57; Greely 1987:7; Park 1961:105–106). The placer deposits near present-day Laguna Dam were discovered in 1859 when Francisco and Antonio Contreras established a store on the banks of the Colorado River within a half mile of the then still visible ruins of Mission San Pedro y San Pablo de Bicuñer. Those on the Arizona side of the Colorado were called the Laguna Placers while the deposits on the California side became known as the Potholes (Figueroa nd).

The caution and hostility that Mexican prospectors developed toward Anglo-Americans after their northern California experiences makes accurate documentation of their activities difficult. They seldom recorded mining claims or discussed their finds with "Gringos" (Renner 1983:121). Their goal was not the development of large underground mines that would produce great wealth, but rather, what could best be described as subsistence mining. Traveling in family groups, they worked high grade surface outcrops and placer deposits rich enough to allow them to make a living. They tended to pursue other means of subsistence during the summer months, returning to placer mining during the cooler months of the year. When the high grade deposits played out they moved on in search of others (Van Wormer and Newland 1996).

As a result, exact documentation of mining activity at Potholes during the period is difficult. The general consensus of various reports is that between 1860 and 1900 the area was worked extensively by Hispanic prospectors. It is estimated that as many as 400 to 500 Mexicans worked the area principally with hand dry washing machines for several decades and extracted an estimated $2,000,000 in gold from the district (Haley 1923:155). The following excerpts are taken from State of California Mineralogists Reports:

> **1894** Pot Holes Mines (Placer). They are on the Colorado River, about half way between Yuma and the California Picacho Mines. In the summer of 1894, Mexicans were hauling the gold bearing dirt to the rivers, there being no water at the mine (Crawford 1894:242).
>
> **1896** These mines are worked in a very desultory way by a few Mexicans (Crawford 1896:344).
>
> **1926** Potholes (Placer). These dry placers are located 10 miles northeast of Yuma and several miles west of Laguna Dam. Elevation 150 feet. The gold bearing gravels of this region were extensively worked in the early days by Indians and Mexicans and the placers have produced considerable gold (Tucker 1926:261).
>
> **1934** Potholes placers are situated 10 miles northeast of Yuma and several miles west of Laguna Dam. This district has reported production of $2,000,000 which it is stated, was recovered by dry washing, some four or five hundred Mexicans having been at work for several years in the past. At present a few small drywashers are operated on these deposits (Tucker and Sampson 1934:245).
>
> **1942** Potholes (Placer). These dry placers . . . were extensively worked in the early days by Indians and Mexicans and are now (still) worked (Sampson and Tucker 1942:125).

In the early years, traditional primitive drywashing methods were used. The Mexican *batea*, or wooden pan method, was one of the most common. Gold bearing sand was scooped into the pan, which was manipulated and shaken with rotary and other motions to settle the heavier material and gather it to the middle while moving the lighter material toward the outer edges. The prospector then blew the lighter material away with the force of his own lungs until only the heavier residue remained which was separated from the gold "by clever manipulation" and then also blown out of the *batea* (Heikes 1913:256). Needless to say, the entire process required incredible skill and a strong diaphragm. The other commonly used method consisted of tossing the sand and gravel on a blanket so that the wind blew the lighter material away. The remaining coarse material was picked over by hand to recover nuggets while the finer gold was caught in the hair of the blanket.

By the 1880s, these methods were replaced by another implement developed by the Mexican prospectors: the bellows drywasher. This machine consisted of a canvas bellows that forced intermittent gusts of air through an angled riffle tray. The air moved the lighter material across the tray while the gold and other heavy material was caught behind the riffles. The bellows was operated by a handle or crank. The apparatus was often referred to as a Mexican drywasher or rocker (Heikes 1913:256). With the adoption of these highly efficient and portable machines Hispanic prospectors extensively worked almost all of the major placer deposits in the southwest including Potholes. The value of gold recovered by a single individual was seldom impressive but many thousands supported themselves with this independent nomadic existence. Their efforts were summarized by geologist Charles Scott Haley who noted in 1923 that:

> Attention must again be called to the fact that, in estimating the gross production … of the desert placers, a large part of the work which has been done there was done when wages averaged about $1.00 a day, in times of industrial depression. Many men, making a bare living at best, preferred to be their own masters while doing it; and the aggregate of all their earnings presents an impressive figure. Considering the length of time involved and the number of men employed, it is no wonder that the aggregate amount, even from very poor diggings, was large.
>
> …
>
> The values won from desert placers have generally been earned by the operations of a multitude of men working independently and that the extent of their operations in point of time covered a great many years.… If this point is borne in mind in considering gold production of the desert, it will be readily understood why it is that to this date not a single large scale operation handling dry placers in Southern California has been successful.… The result of the writers investigation, which has covered every dry placer district of importance in the desert, have on the whole been decidedly disappointing and the writer is forced to the conclusion that these placers have already been exploited by the only practical means: the Mexican with his little hand operated bellows and rocker (Haley 1923:155, 159).

By the mid 1880s, Anglo-American prospectors also began to take an interest in the Potholes area. Unlike their Hispanic colleagues they formed a mining district and began to file claims so that records of their activities can be found in newspapers and mining records. One of the earliest claims was the Colorado Grande Mine, claimed in March 1884 by W.A. Winninger (*San Diego Union* 3-19-1884 3:1). The following year John E. McFee located the Yuma Gravel Claim in August 1885 (*San Diego Union* 8-30-1885; Mining Claims 4:149). On July 20, 1889, J.K. Moore and Geo K. Cusenbary recorded the Golden Nugget. It consisted of forty acres situated "about eighty rods from the bank of the Colorado River on the south side of Volcano Gulch" (Location Notice 1889).

Activity increased during the following decade. In March 1894, the *San Diego Union* reported that "A deed was filed in the county recorders office yesterday conveying from O.F. Townsend and L.J.F. Lager to Jennie L. Vanderwater, for $1,000, a one third interest in the original Pot Hole placer mining claim." The paper noted that this claim "is one of the richest and best known mines on the desert" (*San Diego Union* 3-6-1894 5:1). The year 1896 was especially active with recordings of the Florence Placer Claim by Charles M. Little (*San Diego Union* 10-10-96 5:1; Mining Claims 6:157), the Felix by C. Thiesen, and the Nugget and Weisshun by S. Weisshun, in addition to the Don Carlos, Josephine, and Peacock (*San Diego Union* 3-10-96 5:1). In March 1897, H.A. Hickton filed a location notice for the Lion and Unicorn mines in the "Pot Holes" District (*San Diego Union* 3-16-1897 5:1). In January 1898, C.H. Martin filed notice locating the Myrtle Placer Claim (*San Diego Union* 1-23-98 5:4; Mining Claims 2:72). An 1899 report stated: "There are now eighteen men using drywashers and rockers in the Potholes district twelve miles above Yuma, … with good results." William Boyle and his partner, Reilly, recovered $2,210.00 in coarse gold from what they estimated to be seven yards of dirt (*Wisconsin Weekly Advocate* 7-6-1899). In 1901 L.D. Carter announced plans to erect a plant to work the gravels at Potholes "on a large scale" with his newly invented dry concentrating machine (*San Diego Union* 6-12-1901). This activity appears to have been part of a general expansion of mining interest throughout the region. In December 1897, the *Yuma Sentinel* had reported that rich gold discoveries had recently occurred in the "Castle Dome, ... Pot Holes and other mining districts ..." (*Yuma Sentinel* 12-17-1897). Mexicans still made up a significant portion of the district's population, as witnessed by a party that visited the Dominguez family at Potholes in June 1897 and were treated to a meal of "*tortillas, frijoles,* (and) *carne pieadita pimientado con chiles (sic.) torridos.*" Residents drank water from the river, which was "clear as mud" (*San Diego Union* 6-3-1897).

Although placer mining has always been the major focus of activity at Potholes, hardrock lode mining was also conducted on a small scale. The most prominent lode mines in the district were the Pot Holes or Duncan (later called the Peace Maker, Golden Nugget and Three C's), and Trio. These are both located about a mile northwest of Laguna Dam. The Duncan is the oldest and best known hardrock mine at Potholes. It was developed in the 1890s by Yuma mayor, Ransord Duncan. By 1896, a 150 foot shaft had been sunk into bedrock following a gold bearing quartz vein. E.A. Witmore was superintendent (Crawford 1896:353). In October 1897, the *San Diego Union* reported the sale of the property and plans to expand the operations. Although the sale resulted in ownership of the mine by outside investors Mayor Duncan still retained a major interest. The shaft had been excavated to a depth of 300 feet and several hundred feet of tunnels and crosscuts had also been completed. The ore which was stock piled on the site assayed at values ranging between 12 and 20 dollars a ton. One particular vein had an assay value above $40 a ton and another about 18 inches in width was said to contain as much as $200 a ton in free milling gold. Breaking up and panning samples of this ore had produced numerous large nuggets. Further plans called for installation of an eight stamp mill, steam engine and other machinery, and a pumping plant on the Colorado River 4,500 feet from the mill site (*San Diego Union* 10-12-1897:2, 3–4).

In 1919, W.H. Trenchard acquired the Duncan Mine and relocated the claims as the Peace Maker Group (Potholes Mining Records 24:114). In 1922, they were purchased by a partnership consisting of A.J. and Chester Mitchell of Alhambra, California and W.G. Reese of Yuma Arizona for 25,000 dollars (Deed 1922). Trenchard, however, remained extremely active in Imperial Valley mining, especially in the Potholes Area. In September 1925, a group that included W.H. Trenchard, J.W. Gray, Flora Gray, J.J. Wilson, Hattie Small, Richard Ramsey, C.L. Patterson, and A.P. Small filed a location notice for the Potholes Placers No.1 Placer Mining Claim, which contained 120 acres that consisted of all of the northeast quarter of Section 25, Township 15 South, Range 23 East that was located north of the main irrigation canal from Laguna Dam (Official Records 93:56). At the same time this group located several additional placer claims in the Potholes District that included the Potholes Indian No. 2 in the southwest quarter of Section 25, Potholes No. 1, in the southwest quarter of Section 19, Township 15 South, Range 24 East, the Potholes Placers No. 2, west of and adjacent to the southwest quarter of Section 19, and the Potholes Placers No. 3 in the northwest quarter of Section 30 (15 South 24 East) (Official Records 93:55–58).

In 1928, W.H. Trenchard and his wife Hanna continued to locate lode claims in the Potholes Mining District that included the old Duncan Mine and surrounding area. These consisted of the Charles Jr., the Golden Contact No. 1 and No. 2, the Golden Nugget, also known as the "Duncan Mine", and the Clifford, described as "adjoining the Old Duncan Mine on the south and constituting a spur of the Old Duncan or Golden Nugget Mine" (Official Records 202:349). Evidence of improvements or proof of labor by Trenchard or any of his partners is lacking. They seem to have lost control of most of the claims. Many were worked by the Three C Mines Corporation during the late 1920s and early 1930s.

The Three C Mines Company was a partnership consisting of C.M. Clapp, J.V. Clapp and B.H. Clark. During the last half of the 1920s they gradually took over and developed several of Trenchard's original claims. In 1926, the Trio Group, consisting of the Trio No. 1, No. 2, and No. 3 as well as the Colorado and Hills claims, was located by C.M. Clapp, manager of Trio Mine Incorporated (Official Records 174:373–375; 200:200; 239:58). These claims covered 120 acres in the southwest quarter of Section 19, Township 15 South, Range 24 East and included most of the Potholes No. 1 Placer Claim located by Trenchard and his partners in September 1925. During the physical year ending June 30, 1928 a quartz mill and concentrating table were installed at the Trio Mines (Official Records 200:200). It appears that at the same time the Clapps and Clark began to develop a shaft located on the Potholes Placer No. 1 Claim originally located by Trenchard's group. Land records at the Bureau of Reclamation Office in Yuma indicate that the Wilson's Three C's Mine was located on July 1, 1928 in Section 25, Township 15 South, Range 23 East (Telles 1989). J.J. Wilson was one of the original locators of the Potholes Placer No. 1 Claim with Trenchard in 1925, while the Three C Mines

Company was formed by Clapp, Clapp and Clark in the late '20s (Official Records 93:56; 239:58; 281:74–75).

A location claim or proof of labor notices for Wilson's Three C's Mine could not be found. However, additional evidence exists indicating that the Clapps and Clark continued to develop several of the old Trenchard claims. In June 1930, a Proof of Labor Certificate was filed by the Three C Mines Company for work on the Golden Nugget and Clifford claims originally located by Hanna Trenchard as a relocation of the Old Duncan Mine in 1928. Tasks performed included cleaning out the old shafts, digging a water conversion ditch, and road and bridge repair (Official Records 281:74). That same year the company filed proof of labor for the Trio Lode that recorded the timbering and sinking of a 5 by 7 foot shaft to a depth of 42 feet in addition to road repair work (Official Records 281:175). In 1931, the Golden Nugget and Trio were the only producing hardrock mines in Imperial County. Their production combined with small additional values from drywashing operations at Picacho brought the total gold production in the county for that year to 651 dollars (Heikes 1934:650).

In 1933, the Trio Mining Co. of Los Angeles, with Sam Mosher as president, acquired the Trio groups and Golden Nugget–Duncan mines. The Trio was under development from 1933 until 1935 when the property was acquired by the Bureau of Reclamation for construction of the All American Canal (Sampson & Tucker 1942:119). The Clapps, however, remained active in mining. They regained ownership of the Golden Nugget–Duncan Claim and renamed it the Three C's Mine. They still owned the property in 1977 (Morton 1977:60).

In conclusion, the majority of the gold mining at Potholes consisted of drywashing placer gravels by independent Mexican prospectors during the late 19th and early 20th centuries. Hardrock lode mining played a minor role in the total gold production of the district.

SIGNIFICANCE

The small hardrock mines and "rat hole" tunnels that can still be seen at Potholes represents a class of miners that remain relatively undocumented by western historians: the independent Mexican prospectors and the self employed small mine owner. Scholarly works on western mining deal largely with rich mining capitalists (Peterson 1977) or employees of large mining companies (Young 1976; Lingenfelter 1974). It is felt by many western historians that once placer deposits were exhausted in a district, resulting in development of hardrock mines, self employed prospectors moved on, leaving only mine owners and their employees to populate mining communities (Bough et al. 1970:358, 360; Young 1970:1). This view has resulted in the classification of many works on western mining as either business or labor histories (Paul 1979).

Indeed, the above scenario was often the case, especially in wealthy mining districts such as California's Mother Lode or Nevada's Comstock mines, where large capitol investment became necessary to extract the vast mineral wealth (Bough et al. 1970). The mine shafts and "rat holes" at the Potholes District, however, represent another type of western miner: the independent prospector and the self employed small mine operator. Many made a successful living while working their claims with only one or two other partners. That these self employed miners are not unique to western history is attested to by numerous small mine shafts throughout the western mountains and deserts. Certainly, independent Mexican prospectors as well as small mine owners such as W.H. Trenchard, the Clapp brothers, and B.H. Clark played an important role in western development that has not yet been extensively defined or explored by professional historians.

REFERENCES

Bogue, Allan G., Thomas D. Phillips, and James E. Wright, 1970, Disposing of the Public Domain, in The West of the American People, edited by Allan G. Bogue, Thomas D. Phillips, and James E. Wright. F.E. Peacock Publishers, Itasca, Illinois.

Crawford, J.J., 1894,Twelfth Report (Second Biennial) of the State Mineralogist, Two Years Ending September 15, 1894. California State Mining Bureau, San Francisco.

Crawford, J.J., 1896,Thirteenth Report (Second Biennial) of the State Mineralogist, Two Years Ending September 15, 1896. California State Mining Bureau, San Francisco.

Deed, 1922, W.H. Trenchard of Potholes, California to A.J. Mitchell and Chester Mitchell of Alhambra, California and W.G. Reese of Yuma, Arizona on file at the Arizona Historical Society Century House Museum Research Archives, Yuma, Arizona.

Figueroa, Paul, no date, Father Figueroa Manuscript on file at the Arizona Historical Society Century House Museum Research Archives, Yuma, Arizona.

Greeley, Michael N., 1987, The Early Influence of Mining in Arizona, in History of Mining in Arizona, edited by J. Michael Canty and Michael N. Greeley, Mining Club of the Southwest Foundation, Tucson:13–50.

Haley, Charles Scott, 1923, Gold Placers of California. California

State Mining Bureau Bulletin No. 92. California State Mining Bureau, San Francisco.

Heikes, V.C., 1913, Dry Placer in Arizona in Mineral Resources of The United States–Calendar Year 1913–Part I Metals. United States Geological Survey, Washington D.C.

Heikes, V.C., 1934, Gold, silver, copper, lead, and zinc in California in Mineral Resources of the United States 1931- Part I Metals. United States Geological Survey, Washington D.C.

Kenny, William Robert, 1967, Mexican-American Conflict on the Mining Frontier, 1848–1852. Journal of the West, 6 (Winter):582–592.

Lingenfelter, Richard, 1974, The Hardrock Miners: A History of the Mining Labor Movement in the American West, 1863–1893. University of California Press, Berkeley.

Location Notice, 1889, Location Notice for the Golden Nugget Placer Claim, Potholes Mining District, July 20, 1889, on file at the Arizona Historical Society Century House Museum Research Archives, Yuma, Arizona.

Love, Frank, 1974, Mining Camps and Ghost Towns: A History of Mining in Arizona and California Along the Lower Colorado, Westmoreland Press, Los Angeles.

Mining Claims, 1896–1919 , Mining Claims Books, Imperial County Recorder's Office, El Centro.

Morton, Paul K., 1977, Geology and Mineral Resources of Imperial County. California Division of Mines and Geology, Sacramento.

Officer, James E., 1987, Hispanic Arizona. University of Arizona Press, Tucson.

Official Records, 1925–1935, Imperial County Recorder's Office, El Centro.

Park, Joseph H., 1961, The History of Mexican Labor in Arizona During the Territorial Period, Masters Thesis, University of Arizona.

Paul, W. Rodman, 1979, A Tenderfoot Discovers There Once Was a Mining West. Western Historical Quarterly 10 (1):5–19.

Peterson, Richard H., 1977, The Bonanza Kings: The Social Origins and Business Behavior of Western Mining Entrepreneurs, 1870–1900. University of Oklahoma Press, Norman.

Potholes Mining Records, 1919, San Diego County, Office of Recorder

Renner, Pamela, 1983, La Paz: Gateway to Territorial Arizona, Journal of Arizona History, 24 (Fall) 119–144.

Robrock, David P., 1991, Argonauts and Indians: Yuma Crossing 1849. The Journal of Arizona History, 32 (Spring): 21–40

Rodriguez, Antonio y Martha Ortega Soto, 1985, Capitolo III: Entrada de Colonos Espanoles En Sonora Durante El Siglo XVII, in Historia General de Sonora Vol. II, De La Conquista al Estado Libre de Sonora, Segio Ortega Noriega y Ignacio Del Rio, editors. Gobenador del Estado de Sonora, Hermosillo, Mexico.

Roske, Ralph J., 1963, The World Impact of the California Gold Rush 1849–1857. Arizona and the West, 5 (Summer):187–232.

Sampson, R.J. and W.B. Tucker, 1942, Los Angeles Field District: Current Mining Activity in Southern California. California Journal of Mines and Geology 36(1):9–31.

San Diego Union, 1885–, Various issues cited in text. San Diego State University Library.

Telles, Carol (Staff Archaeologist, Bureau of Reclamation, Yuma), 1989, Telephone Communication to Stephen R. Van Wormer.

Trafzer, Clifford E., 1980, Yuma: Frontier Crossing of the Far Southwest. Heritage Books, Wichita.

Tucker, W.B., 1926, Los Angeles Field Division, Imperial County. Report XXII of the State Mineralogist 22(2):248–261.

Tucker, W.B. and R.J. Sampson, 1934, Los Angeles Field District: Current Mining Activity in Southern California. California Journal of Mines and Geology 30(4):310–315.

Van Wormer, Stephen R. and James D. Newland, 1996, The History of Hedges and the Cargo Muchacho Mining District, Part I: A Case Study of Mexican Mineros and Corporate Mining in the Southern California Desert. Journal of San Diego History 42 (2): 1–30.

Wisconsin Weekly Advocate, 1899, Issue of July 6 cited in text. Online at GenealogyBank.com.

Young, Otis E., 1970, Western Mining: An Informal Account of Precious-Metals Prospecting, Lode Mining, and Milling on the American Frontier from Spanish Times to 1893. University of Oklahoma, Norman.

Young, Otis E., 1976, Black Powder and Hard Steel: Miners and Machines on the Old Western Frontier. University of Oklahoma Press, Norman.

Yuma Sentinel, 1897, issue cited in press. On microfilm at the Yuma City Library.

OLD MINING DISTRICT TO BE REVIVED

New Company Operating Senator Mine North of Yuma on Colorado

YUMA (Ariz.) June 5. (Exclusive)—The Pioneer Potholes district, north of Yuma, on the Colorado River, is coming back to activity, after years of idleness. A new corporation, the Skelton Mining Company, has resumed work on the Senator mine, which has been developed to 300 feet, with 150 feet of drifting at that level, now flooded.

The same amount of work has been done at 200 feet demonstrating the existence of about thirty feet of ledge matter, said to average around $10 gold. The ore is free milling, in iron-stained quartz, with 1000 tons already on the dump.

In Los Angeles the Kenemott Iron Works is constructing a specially designed ball mill for the property, with power to be furnished bv a fifty-horsepower Fairbanks-Morse engine. In charge of operation is Pearl Skelton of Yuma, who, with Ward H. Preston of Shortville, N. Y., purchased the property from W. T. Trenchard.

In the same section is the old Duncan mine, discovered in 1870 by Mexican placer miners and, now being worked by the Clapp brothers, who are reported to have ore of remarkable richness, with much free gold. The district first was known for its placer deposits, first operated about 1860. In the same section is the Laguna dam, from which water is diverted for irrigation around Yuma.

From Los Angeles Times, *June 6, 1927.*

Geologic and Geochemical Study of the Picacho Gold Mine, California: Gold in a low-angle normal fault environment

STEVEN LOSH, DAN PURVANCE, ROSS SHERLOCK, AND E. CRAIG JOWETT

S. Losh, Department of Earth and Atmospheric Sciences, Cornell University, Ithaca, NY 14853, USA E-mail: losh@geology.cornell.edu

D. Purvance, Glamis Gold Inc, Picacho Mine, 1891 Rail Avenue, Yuma, AZ 85365, USA

R. Sherlock, SRK Consulting, Suite 800, 580 Hornby Street, Vancouver, BC, Canada V6C 3B6

E. C. Jowett, Waterloo Centre for Groundwater Research, University of Waterloo, Waterloo, ON, Canada N2L 3G1

ABSTRACT

The Picacho gold deposit, located in south-easternmost California, is a low-grade gold deposit in a nearly flat-lying denudational fault of regional extent and probable Oligocene age. The deposit is hosted by intensely fractured and faulted Mesozoic leucogranite and by chloritic augen gneiss and schist, and is overlain unconformably and in fault contact by unmineralized late Oligocene Quechan volcanic rocks. The deposit is structurally characterized by normal and normal-oblique faults of low to high dip at shallow depths in the mine, merging downward with a synchronous, low-dipping ore-stage extensional fault system (the Chocolate Mountains/Gatuna Fault) of probable Oligocene age in deeper portions of the deposit. The fault system was infiltrated during much of its active life by hot, dilute, highly exchanged meteoric water having temperatures of 170°–210° C, salinity <2 wt% NaCl equivalent and calculated $\delta^{18}O$ fluid between -2.6‰ and 5.2‰.This main-stage fluid precipitated quartz, pyrite, and specular hematite, accompanied by silicification and sericitization. Auriferous ore-stage pyrite was precipitated late in the fault evolution probably by mixing of reducing ore fluid with relatively oxidized main-stage fluid during regional Oligocene extension on the Chocolate Mountains/Gatuna Fault. The Picacho deposit is characterized by a gold–arsenic–antimony geochemical signature consistent with bisulfide complexing of gold in reducing fluid, in contrast with typical denudation fault-hosted base-metal-rich deposits associated with high-salinity fluids elsewhere in the southwestern United States. The deposit is overprinted by Miocene normal faults having a wide range of dips. These postore faults are associated with red earthy hematite precipitation, pyrite oxidation, and supergene enrichment of gold.

INTRODUCTION

IN THE 1980S, SEVERAL LARGE GOLD DEPOSITS in the southwestern United States were discovered and linked to newly described middle Tertiary low-angle normal faults, variously known as denudational faults or detachment faults (for example, Seedorff 1991; Maldonado 1991; R. Benson, unpublished data). These faults are associated with regional extension and are commonly but not always associated with footwall metamorphic core complexes (works in Crittenden et al. 1980). This type of fault was first recognized in the southwestern United States more than 30 years ago (Anderson 1971), and has since been widely described in the US Basin and Range Province and throughout the world. Movement on low-angle normal faults has exhumed their footwalls, and the faults are presently exposed as low-dipping to grossly horizontal undulating or domal surfaces that cover areas

Reprinted from *Mineralium Deposita*, Volume 40, Number 2, March 2005, p. 137–155.

of tens to as much as thousands of square kilometers. In many instances, low-angle normal faults initially had moderate to steep dips, then rotated to low dip; in other instances, normal faults had low dip throughout their movement history.

"Detachment faults" were found to host gold deposits, and became an attractive exploration target (e.g., Johnson 1987). Many relatively small "detachment-hosted" gold deposits in the southern Basin and Range province (e.g., Arizona, southeastern California) are characterized by an epithermal Au–Cu–Mn–F geochemical assemblage and by high-salinity fluid inclusions (Wilkins et al. 1986), in contrast to large bonanza-type epithermal volcanic rock-hosted gold deposits, characterized by low-salinity fluids, of the same age and in the same region. Kerrich and Reynolds (1987) postulated that gold in these detachment fault-hosted deposits was precipitated due to mixing of oxidized and reduced fluids. The difference between volcanic-hosted and detachment-hosted gold deposits in terms of both chemistry and size raises questions that bear not only on "detachment-hosted" deposits but on epithermal gold deposits, in general: What controls the chemistry of the detachment-hosted gold deposits? What factors contribute to the formation of moderate-sized to large "detachment-hosted" gold deposits?

In the 1980s, several workers described the Picacho deposit, in southeasternmost California, as being associated with a low-angle normal fault (Van Nort and Harris 1984; Drobeck et al. 1986: Liebler 1988). The Picacho deposit produced about 18.6 metric tons (600,000 troy oz.) of gold from faulted and fractured Mesozoic leucogranite and schist as well as from Tertiary postore conglomerate that was derived from the mineralized crystalline rocks. Notably, the deposit is characterized by low-salinity fluid inclusions (Liebler 1988) and a Au–As–Sb geochemical assemblage similar to that of many volcanic-hosted epithermal gold deposits, making it of particular value in addressing the questions posed above. Drobeck et al. (1986) noted that rocks in the Picacho deposit were geochemically and mineralogically similar to rock from a low-angle normal fault (their "Chocolate Mountains Detachment Fault"), exposed about 1 km north of the deposit. Liebler (1988) noted that the ore bodies were roughly flat-lying in highly brecciated rocks and that gold grades increased upward, approaching the nearly horizontal "Chocolate Mountains Detachment Fault," which separated mineralized Mesozoic crystalline rocks from unmineralized Oligocene volcanic rocks at the mine. Based on these lines of evidence, these workers concluded that the Picacho deposit is genetically related to that fault. However, the setting of the deposit in a detachment fault of regional extent was not unequivocally established, perhaps owing to the limited amount of exposure of the deposit at the time of those earlier studies.

Since these early works, mining has exposed much more of the deposit, allowing it to be investigated in greater detail. The purpose of this paper is to document the structural setting of the Picacho deposit and present geochemical data that bear on how it formed. In addition to work in the mine, sampling was carried out at fault exposures outside the mine to further evaluate the relationship between the Picacho deposit and nearby low-angle normal faults. This work documents two main stages of low-angle normal faulting in the Picacho deposit. Gold mineralization in the mine is associated with late stages of the older extensional event, which is correlated with regional extension and tectonic exhumation during the Oligocene and early Miocene epochs. Geochemical data indicate the deposit probably formed by mixing of oxidized and reduced low-salinity fluids of meteoric origin that interacted with different rock types at different temperatures along their respective pathways.

Mine setting and history

The Picacho deposit is located in the historic Picacho mining district, 29 km north of Yuma, Arizona (Fig. 1). It is situated in Sonoran-type sparsely vegetated low hills near the base of Picacho Peak. Temperatures at the mine commonly exceed 43°C in the summertime, and annual rainfall is about 6.5 cm. In 1862, Jose Maria Mendivil discovered the Picacho lode and staked the original claims. Primarily using underground mining methods, hundreds of miners worked in the narrow high-grade veins that outcropped along the flanks of the washes. Through the 1940s, approximately $15 million worth of gold was extracted from the Picacho Mine district. In 1980, Chemgold, a wholly owned subsidiary of Glamis Gold, Inc., began to mine the deposit in four shallow open pits (Dulcina, Apache, San George, and Diablo). Since 1992, mining focused on the deeper downdip extent of the Dulcina ore body referred to as the Dulcina East and Dulcina Extension pits (Fig. 1). By 1999, all Picacho deposit ore reserves were exhausted and mining ceased, and by 2002, all open pits were backfilled. Over the 20year mine life, Chemgold mined 15.5 Mt of ore at an average grade of 1.0 g/ton (0.033 oz/ton), and recovered 12.1 Mt (390,000 troy oz.) of gold. Including historical gold production, total gold production from the deposit is estimated at over 18.6 Mt (600,000 troy oz).

Methods

Mapping and sampling were carried out in the mine and at several locations outside the mine. All samples were examined petrographically, and selected samples were further analyzed in terms of their geochemical (bulk and stable isotope) and fluid inclusion properties.

Microthermometric analysis of fluid inclusions was conducted on a Fluid Inc. adapted USGS gas flow heating–freezing stage. The accuracy of the Trendicator thermocouple is better than 0.4°C over the temperature range involved in this study. All of the inclusions in demonstrably cataclasite-stage quartz were between about 2 and 10 μm in diameter. They consisted of two phases, liquid (90%) and vapor (10%) water, and homogenized in the liquid phase. Primary fluid inclusions, as defined by Roedder (1984), were found in quartz and calcite hosts as isolated single inclusions or groups of inclusions. Calcite-hosted inclusions were from veins that predate specular hematite (94PIC16) as well as veins that postdate it (P130). Secondary inclusions were found only in quartz and were located along late fractures.

Thirty-four samples were analyzed for Au, Ag, As, Sb, Hg, Cu, Bi, Zn, and Pb. Trace-element geochemical analysis was performed by X-ray Assay Laboratory of Don Mills, Ontario and American Assay Labs, Reno, Nevada. Gold was analyzed by fire assay preconcentration followed by directly coupled plasma spectroscopy (DCP); silver, lead, zinc, and copper by DCP; arsenic, antimony, and bismuth by atomic absorption; and Hg by wet chemical methods and cold fusion AA. Twenty-three samples were collected within the mine, and eleven samples were taken from faults at Little Picacho Wash and Copper Basin (Fig. 1).

Oxygen isotope data were collected from rocks in and near the Picacho mine to evaluate the source of hydrothermal fluid and the extent of water-rock interaction during main-stage faulting and to ascertain whether fluids involved in main-stage deformation in the Picacho deposit were had different exchange histories or pathways compared to fluids that circulated in the low-angle normal fault(s) outside the deposit. Mineral separates for

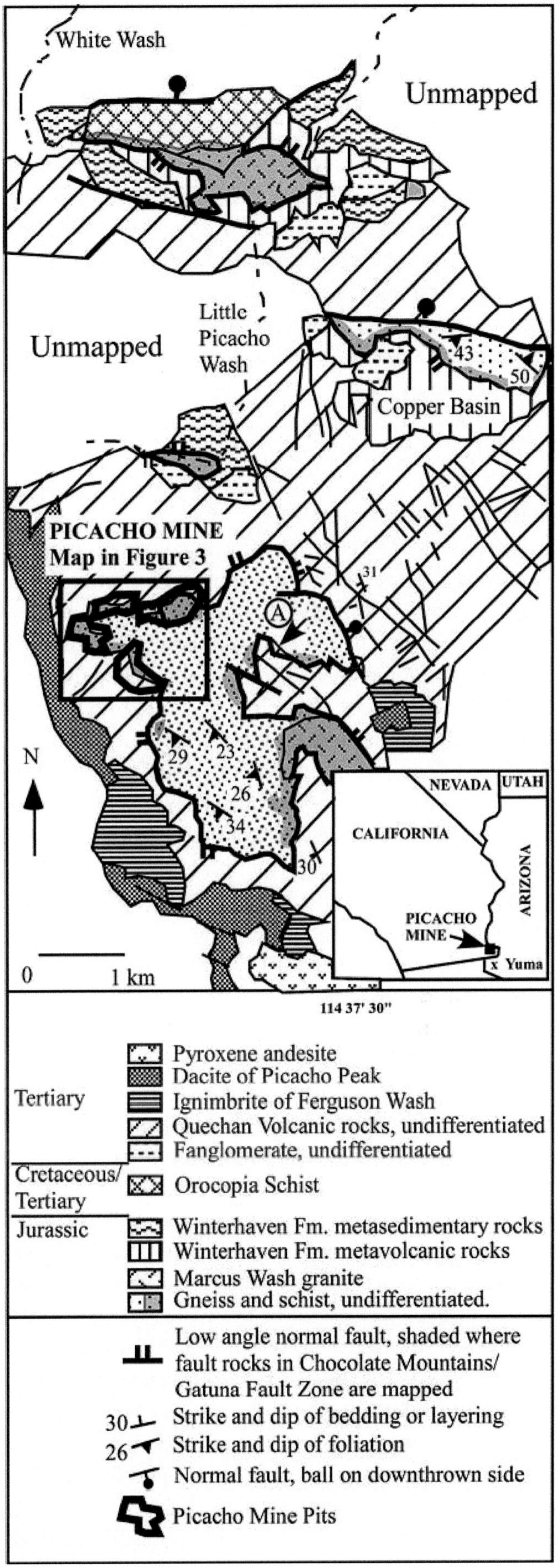

Figure 1. *Geologic map of vicinity of Picacho Mine, modified from Drobeck et al. (1986) in the southern portion and Haxel et al. (1985) in the northern portion. Low-angle normal fault trace in the mine area was taken from Drobeck et al. (1986) as were locations of extensively brecciated crystalline rocks herein assigned to the Chocolate Mountains/Gatuna Fault south and east of the mine. Shaded areas correspond to brecciated or faulted lower plate rocks in the Chocolate Mountains/Gatuna Fault. Location "A" on the map corresponds with an unconformity between schist and Quechan volcanic rocks*

isotopic analysis were made using conventional heavy liquid, magnetic, and hand-picking techniques. Silicate oxygen isotopic analysis was performed on a SIRA, triple collecting, isotope ratio, gas-source mass spectrometer at the University of Western Ontario. Standards of NBS-28 were run with each batch of six samples; results were 9.7±0.1‰ relative to V-SMOW. Calcite carbon and oxygen isotope analyses were obtained by extracting CO_2 from calcite using phosphoric acid (McCrea 1953) and analyzing the gas in a Micro-mass 903 triple collecting gas-source mass spectrometer at the University of Waterloo. The lab reports uncertainties of 0.1‰ for both oxygen and carbon.

RESULTS

Regional geologic setting

Rock types exposed in the Picacho mine area comprise Jurassic to Cretaceous metamorphic and granitic rocks as well as Jurassic clastic sedimentary rocks and andesite. These are overlain unconformably and in fault contact by Oligocene fanglomerate and volcanic rocks, in particular, the andesitic to dacitic Quechan volcanic rocks, dated at 24–26 Ma (Spencer et al. 1995) (Figs. 1, 2). Younger Miocene volcanic rocks overlie the Quechan volcanic rocks and define much of the rugged skyline of the area.

Rocks in the Picacho mine area have been affected by several episodes of deformation (Table 1). Regional Mesozoic compression and prograde metamorphism gave way in the early Tertiary to extension and localized retrograde metamorphism. At this time, the major Mesozoic compressional structure in the area, the Chocolate Mountains Thrust Fault, was locally reactivated as an extensional fault (e.g., Simpson 1990).

Oligocene extension Regional extension and substantial tectonic exhumation took place on two subparallel faults that are well exposed in the Gavilan Hills, about 7 km northwest of the Picacho Mine (Oyarzabal et al. 1997; Jacobson et al. 2002). These faults currently have low to moderate dips. Jacobson et al. (2002) concluded that early extension in the Gavilan Hills was accommodated on the ductile Chocolate Mountains Fault in the middle crust, and later extension took place on the nearby primarily brittle Gatuna fault in the upper crust. For the purposes of this paper, the two faults are considered together and are here termed as the "Chocolate Mountains/Gatuna Fault." As described in a subsequent section, the Chocolate Mountains/Gatuna Fault, or its equivalent, is exposed in the Picacho Mine and the surrounding area (Fig. 1). These exposures may all represent a single fault zone, or there may be a linked system of listric extensional faults in the area, accommodating progressive Tertiary exhumation of rocks from midcrustal depths (e.g., Richard and Haxel 1991; Haxel et al. 1988).

Rapid tectonic denudation of the footwall along the Gatuna Fault in the Gavilan Hills began shortly before eruption of the 24–26 Ma Quechan basal volcanic flows (age from Spencer et al. 1995) and ended at about the same time they erupted. As shown in a subsequent section of this paper, Quechan volcanic rocks unconformably overlie rocks deformed in the Chocolate Mountains/ Gatuna Fault at the Picacho Mine, demonstrating that rocks in the footwall of the Chocolate Mountains/Gatuna Fault were exposed at the surface in the mine area by 24–26 Ma. This timing constraint corroborates the 24–28 Ma cooling ages for feldspars in footwall rocks in the Gavilan Hills (Jacobson et al. 2002).

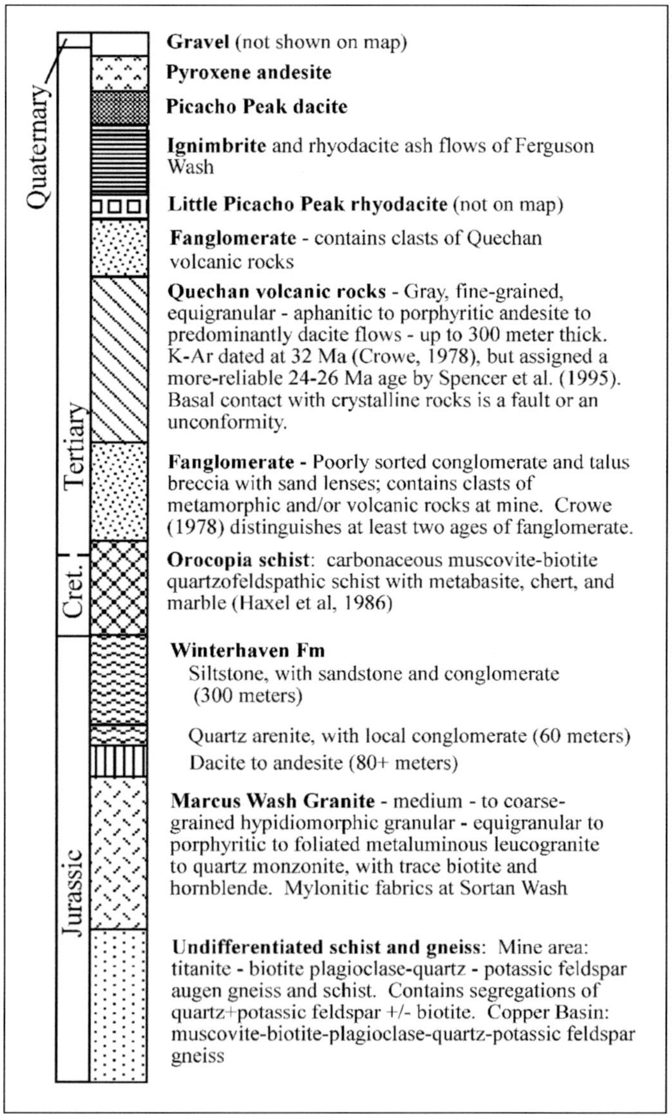

Figure 2. *Stratigraphic column, Picacho Mine area, compiled from Drobeck et al. (1986) and Haxel et al. (1985)*

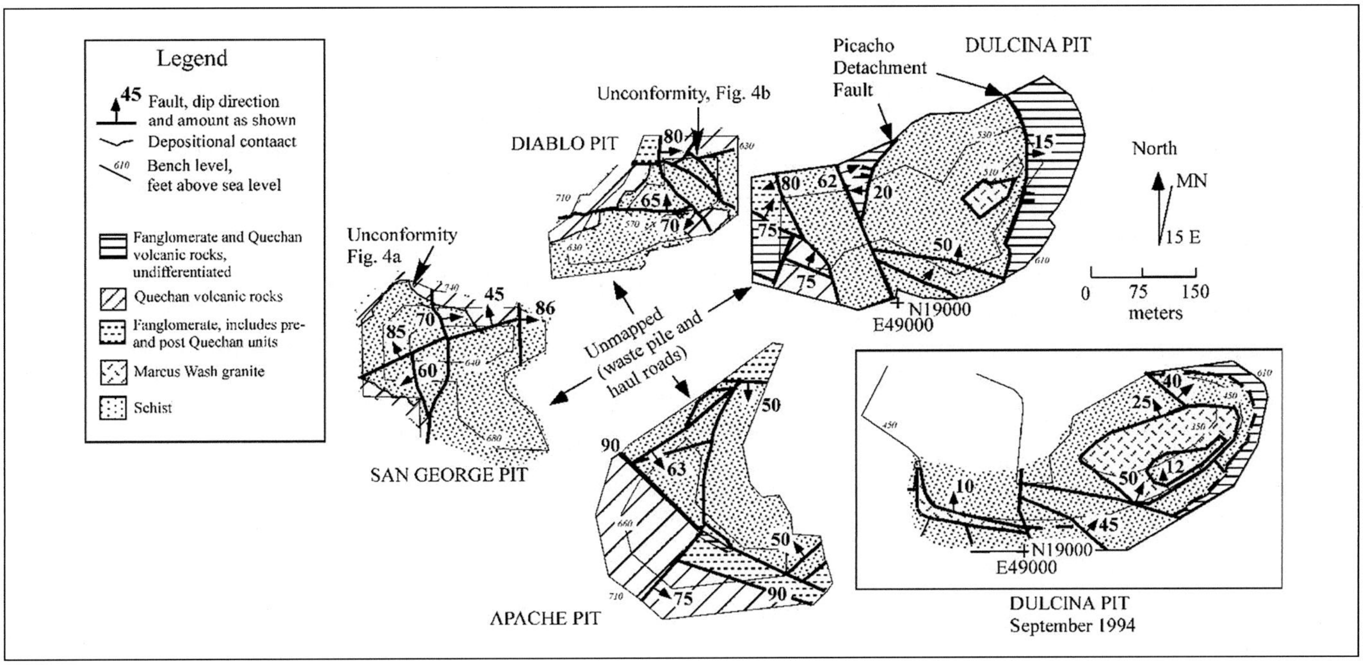

Figure 3. *Geologic map of mine pits as of December 1988; inset shows map of lower portions of the expanded Dulcina Pit (Dulcina Extension) in September 1994. Only major faults are shown. Upper part of Dulcina Pit was inaccessible in 1994, and only features that could be confidently mapped from the bottom of the pit are shown on the map. See Figure 1 for location of pits. All of the crystalline rocks in the mine lie within the Chocolate Mountains/Gatuna Fault zone, as shown in Figure 1. Low-angle fault contact between crystalline rocks and Quechan volcanic rocks/fanglomerates in the Dulcina pit (1988 map) corresponds to the "Picacho Fault." High-angle faults cutting low-angle fault in 1994 Dulcina pit map contain red, earthy hematite-stained gouge and are post-mineral. Area between pits was not mapped, as it was the site of haul roads and waste dumps*

Miocene extension

At Little Picacho Wash about 1 km north of the mine, a well-exposed low-dipping (8°NE) normal fault separates footwall Mesozoic schist from a highly normal-faulted hangingwall that includes tilted Quechan volcanic rocks (Drobeck et al. 1986). These hangingwall faults are kinematically linked to movement on the low-angle fault at which they end. Drobeck et al. (1986) named this low-angle fault the "Chocolate Mountains Detachment Fault." The latest episode of movement on this fault is constrained to be younger than the age of the tilted hangingwall Quechan volcanic rocks, or 24–26 Ma, but older than nearly flat-lying basalt flows, dated at ~20 Ma, that overlie moderately tilted Miocene volcanic rocks in the nearby Castle Dome Mountains (Spencer et al. 1995). In the mine area, the Picacho fault generally reactivates the older Chocolate Mountains/Gatuna Fault. Where untilted Quechan volcanic rocks unconformably overlie faulted Mesozoic schist and granite in the mine, the Picacho Fault is constrained to lie entirely within Quechan volcanic rocks above the unconformity. As the low-angle normal fault having post-Quechan movement is not exposed in the Chocolate Mountains (at least 35 km northwest of the mine) but is instead found primarily in the area around the Picacho Mine and to the north and east, we propose that the "Chocolate Mountains Detachment Fault" of Drobeck et al. (1986) be renamed the "Picacho Fault," as the latter name more accurately describes its geographic extent.

GEOLOGY AND GEOCHEMISTRY OF THE PICACHO DEPOSIT

Geologic setting of Picacho deposit

The Picacho deposit (Fig. 3) is hosted primarily by faulted and fractured Mesozoic schist and leucogranite.

In the Apache and Dulcina pits, ore also consists of Tertiary fanglomerate that contains detrital pyrite and clasts of mineralized granite and schist (also see Drobeck et al. 1986); approximately one-third of the ore mined at Picacho consisted of fanglomerate. Barren Quechan volcanic rocks unconformably overlie specular hematite + pyrite ± gold-bearing cataclasite in the San George and Diablo pits (Figs. 3, 4), indicating that the Quechan volcanic rocks are younger than the hydrothermal system responsible for the Picacho deposit. Quechan volcanic rocks were reported by Van Nort and Harris (1984) to be locally mineralized along unconformable contacts with underlying crystalline rocks, but we (DP) have seen no mineralized volcanic rocks in several years of examining ore at the Picacho Mine. Overall, the pyrite and specular hematite deposition and gold mineralization at the Picacho Mine is stratigraphically bracketed between the Jurassic Marcus Wash granite and the Oligocene fanglomerate and Quechan volcanic rocks.

In the deposit, fault geometry varies with depth. At shallow levels in the deposit, faults in the crystalline rocks typically dip moderately to steeply to the northeast (Fig. 5i). Where measurable, fault displacements are in the order of centimeters to meters, with normal to oblique slip. In contrast, low-angle faults are predominant at deeper levels in the Dulcina Pit (350–450 feet elevation) (Figs. 3, 5ii). Throughout the mine, specular hematite and/or pyrite veins are generally parallel to the nearest faults. Near the base of the pit, white cataclasite and green to gray fault gouge bound a gently-dipping tabular body of Marcus Wash granite generally between 6 to 12 meters thick. The granite is transected by numerous gently to moderately dipping faults that merge with the bounding faults (Fig. 6), forming a small-scale extensional duplex. The sill is the deepest occurrence of Marcus Wash granite in the mine, and its base defines the lower limit of ore. High-angle down-to-the-north listric faults merge with the low-angle fault above the granite body in at least two locations (ex. Fig. 7). This linked fault system has been disrupted by translation and rotation associated with later, high-angle faults (e.g., Fig. 3 inset).

A low-angle normal fault was also mapped in the shallow portion of the Dulcina Pit in 1988 (Fig. 3), where it separates mineralized crystalline rocks in the footwall from fanglomerate and unmineralized Quechan volcanic rocks in the hangingwall. Where mapped, it comprised a several-centimeter-thick zone of red clay gouge that incorporates clasts of mineralized, cataclastically deformed leucogranite. Like the deeper low-angle fault system, this shallower, later fault has also been disrupted by late, high-angle faults.

There are several types of fault rock in the Picacho Mine, each corresponding to different stages of deformation (Table 2). (1) Protomylonite is the oldest fault product in the mine and lies only below the Marcus Wash granite sill in the Dulcina pit. It deformed under lower greenschist facies conditions and contains ductilely deformed calcite veins as well as later undeformed specular hematite and pyrite veinlets. Just below the contact with overlying Marcus Wash granite, protomylonite foliation is generally parallel to the gouge zone at the base of the Marcus Wash granite sill. (2) Cataclasite is commonly developed

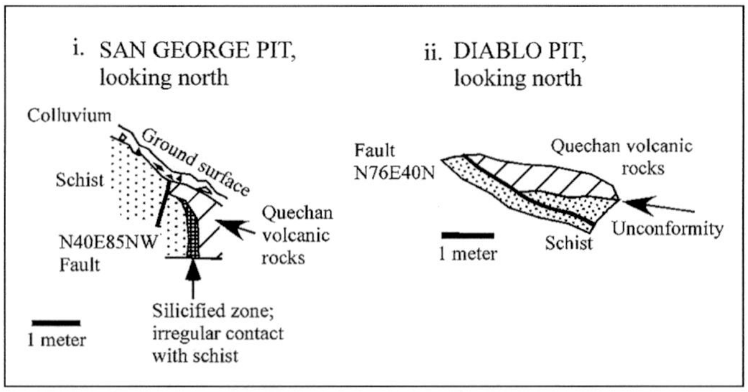

Figure 4. *(i) Sketch of unconformity, exposed in a vertical pit wall, between mineralized, faulted chlorite schist and unmineralized Quechan volcanic rocks. 740 level, San George Pit. (ii) Sketch of unconformity between mineralized schist and Quechan volcanic rocks in pit wall, Diablo Pit*

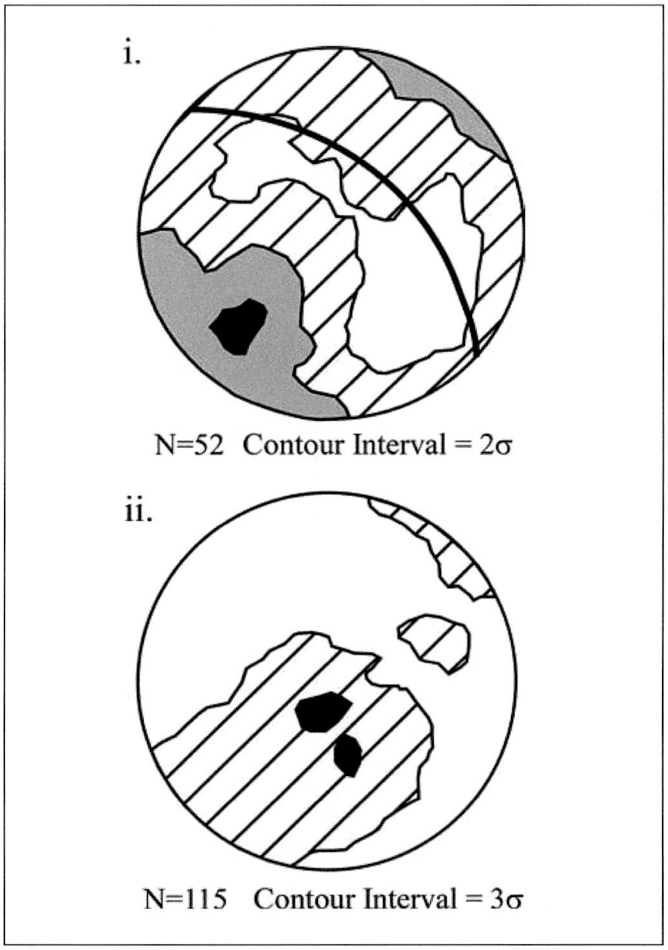

Figure 5. *(i) Kamb contour plot of poles to syn-ore faults at levels above the 510 level, Picacho mine; data are from all four pits. Maximum corresponds to fault orientation of 315/61 E. (ii) Kamb contour plot of poles to faults in the Dulcina Extension pit, 390–450 levels. Maxima correspond to fault orientations of 281/15 N and 253/37 N. Kamb contour plots are more appropriate to small data sets than are standard contour plots of poles/1% area (Kamb 1959). Significant clustering is defined as 3r or greater. Stereonet program of Rick Allmendinger, Cornell University*

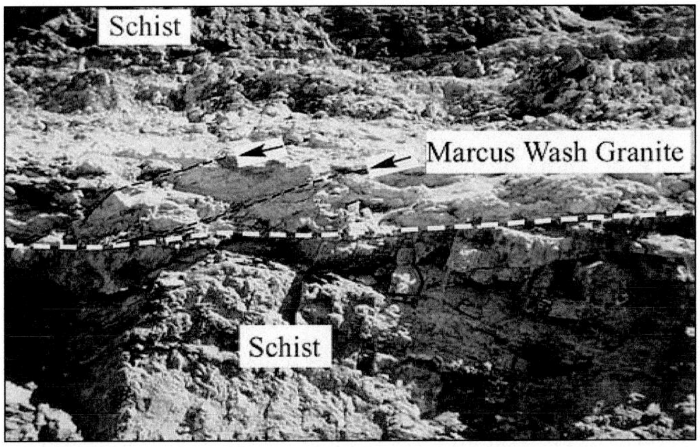

Figure 6. *Faults comprising extensional duplex in slice of Marcus Wash granite. Low-angle bounding fault runs left to right across middle of photo. Higher-angle faults above it (arrow) merge at their lower ends into the bounding fault shown, and at their upper ends into the upper bounding fault, which traverses the top of the photo. Specular hematite ± pyrite are found within the faults and in adjacent veinlets. Photo taken at the 390 level, Dulcina Pit, looking southeast; field of view approx. 3m.*

Figure 7. *Low-angle normal fault separates Marcus Wash granite (bottom, light colored) from hanging wall chlorite schist (top, darker gray). Two listric faults in the schist (arrows) merge into the low-angle fault and are geometrically consistent with down-to-the-north movement. Photo taken September 1994 in Dulcina Pit, looking west; height of wall is about 45m.*

Figure 8. *(i) Cataclasite ("cata1") overgrown by specular hematite ("sp"), crosscut by later cataclasite ("cata2") (P63). Elsewhere in slide, "cata2" is overprinted by specular hematite and gold-bearing oxidized pyrite, and is crosscut by yet another mineralized cataclasite. Each stage of mineralization in the sample is crosscut by later cataclasite, indicating repeated cycles of cataclasis and growth of specular hematite ± pyrite. (ii) Reflected light view of same field as 8i. Contact between "cata1" and "cata2" traverses photo about halfway up. (iii) Specular hematite ("sp1") overgrown on cataclasite. Bottom half of photo is of brecciated "sooty" specular hematite ("sp2") in later cataclasite. Elsewhere in the slide, cataclasite containing "sooty" specular hematite constitutes clasts in breccia having groundmass of red earthy hematite (P57). (iv) Red earthy hematite-bearing breccia (P91) incorporates clasts containing specular hematite ("sp," left). Complex aggregate of oxidized pyrite grains ("py") overgrows breccia fabric on right. (v) Goethite pseudomorph after pyrite ("goet") overgrows breccia containing red earthy hematite (P92). Similar textures without red earthy hematite are also common in deeper parts of the Dulcina pit. Red earthy hematite postdates pyrite, and probably was precipitated at the same time pyrite was oxidized. (vi) Gold (Au) in oxidized pyrite that overgrows breccia (P91)*

in leucocratic rocks and to a lesser extent in schist, and is present throughout the mine. As shown by fluid inclusion data from quartz veins associated with this faulting event (next section), these cataclasites formed at temperatures on the order of 200°C. Specular hematite and some of the pyrite in the deposit was precipitated during multiple episodes of cataclastic deformation (Fig. 8i, ii). Brecciation of specular hematite, which is normally well crystallized in these rocks, produces a sooty-appearing groundmass in which specular hematite grain size is much reduced (Fig. 8iii). (3) Postcataclasite breccias incorporate angular clasts of specular hematite-bearing cataclasite. Microbreccia zones of this age crosscut cataclasite-stage quartz veins and incorporate clasts of brecciated specular hematite. Gold-bearing pyrite overgrew these breccias and microbreccias (Fig. 8iv, v, vi). Except where cataclasites and angular-clast breccias are near post-Quechan faults, they do not contain the red earthy hematite that is typical of fault breccias formed during Picacho Fault-stage and later, high-angle faulting. (4) Postmineral faults cut Quechan volcanic rocks, and are characterized by poorly cemented breccia, calcite veins, and abundant clay + red earthy hematite matrix. In contrast to the cataclasites that are distributed throughout the deposit, these late faults are relatively narrow and widely spaced.

Alteration and veins in the mine

Schist exhibits sericitization and calcite alteration of plagioclase and chloritization of biotite (Table 2). Chloritization of biotite is widespread. Where biotite remains, it is locally altered to specular hematite, titanite, and/or epidote. Chlorite and epidote are incorporated into ductile shear fabrics, and formed early in the deformation history. Later cataclasis was typically accompanied by calcite precipitation in veins and cataclasite matrix and calcitization of plagioclase (discussed below). Calcite + (barren) pyrite veins up to 5 mm across are common in schist in deep portions of the Dulcina pit. Where calcite veins are present in schist, wallrock plagioclase is also partially altered to calcite. Calcite veinlets branch off the calcite + pyrite veins and crosscut coarse sericite in wallrock schist, indicating that calcite veins and, by extension, calcitization of plagioclase postdate sericitization. In rare instances, specular hematite is also intergrown with vein calcite in the deep parts of the Dulcina Pit. Calcite is common in cataclasite groundmass in schists, where it is intergrown with porphyroclastic quartz and feldspar and may constitute up to 25% of the cataclasite by volume. Clear and para-genetically late calcite veinlets are common throughout the deposit. In one instance (sample 94PIC16), clear calcite predates specular hematite veins. Syntaxial quartz veins that crosscut cataclastic fabric are rare in the schist, and locally contain pyrite that is barren of gold. These veins are transected by later calcite veins. Silicification of schist is more common in deeper parts of the deposit but is weak and spotty. Titanite, intergrown with calcite in portions of veinlets, is also increasingly abundant in the deeper portions of the Dulcina Pit: locally, it constitutes up to 2% of the rock.

Leucogranite in the mine is commonly weakly silicified, with quartz partially replacing plagioclase and potassic feldspar. As in the schists, calcite is common but not ubiquitous in the groundmass of cataclasite derived from leucogranite. However, calcite alteration of leucogranite plagioclase is absent and sericitization of plagioclase is rare, even where leucogranite is cataclastic or incorporated into angular breccia. Quartz veins up to 5-mm-wide are present in cataclastically deformed leucogranite and commonly contain specular hematite. Euhedral blades of specular hematite to 2 mm long have also overgrown cataclastic fabric in many leucogranite samples.

In both faulted and jointed schists and leucogranite, red, earthy hematite stain is common at higher levels in the mine and within and near postore faults throughout the mine. It is commonly present where pyrite is altered to goethite and is commonly, but not always, found in the matrix of breccias that contain specular hematite-bearing clasts. Its presence in postore fault gouge is consistent with activity of those faults during oxidative supergene alteration.

Quechan volcanic rocks at the mine display weak clay mineral alteration of plagioclase and red earthy hematite stain. These rocks are pervasively stained yellow by the application of potassium ferricyanide after HF etching, implying that they have experienced potassium metasomatism. Roddy et al. (1988) describe pervasive potassium metasomatism similar to this that took place at low temperatures (80–110°C) in volcanic rocks near extensional faults in the southern Basin and Range. Calcite veins and amygdaloidal fillings are common in Quechan volcanic rocks as well as in agglomerates and basalts in the mine area.

Mineralization in the Picacho deposit

Ore at the Picacho Mine is mineralogically simple (also see Van Nort and Harris 1984; Drobeck et al. 1986; Liebler 1988). Gold is associated with pyrite that has overgrown finely brecciated ("microbreccia") zones in cataclasite in either schist or leucogranite (Table 2). These microbreccia zones crosscut specular hematite-bearing cataclasite as well as cataclasite-stage quartz veins. Auriferous pyrite has also overgrown breccia that contains clasts of bladed specular hematite and cataclasite. In some samples from the Dulcina Extension pit, auriferous pyrite has overgrown cataclasite that contains calcite in its groundmass; in some instances, the pyrite is fractured and crosscut with clear calcite veinlets. Auriferous pyrite is either in the form of isolated cubes or aggregates of subhedral to euhedral grains having a total length of up to 1 cm. Gold is not visible in pyrite under 80· magnification. It is likely in solid solution or in submicron grains intergrown with pyrite. In samples where pyrite was oxidized to goethite, gold is detectable as grains up to 0.1 mm across within and next to the former pyrite grains (Fig. 8vi), and is rarely visible to the naked eye. From textural relationships, gold-pyrite precipitation postdated cataclasis and associated silicification, calcite precipitation, and specular hematite and quartz veins. In no case is auriferous pyrite inter-grown with gangue mineralization. Gold-bearing pyrite is not brecciated, and thus was precipitated at the latest stages of widespread faulting in the deposit. Throughout the mine, gold grades in crystalline rocks are typically higher near high-angle listric faults. In the Dulcina Extension pit, grades are particularly high (up to 17 g/t [0.55 oz/ton], sample C–D Ext; Table 3) near the intersection of listric faults with the low-angle fault system atop the Marcus Wash granite sill.

Although the Quechan volcanic rocks are highly jointed and should therefore be susceptible to mineralization if hydrothermal fluids were present in them, they are essentially barren of mineralization at Picacho. The volcanic rocks show only low-temperature alteration in the mine (e.g., Roddy et al. 1988). The large bulk, if not all, of gold mineralization at Picacho is associated with late stages of movement on Chocolate Mountains/Gatuna Fault–related structures that predate the Quechan volcanic rocks.

Fluid inclusions in mine samples

Several samples of quartz and calcite veins suitable for fluid inclusion analysis were collected from the Dulcina, San George, and Apache Pits (Table 4, Fig. 9; descriptions in Appendix). Although auriferous pyrite is not intergrown with gangue minerals that could directly yield mineralization temperatures and salinity of ore fluid, determination of these parameters for main-stage cataclasis at Picacho is essential for evaluating the geologic context in which gold mineralization took place.

Final ice melting temperatures (Fig. 9i) indicate the fluid inclusion salinity ranges from 0.5 to 2.2 weight percent NaCl equivalent (Bodnar 1993). These measurements, though few in number owing to the scarcity and small size of the inclusions, corroborate the salinities of 0.5–0.7 wt% NaCl reported by Liebler (1988) for samples from the Apache Pit. No systematic relationship exists between salinity and paragenetic stage of the host mineral. Homogenization temperatures for the primary inclusions show a well-defined range from 170° to 215°C (Fig. 9ii), consistent with Liebler's (1988) results. Secondary fluid inclusions in quartz yield nearly the same homogenization temperatures as primary inclusions in calcite veins of both pre-and post-specular hematite age: 132°–160°C. Thus, fluid temperatures appear to have fluctuated by 50°C during calcite veining and specular hematite and pyrite mineralization.

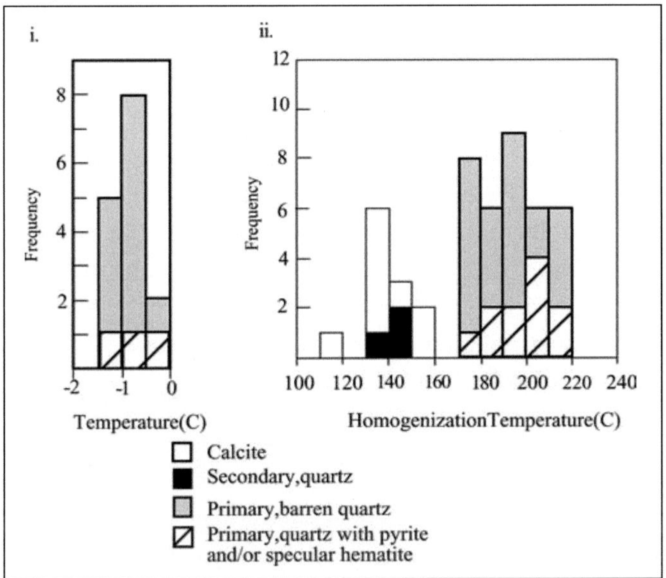

Figure 9. *Histogram of (i) final ice melting temperatures and (ii) fluid inclusion homogenization temperatures. Data are presented in Table 4.*

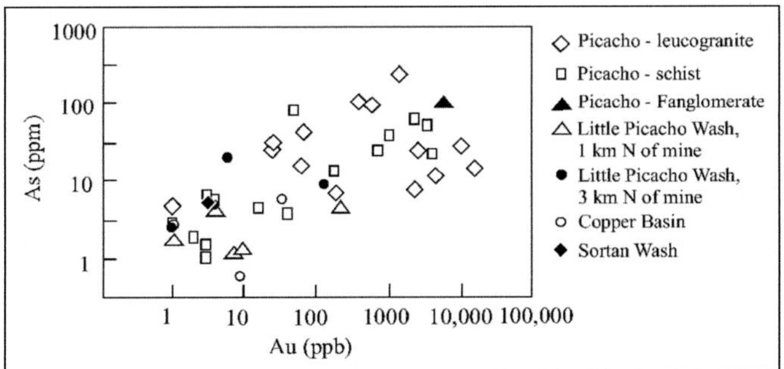

Figure 10. *Log–log plot of As versus Au concentrations for samples from mine and from low-angle fault breccias outside of mine. Data are presented in Table 3.*

GEOCHEMISTRY OF THE PICACHO DEPOSIT

Trace elements

Samples from the mine are characterized by a gold–arsenic–antimony–mercury metal association and are low in copper and silver (Table 3). Although there is appreciable scatter, the correlation of arsenic with gold (Fig. 10) indicates that the two were coprecipitated.

Stable isotopes

Silicate samples from the mine (Table 5) consist of quartz from veins associated with cataclasites and mineral separates made from silicified d18 rocks and from schist. $\delta^{18}O$ values of samples from quartz veins in the mine range from 10.2‰ to 15.4‰ VSMOW, essentially the same as the range of quartz from silicified rocks, whereas schist quartz $\delta^{18}O$ ranges from 9.5‰ to 11.1‰. Feldspar separates consist of approximately 1:2 mixtures of potassic feldspar and oligoclase (~An30). For purposes of this study, an average composition of An10 is used for computing oxygen isotope fractionation (O'Neil and Taylor 1967) between feldspar and water. Feldspar $\delta^{18}O$ values range from 8.5‰ to 13.9‰, the higher $\delta^{18}O$ values being associated with silicified cataclasites. These data are discussed in a subsequent section.

Calcite vein samples (Table 5) from the mine yield $\delta^{18}O$ values between 9.5‰ and 14.5‰ V-SMOW, and $\delta^{13}C$ between −2.4‰ to −13.3‰ PDB. Calcite associated with specular hematite and pyrite and calcite that postdates sericitic retrograde alteration has $\delta^{13}C$ lower than −9.5‰, and is thus distinguished from higher −$\delta^{13}C$ vein calcite that is not associated with opaque minerals (Fig. 11).

ALTERATION, MINERALIZATION, AND GEOCHEMISTRY OF FAULT ZONES IN THE MINE AREA

Alteration and mineralization

Low-angle normal faults that separate schist from volcanic and sedimentary rocks of the Winterhaven Formation at Copper Basin and Little Picacho Wash, respectively, contain schist and leucogranite that have undergone retrograde lower greenschist facies alteration and are locally protomylonitic. Fault breccias at both of the localities noted above are commonly silicified and contain calcite, specular hematite, and pyrite, much of which has been altered to goethite. Furthermore, Marcus Wash granite below a low-dipping fault contact with volcanic rocks of the Winterhaven Formation in Sortan Wash, 7 km northeast of the mine, shows sericitization of plagioclase and K-feldspar. Retrograde white mica and dynamically recrystallized quartz locally define a strong S–C fabric in this shear zone. Extensively developed cataclasites there contain calcite ± specular hematite ± pyrite. In all, the style of faulting and the type and extent of alteration and mineralization in these low-angle normal faults outside the mine are similar to the type of faulting, alteration, and mineralization in the mine and to the Chocolate Mountains/Gatuna Fault, as identified by Oyarzabal et al. (1997). Thus, these fault exposures are also correlated with the Chocolate Mountains/Gatuna Fault.

Geochemistry

Although concentrations of gold and other analyzed elements are lower in rocks from low-angle faults outside the mine than in samples from the mine, fault rocks from outside the mine show the same metal assemblage of Au–As–Sb–Hg, with low Cu and Ag, that characterizes the mine samples (also noted by Drobeck et al. 1986). Correlations between these elements in samples from outside the mine are similar to mine samples; for example, the ratios of gold to arsenic are similar to one another, regardless of location (Fig. 10). Other element pairs show similar patterns, indicating the trace-element geochemical "fingerprint" of the fluid represented by these samples was largely the same throughout the sampled fault system.

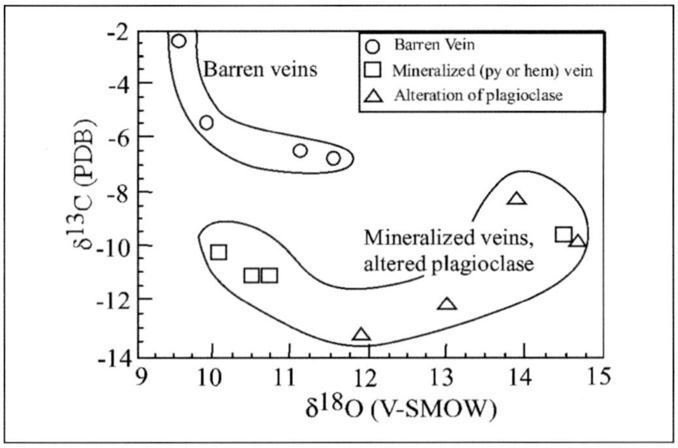

Figure 11. *$\delta^{13}C$ versus $\delta^{18}O$ for calcite veins not associated with pyrite or specular hematite ("barren veins"), veins associated with opaque minerals, and calcite replacement of plagioclase in altered rock.*

Quartz from strongly silicified cataclasite and quartz + pyrite veins in and near low-angle faults described above have $\delta^{18}O$ values ranging from 9.2‰ to 13.5‰ (Table 5), largely overlapping the range of $\delta^{18}O$ values of hydrothermal quartz in the mine. A single calcite sample from cataclastic Marcus Wash granite at Sortan Wash (Table 5) yielded a $\delta^{18}O$ value of 13.5‰ and $\delta^{13}C$ of -8.1‰, virtually the same as calcite associated with pyrite from the Dulcina Extension pit. Assuming that temperature during deformation at these locations was the same as that in the mine, stable isotopic compositions of fluids in the low-angle faults outside the mine are the same as fluids that circulated during main-stage cataclasis in the Picacho deposit.

DISCUSSION

Structural setting of the Picacho deposit

The similarities between fault products and associated mineralization at shallow and deep levels in the mine, as well as exposed connections between the shallow and deep-level fault systems, indicate that the low-angle fault system in the Dulcina pit and the steeper-dipping ore-stage faults at shallower levels are the same age and are thus kinematically related (e.g., Fig. 7). The field evidence presented in this study indicates that the Picacho deposit is controlled by low-angle and subsidiary high-angle normal faults. Supergene redistribution of gold is minor at most. We furthermore conclude that these faults are correlative with Chocolate Mountains/Gatuna — age low-angle normal faults exposed outside the mine for the following reasons: (1) The deposit is characterized by low-to high-angle normal to normal-oblique faults, including a throughgoing, ore-stage low-angle normal fault deep in the deposit. The structural style of faulting in the mine is similar to many upper-or middle-plate settings associated with low-angle normal faults (Rehrig and Reynolds 1980; Kerrich and Rehrig 1987; Gross and Hillemeyer 1982). The faults exposed outside the mine are also extensional in nature. (2) The style of faulting within the mine is very similar to that in Chocolate Mountains/Gatuna-age faults outside the mine: retrograde lower-greenschist facies mylonite-series rocks are overprinted by cataclastic fabrics, indicating similar conditions during faulting at all locations. (3) The faulting and attendant mineralization in the mine are stratigraphically constrained to a time interval consistent with that of Chocolate Mountains/Gatuna-age low-angle normal faulting outside the mine. (4) Trace-element geochemistry, alteration and mineralization assemblages, and stable isotope data indicate that the fluid responsible for mineralization in the Picacho deposit was very similar to fluid that was present in nearby exposures of Chocolate Mountains/Gatuna-age faults (also noted by Drobeck et al. 1986). Such similarities would be expected if the mineralization at the Picacho mine and fluid circulation in the nearby low-angle normal fault system took place at the same time and under similar conditions. Based on these lines of evidence, we conclude that the Picacho deposit is genetically related to pre-Quechan, Chocolate Mountains/Gatuna-age extensional faulting, and that the area of the mine has been overprinted by post-Quechan low-angle (Picacho Fault) and high-angle faults (Fig. 12). Oyarzabal et al. (1997) proposed that the early regional extension was related to processes that operated in an accretionary wedge complex. Crustal rheology may have been rather different from that which characterized the formation of metamorphic core complexes in the Miocene, leading to a different geometry of extensional faulting. However, it is clear that the Chocolate Mountains/Gatuna Fault system shares many features with metamorphic core complexes in terms of its regional extent, rapid footwall denudation, and fault product evolution. It was either active at low dip, or was intimately associated with a low-dipping "detachment" fault into which it soled.

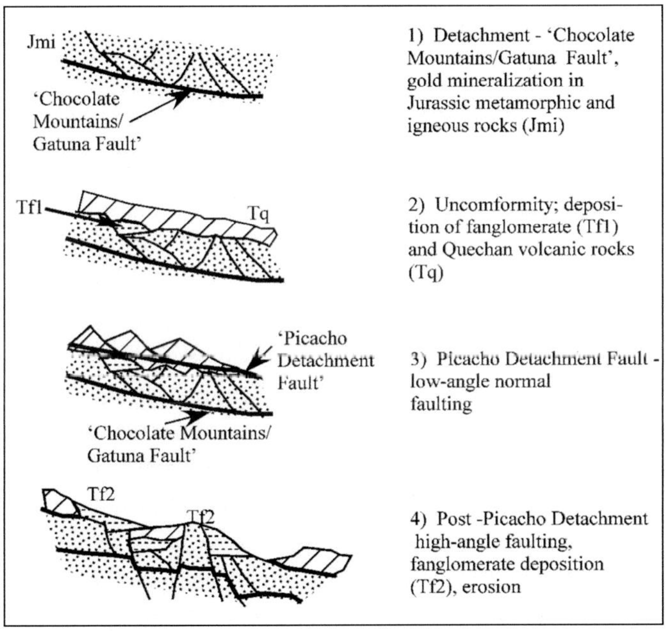

Figure 12. *Interpreted structural history of Picacho deposit. Sections are schematic. Early low-angle normal faulting ("Chocolate Mountains/Gatuna"-stage) was synchronous with mineralization, followed by erosion, deposition of fanglomerate and Quechan volcanic rocks, a later episode of low-angle normal faulting (Picacho Fault), and still-later high-angle faulting*

SOURCE OF FLUIDS IN THE PICACHO DEPOSIT AND DEPOSITION OF GOLD

Calculated stable isotopic composition of water based on either calcite or hypogene quartz isotope data shows a significant range of values. Main-stage vein and replacement quartz are in O-isotopic equilibrium with fluids having $\delta^{18}O$ between −2.6‰ and 5.2‰, based on fluid inclusion homogenization temperatures of 170°–215°C (Table 5; fractionation equation of Matsuhisa et al. 1979). Feldspar in silicified rocks is calculated at or near isotopic equilibrium with these fluids. Calculated $\delta^{18}O$ of water in equilibrium with calcite associated with opaque minerals is between 1.3‰ to 5.7‰, whereas that of water in isotopic equilibrium with barren calcite ranges from −3.1‰ to −0.4‰ (fractionation equation of O'Neil et al. 1969). Incorporation of a pressure correction of 20°C (e.g., Potter 1977), representing a fluid pressure of 200 bars (2 km of hydrostatic head), would increase the calculated fluid $\delta^{18}O$ values by about 0.5‰. The calculated carbon isotopic composition of fluids that precipitated calcite also varies according to the paragenetic stage of calcite. Fluids in C-isotopic equilibrium with pyrite-or specular hematite-bearing calcite have calculated $\delta^{13}C$ values between -9.6‰ to -13.2‰, whereas fluids in equilibrium with barren calcite have $\delta^{13}C$ from -3.4‰ to -7.7‰ (fractionation equation of Ohmoto and Rye 1979). Based on the above computations, the fluids in the Picacho deposit are interpreted as isotopically exchanged meteoric waters that interacted to varying extent with rock along their flow paths and, in the case of calcite-precipitating fluids, a ^{13}C-depleted reservoir.

As demonstrated here, as well as by Drobeck et al. (1986) and Liebler (1988), Picacho is a gold-arsenicantimony system characterized by low-salinity fluid inclusions, whereas most documented low-angle normal fault-related mineralization in the lower Colorado River trough is copper-or manganese-dominated (ex. Wilkins and Heidrick 1982; Spencer and Welty 1985; Roddy et al. 1988; works edited by Frost and Martin 1982; and Johnson 1987), and is characterized by briny fluid inclusions (Wilkins et al. 1986). The high fluid salinity of fluid inclusions in these deposits reflects incorporation of evaporitic basinal brines, with the metals likely having been transported as chloride complexes. The extensional faults associated with the Picacho deposit evidently were not infiltrated by evaporitic brines, implying that large syntectonic evaporitic basins that were present in upper plates of Miocene metamorphic core complexes throughout the lower Colorado River trough were absent in this setting.

Stable isotope data point to the involvement of two different hydrothermal fluids in main-stage cataclasis in the Picacho deposit. The near-coincidence of quartz $\delta^{18}O$ values of schist porphyroclasts and some of the hypogene quartz veins (Table 5) points to local rock buffering of isotopic composition of some of the fluid at Picacho, and indicates oxygen isotopic exchange between fluid and rock at a low cumulative water/rock ratio or at the rock-dominated end of a fluid exchange path (e.g., Bowman et al. 1994). High-$\delta^{18}O$ vein and replacement quartz (samples P63, P212) was deposited by water having $\delta^{18}O$ of 4–6‰. These waters could not have acquired their ^{18}O-enriched isotopic composition by local interaction with host rock (~9‰) at 200°C. These fluids must have either interacted with Jurassic biotite schist (or other rock having quartz $\delta^{18}O$ of 10‰) at temperatures of 300–350°C, considerably higher than temperatures in the deposit, or with rock having $\delta^{18}O$ significantly higher than 10‰. In either case, these isotopically-heavy fluids exchanged with rock at some distance from the deposit. As these two isotopically different sets of veins belong to the same general para-genetic stage, it is evident that there were two different fluids that deposited quartz in the fault zone during main-stage cataclasis.

Carbon isotope data indicate the presence of different calcite-depositing fluids in the fault as well as during main-stage cataclasis. $\delta^{13}C$ of vein calcite that predates specular hematite (sample 94PIC16, Table 5) is similar to that of barren vein calcite that postdates specular hematite elsewhere in the mine, and is significantly higher than that of calcite associated with opaque minerals. The most feasible source of low-$\delta^{13}C$ carbon in the vicinity of the Picacho deposit is the Orocopia schist, which contains significant amounts of carbonaceous material (Haxel et al. 1986). This unit also contains marble layers that may have served as a source for the abundant calcite associated with Chocolate Mountains/Gatuna-age faults. Although the existence of two carbon isotopic populations of calcite in the deposit does not in itself demonstrate fluid mixing, in that the sampled veins are not all known to be contemporaneous, the variability in fluid $\delta^{13}C$ indicates that the fluid pathways to the deposit changed during faulting. In particular, fluids associated with pyrite interacted more extensively with low-$\delta^{13}C$ rocks along their flow path than did fluids that precipitated the barren calcite veins.

As direct measurements of ore fluid chemistry could not be made due to the lack of gangue minerals, the causative mechanism of gold precipitation can only be inferred from paragenetic information. The lack of genetic association of gold-bearing pyrite with hypogene quartz, which is widely distributed throughout the mine, implies that cooling of fluid was not an important control on auriferous pyrite deposition (e.g., Fournier 1985). Similarly, the abundance of pyrite in leucocratic, iron-deficient rocks indicates that sulfidation of preexisting iron-bearing minerals is not a significant factor in pyrite deposition at Picacho. In numerous instances, specular hematite immediately next to later pyrite is unaltered, also indicating that iron-bearing minerals were not sulfidized by ore-stage fluids. The ore fluid was of sufficiently high pH (or [K+]) that K-feldspar was not sericitized or argillized during gold deposition. Boiling is precluded by the scarcity of silica and argillic alteration. The occurrence of solitary ore-stage pyrite throughout the deposit implies that oxidation of fluid, probably related to fluid mixing, may have been the primary cause of gold mineralization at Picacho. The source of the ore fluid may be addressed through lead and sulfur isotope studies on auriferous pyrite, integrated into a regional study incorporating possible gold sources. The latest pulse of hydrothermal fluid, carrying gold as a sulfur ligand, may have encountered the oxidizing, specular hematite-precipitating fluid that was present during main-stage deformation in the Chocolate Mountains Fault zone at Picacho, resulting in gold mineralization (e.g., Templeman-Kluit 1984).

Fluid circulation in the fault was probably driven by a magmatic heat source that was ultimately associated with the Quechan and younger volcanic rocks in the area. Spencer et al. (1995) inferred that an eruptive center for these rocks existed several kilometers north of the mine. The Chocolate Mountains/Gatuna Fault was active almost up to, or overlapped with, the time of eruption of the Quechan lavas, and served as a conduit for meteoric fluids that flowed through and interacted with Mesozoic rocks for as much as several kilometers before arriving at the site of the Picacho deposit. There, the fluids mixed with oxidizing waters and precipitated gold (Fig. 13). Mixing of relatively reducing and oxidizing fluids was enhanced in the >100-m thick zone of highly faulted and fractured rock, which may represent a footwall ramp or middle plate setting into which near-surface fluids were also focused.

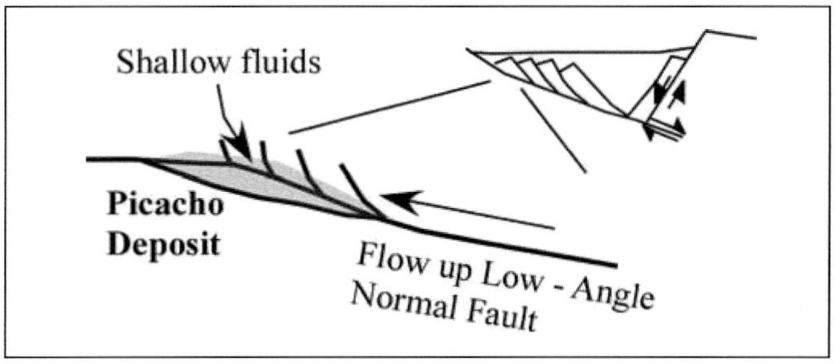

Figure 13. *Schematic fluid flow model for the low-angle normal fault hydrothermal system, including the Picacho deposit. In this interpretation, meteoric water enters the system at topographically elevated recharge points, flows downward along upper plate faults, and up dip along the low angle normal fault, cooling as it ascends. The deposit is in a thick fractured zone that represents a site where fluid may have flowed at high angles to isotherms and also mixed with other fluids that infiltrated downward through the upper plate. Structurally, the mine may be in a middle plate between two low-angle normal faults*

SUMMARY AND EXPLORATION IMPLICATIONS

The Picacho deposit is structurally controlled by a regionally exposed low-angle normal fault system of early to middle Tertiary age. Fault orientations within the deposit and the intimate relationship between Chocolate Mountains/Gatuna-age faulting and mineralization are consistent with introduction of ore-stage fluids during this important extensional event. The structural and chemical similarities between rocks in the Picacho deposit and those in low-angle Chocolate Mountains/Gatuna-age normal faults outside the mine give strong support to the interpretation that mineralization at Picacho and fluid circulation in the low-angle normal fault system were synchronous. Oxygen isotope and fluid inclusion data indicate that the cataclasite-stage hydrothermal fluid was a low-salinity, moderately to highly exchanged meteoric water at about 200°C, and that cumulative water/rock ratios in the exchange reservoir were relatively low. Auriferous pyrite was precipitated at the latest stage of regional extension along the Chocolate Mountains/Gatuna Fault.

In many places in the southwestern US, mineralization in low-angle extensional faults is widespread, with spotty areas of elevated gold grades. This style of mineralization may be attributable to fluids that ascended the fault at low angles to isotherms without mixing with descending, oxidizing waters in a concentrated area. The moderately large size of the Picacho deposit was a product of a favorable "plumbing system" in which fluid circulation through a large volume of rock in a regionally extended area was invigorated by syn-extensional magmatic heating, and where mixing was facilitated by the thick, structurally complex zone of fractured and faulted rock at the mine. The lack of high-salinity fluids, which may be due to the absence of playa lakes in the upper plate, may have favored bisulfide complexing of gold, making it more amenable to precipitation in a restricted area by mixing with oxidized fluids. Thus fluid chemistry may play a role in the ultimate size of gold deposits formed in this environment.

ACKNOWLEDGEMENTS

The Waterloo Centre for Groundwater Research funded this study. Picacho Mine personnel, particularly Mike Fitzgerald, Kevin Sur, and Steve Baumann, helped greatly by providing access to the mine, much logistical support, gold assays, and stimulating discussions during early stages of this study. We thank Dick Tosdal, Kevin Cassidy, Eric Seedorf, Dave Craw, and anonymous reviewers for very constructive and useful critiques of earlier versions. We thank Bud Hillemeyer for unpublished large-scale geologic maps of the area.

APPENDIX

Descriptions of samples used for geochemical analysis

P3 Little Picacho Wash, about 1 km north of mine. Silicified cataclasite with trace pyrite, specular hematite, 15 cm below microbreccia ledge.

P6 Little Picacho Wash, silicified cataclasite 1.5 m below P3.

P12 Little Picacho Wash, silicified cataclasite, about 1.5 m below microbreccia ledge and 15 m west of P6.

P16 Little Picacho Wash — quartz from iron-stained microbreccia ledge, about 30 m south of P3.

P22 Little Picacho Wash, silicified cataclasite, brecciated and veined with hematite and quartz, about 3 m below microbreccia ledge and 150 m east of P12.

P25 Little Picacho Wash, mylonitic gneiss, weakly sericitized, with calcite alteration of plagioclase, about 9 m from P22.

P47 Copper Basin — silicified Winterhaven volcanic rock about 1.5 m above detachment breccia zone.

P50 Copper Basin — mylonitic gneiss, with quartz ± specular hematite veins.

P56 Copper Basin — muscovite gneiss, about 45 m below projection of detachment fault.

P57 San George Pit 740 level — Silicified gneiss with quartz + cataclastically — deformed specular hematite veins.

P58 San George Pit 740 level — gneiss with specular hematite veinlets.

P63 San George Pit, stockpile on 660 level — silicified leucogneiss cataclasite with quartz + specular hematite. Oldest cataclasite (Cata I) contains no specular hematite. It is transected by two generations of cataclasite which were each overgrown by specular hematite. Later gold-bearing pyrite overgrew postspecular hematite microbreccia zone in which specular hematite is brecciated. Pyrite is oxidized.

P65 Same location and rock type as P63.

P73 San George Pit 640 level — brecciated sericitized schist, weakly cataclastic, calcite ± pyrite veinlets postdate cataclasite.

P76 San George Pit — biotite-chlorite schist — least altered in vicinity.

P85 Diablo Pit 590 level — biotite schist with specular hematite veinlets about 45 cm above moderately-dipping fault.

P88 Diablo Pit 590 level — biotite schist, biotite partially altered to chlorite.

P91 Diablo Pit 550 level — silicified cataclastic leucogneiss, brecciated with red earthy hematite matrix.

P92 Diablo Pit 550 level — weakly silicified leucogneiss, brecciated, abundant (2–3%) specular hematite.

P94 Dulcina Pit 530 level — schist, chloritized, with 1/2 inch wide quartz + pyrite vein, crosscut by later calcite veinlets.

P94a Dulcina pit 530 level — silicified leucogneiss, strong red earthy hematite stain, with calcite veins.

P96 Dulcina Pit, east part, 510 level — 30 m west of P94a — leucogneiss, about 2% specular hematite veinlets, minor pyrite, unoxidized.

P97 Dulcina Pit, 510 level — Leucogranite, cataclastic, specular hematite in bands, brecciated in younger cataclasite, then crosscut by plain calcite vein-lets.

P99 Dulcina Pit, 510 level— Leucogranite — cataclastic, crosscut by pyrite + calcite (± epidote) veins, which are crosscut by calcite + specular hematite veins. Pyrite locally brecciated where crosscut by specular hematite veinlets.

P104 120 m west of main road leading to Picacho State Park, just east of mine. Leucogranite, brecciated, quartz veinlet associated with oxidized pyrite.

P107 60 m east of main road leading to Picacho State Park, just east of mine, quartz + K-feldspar + biotite (trace) segregation in schist.

P109 Same location as P107, quartz + K-feldspar + biotite segregation in schist.

P110 Same location as P107, Vein quartz in schist, no associated opaque minerals.

P129 Western Dulcina Pit, quartz veinlet in schist.

P130 Western Dulcina Pit, barren calcite vein in leucogranite, crosscuts specular hematite mineralization.

P135 Western Dulcina Pit–chloritized schist.

P170 Dulcina Pit, east part-strongly oxidized biotite gneiss adjacent to fault.

P171 Dulcina Pit — same rock type as P170 1/3 m from P170.

P172 Dulcina Pit, same rock type as P170, 1/2 m horizontally from P170.

P212 Apache Pit, 680 level. Leucogranite, cataclasite, with 5-mm wide quartz vein; crosscut by microbreccia zone which is overgrown by gold-bearing pyrite.

P214 Dulcina Pit, 530 level — leucogneiss, 2% specular hematite veinlets, minor pyrite. Calcite veinlets postdate cataclasite.

P215 Dulcina Pit, 530 level — leucogneiss, cataclastic, with veinlets containing quartz + pyrite, and specular hematite and calcite.

P228 Little Picacho Wash about 3 km north of mine, from moderately south-dipping highly altered breccia zone separating biotite schist from Winterhaven volcanic rocks -biotite schist with quartz + pyrite veins about 3 m below projection of fault zone.

P234 Little Picacho Wash about 3 km north of the mine and about 1000 feet west of P228 within same brecciated zone — strongly silicified rock about 1.5 m north of 35-degree south dipping shear zone between schist and Winterhaven andesites. Protolith unknown but probably biotite schist, with about 2% disseminated medium grained euhedral to subhedral pyrite.

P236 Little Picacho Wash about 3 km north of the mine and 30 m north of P234 (in footwall)—biotite gneiss with quartz + oxidized pyrite veins.

P240 Little Picacho Wash about 3 km north of the mine and about 45 m south of P234—Winterhaven volcanic rock (andesite)-gray, fractured.

PIC7 Dulcina Pit, 450 level — Chlorite schist with foliated and vein calcite, 15 cm below low-angle fault juxtaposing Marcus Wash granite with schist. Sample is from foliated calcite in schist.

PIC11 Dulcina Pit, 430 level — Chlorite schist — cataclasite, with calcite+specular hematite+pyrite vein.

PIC12a Dulcina Pit, 430 level — Chlorite schist, protomylonite fabric, overprinted by cataclasite, calcite+pyrite veins.

PIC13 Dulcina Pit, 430 level — Chlorite schist—1.5 m east of PIC12a, largely undeformed, with calcite+quartz vein crosscutting chlorite + fine-grained quartz vein.

PIC14 Dulcina Pit, 430 level— Chloritic schist, cataclastic. Specular hematite overgrew cataclasite in veinlets, then was later brecciated. Late calcite veins are also brecciated.

PIC16 Dulcina Pit, 410 level — Chlorite schist — specular hematite+calcite vein in fault. This vein crosscuts a barren calcite vein, which was sampled for isotopic and fluid inclusion analysis.

PIC23 Dulcina Pit, 410 level — Chlorite schist — pyritic, weak protomylonite fabric, 15 cm from N8W57E fault zone, schist incorporated into cataclasite with calcite in groundmass with pyrite. This cataclastic fabric is crosscut by later fine-grained cataclastic zones with no calcite; pyrite is brecciated.

PIC24 Dulcina Pit, 390 level — Chlorite schist, 30 cm from N55W72SW fault zone. Weak protomylonite fabric (ribboned, sutured quartz) incorporated into cataclasite, calcite + pyrite vein postdates cataclasite. Feldspars show crack-seal texture, with quartz filling fractures; calcite crosscuts quartz. Titanite in one vein.

PIC27 Sortan Wash, about 10 km northeast of mine, about 90 m south of where "Sortan Fault" crosses Sortan Wash — cataclastic rock derived from Marcus Wash granite, contains calcite; from N80E15N fault zone.

T.B. Dulcina Pit, upper benches. Talus breccia (fanglomerate).

330W Dulcina Pit, northwest extension, 330 level. Marcus Wash granite, brecciated, with scattered pyrite; oxidized.

C-D Ext.—Dulcina Pit, eastern part, lower bench (450 level). Marcus Wash granite, containing calcite-cemented cataclasite overgrown by euhedral pyrite.

C-Dul Dulcina Pit, northwest extension, 450 level. Leucogranite containing calcite — and specular hematite-bearing cataclasite, overgrown by pyrite.

C-7 Dulcina East Pit, 490 bench. Leucogranite, cataclasite, scarce calcite. Cataclastic fabric is overgrown by clots of euhedral pyrite.

C-8 Dulcina Pit, northwest side, 450 level. Schist, locally incorporated into calcite-bearing cataclasite, crosscut by many clear calcite veinlets. Biotite is unchloritized, sericite is absent.

C-9 Dulcina Pit, same location as C-8. Schist, locally cataclastically deformed, with calcite in ground-mass. Pyrite overgrew cataclasite. Biotite partially chloritized.

REFERENCES

Anderson R.E., 1971, Thin skin distension in Tertiary rocks of southestern Nevada. Geol. Soc. Am. Bull., 82:43–58.

Bodnar R., 1993, Revised equation and table for determining the freezing point depression of H_2O–NaCl solutions. Geochim Cosmochim Acta 57:683–684.

Bowman J., Willett S., Cook S., 1994, Oxygen isotope transport and exchange during fluid flow: one dimensional models and applications. Am J Sci 294:1–55

Crittenden Max D., Jr., Coney Peter J., Davis George H. (eds.), 1980, Cordilleran metamorphic core complexes. Memoir — Geological Society of America, Boulder, 153:490.

Crowe B., 1978, Cenozoic volcanic geology and probable age of inception of basin-range faulting in the easternmost Chocolate Mountains, California. Geol. Soc. Am. Bull., 89:251–264.

Dillon J., Haxel G., Tosdal R., 1990, Structural evidence for northeastward movement on the Chocolate Mountains Thrust, southeasternmost California. J. Geop. Res. B., 95:19953–19971.

Drobeck P., Frost E., Hillemeyer B., Liebler G., 1986, The Picacho Mine: a gold mineralized detachment in southeastern California. Ariz. Geol. Soc. Digest., 16:187–221.

Fournier R., 1985, Silica minerals as indicators of conditions during gold deposition. In: Tooker E (ed) Geologic characterization of sediment-and volcanic-hosted disseminated gold deposits — search for an occurrence model. U.S. Geol. Surv. Bull., 1646:15–26.

Frost E., Martin D. (eds)., 1982, Mesozoic–Cenozoic tectonic evolution of the Lower Colorado River Trough, Arizona and California. Cordilleran Publishers, San Diego.

Gross W., Hillemeyer B., 1982, Geometric analysis of upper plate fault patterns in the Whipple-Buckskin detachment terrane, California and Arizona. In: Frost E., Martin D (eds) Mesozoic– Cenozoic tectonic evolution of the Lower Colorado River Trough, Arizona and California. Cordilleran Publishers, San Diego.

Haxel G., Tosdal R., Dillon J., 1985, Tectonic setting and lithology of the Winterhaven formation: a new Mesozoic stratigraphic unit in southeasternmost California and southwestern Arizona. U.S. Geol. Surv. Bull., 1599:19.

Haxel G., Tosdal R., Dillon J., 1986, Field guide to the Chocolate Mountains thrust and Orocopia Schist, Gavilan Wash area, southeasternmost California. Ariz. Geol. Soc. Digest, 16:282–293.

Haxel G., Jacobson C., Tosdal R., 1988, Multiple low-angle faults atop the Orocopia Schist, southwest Arizona and southeast California. Geol. Soc. Am. Abstr. w/Prog. Annual Mtg., 20:A383.

Jacobson C., Grove M., Stamp M., Vucic A., Oyarzabal F., Haxel G., Tosdal R., Sherrod D., 2002, Exhumation history of the Orocopia Schist and related rocks in the Gavilan Hills area of southeasternmost California. In: Barth A (ed) Contributions to crustal evolution of the Southwestern United States. Geol. Soc. Am. Spec. Pap., 365:129–154

Johnson J (ed)., 1987, Bulk mineable precious metal deposits of the western United States: Guidebook for field trips. In: Geological Society of Nevada symposium, Reno, 1–12 April 1987.

Kamb W., 1959, Ice petrofabric observations from Blue Glacier, Washington, in relation to theory and experiment. J. Geop. Res., B 64:1891–1910.

Kerrich R., Rehrig W., 1987, Fluid motion associated with Tertiary mylonite and detachment faulting: 18O/16O evidence from the Picacho metamorphic core complex, Arizona. Geology, 15:58–62.

Liebler G., 1988, Geology and gold mineralization at the Picacho Mine, Imperial County, California. In: Schafer R., Cooper J., Vikre P., (eds), Bulk mineable precious metal deposits of the Western US: symposium proceedings. Geological Society of Nevada, Reno, pp 453–472.

Maldonado F., 1991, Structure, stratigraphy, and mineralization of the upper plate of the Bullfrog Hills detachment system, Bullfrog Hills area, with emphasis on the geology of Bullfrog Mountain, Nye County, Nevada. In: Buffa R., Coyner A., (eds), Geology and ore deposits of the Great Basin field trip guidebook. Geological Society of Nevada, Reno, pp 1062–1075.

Matsuhisa Y., Goldsmith J., Clayton R., 1979, Oxygen isotopic fractionation in the system quartz-albite-anorthite-water. Geochim Cosmochim Acta, 43:1131–1140.

McCrea J. G., 1953, Isotopic chemistry of carbonates and a paleotemperature scale. J. Chem. Phys., 18:849–857.

Ohmoto H., Rye R., 1979, Isotopes of sulfur and carbon. In: Barnes H (ed) Geochemistry of hydrothermal ore deposits, 2nd ed. Wiley, New York, pp 509–567.

O'Neil J., Taylor H., 1967, The oxygen-isotope and cation exchange chemistry of feldspar. Am. Mineral, 52:1414–1437.

O'Neil J., Clayton R., Mayeda T., 1969, Oxygen isotope fractionation in divalent metal carbonates. J. Chem. Phys., 51:5547–5558.

Oyarzabal F., Jacobson C., Haxel G., 1997, Extensional reactivation of the Chocolate Mountains subduction thrust in the Gavilan Hills of southeast California. Tectonics, 16:650–661.

Potter R., 1977, Pressure corrections for fluid inclusion homogenization temperatures based on the volume properties of the system NaCl–H_2O. J. Res. U.S. Geol. Surv., 5:603–607.

Rehrig W., Reynolds S., 1980, Geologic and geochronologic reconnaissance of a northwest trending zone of metamorphic core complexes in southern and western Arizona. In: Crittenden M., Coney P., Davis G (eds) Cordilleran metamorphic core complexes. Geol. Soc. Am. Memoir., 153:131–158.

Richard S (ed)., 1992, Deformation associated with the Neogene Eastern California shear zone, Southeast California and northwest Arizona. San Bernardino Co. Museum Spec. Publ. 92–1, Redlands.

Richard S., Haxel G., 1991, Progressive exhumation of the Orocopia and Pelona schists along a composite normal fault system, southeast California and southwest Arizona. Geol. Soc. Am. Abstr. w/ Prog. Cordilleran Sect. 23:A92.

Roddy M., Reynolds S Smith B., Ruiz J., 1988, K-metasomatism and detachment-related mineralization, Harcuvar Mountains, Arizona. Geol. Soc. Am. Bull., 100:1627–1639.

Roedder E., 1984, Fluid inclusions. Mineral Assoc. Am. Rev. Mineral 12:251–290.

Seedorff E., 1991, Magmatism, extension, and ore deposits of Eocene to Holocene age in the Great Basin — Mutual effects and preliminary proposed genetic relationships. In: Raines G., Lisle R., Schafer R., Wilkinson W (eds) Geology and ore deposits of the Great Basin, symposium proceedings, Geol. Soc. Nevada, Reno, pp 133–178.

Simpson C., 1990, Microstructural evidence for northeastward movement on the Chocolate Mountains Fault zone, southeastern California. J. Geophys. Res., 95:529–538.

Spencer J., Welty J., 1985, Possible controls of base-and precious metal mineralization associated with Tertiary detachment faults in the lower Colorado River Trough, Arizona and California. Geology, 14:195–198.

Spencer J., Richard S., Reynolds S., Miller R., Shafiqullah M., Gilbert W., Grubensky M., 1995, Spatial and temporal relationships between mid-Tertiary magnetism and extension in southwestern Arizona. J. Geop. Res. B., 100:10321–10351.

Templeman-Kluit D., 1984, Meteoric water model for gold veins in a detached terrane. Geol. Soc. Am. Abstr. w/Prog., 16:674.

Tosdal R., 1986, Mesozoic ductile deformations in the southern Dome Rock Mountains, northern Trigo Mountains, Trigo Peak and Livingston Hills, southwestern Arizona, and Mule Mountains, southeastern California. Ariz. Geol. Soc. Digest, 16:62–71.

Tosdal R., Haxel G., 1987, Late Jurassic regional metamorphism and deformation, southeastern California and southwestern Arizona. Geological Society of America Abstracts with Programs, 19:870–871.

Van Nort S., Harris M., 1984, Geology and mineralization of the Picacho gold prospect, Imperial County, California. Ariz. Geol. Soc. Digest, 15: 175–183.

Wilkins J., Heidrick T., 1982, Base and precious metal mineralization related to low-angle tectonic features in the Whipple Mountains, California, and Buckskin Mountains, Arizona. In: Frost E., Martin D (eds) Mesozoic–Cenozoic tectonic evolution of the Lower Colorado River Trough, Arizona and California. Cordilleran Publishers, San Diego, pp 182–204.

Wilkins J., Beane R., Heidrick T., 1986, Mineralization related to detachment faults: a model. Ariz. Geol. Soc. Digest, 16:108–117.

Table 1 *Deformation events in the Picacho Mine area*

Deformation event/fault	Age	Fault rocks	Comments	Reference
Regional metamorphism	Jurassic		Includes augen schists in Picacho Mine area; foliation dips southwest	Tosdal (1986), Tosdal and Haxel (1987)
Chocolate Mountains Thrust	Late Cretaceous–early Tertiary	Prograde lower amphibolite facies mylonites	Accommodated subduction; exposed 3 km north of mine; at least 300 m below surface at mine. Exposed 2 km north of Picacho Mine	Dillon et al. (1990), Simpson (1990), Oyarzabal et al. (1997), Jacobson et al. (2002)
Chocolate Mountains Fault	50–40 Ma footwall cooling ages from 500C to 350C, due to movement on fault.	Retrograde greenschist facies mylonitic rocks	Low-angle normal fault subparallel to Chocolate Mountains Thrust	Oyarzabal et al. (1997), Jacobson et al. (2002)
Gatuna Fault	28–24 Ma footwall cooling from 350C to 150C due to fault movement. Predates/ synchronous with 26–24 Ma Quechan volcanic rocks	Cataclastic rocks, locally silicified, with calcite veins, specular hematite, pyrite	Low-angle normal fault subparallel to Chocolate Mountains Fault in Gavilan Hills, 10 km NW of Picacho Mine. Separates footwall Jurassic schist, gneiss, and granite from hanging wall Jurassic Winterhaven Fm	Haxel et al. (1985), Simpson (1990), Dillon et al. (1990), Oyarzabal et al. (1997), Jacobson et al. (2002)
Picacho Fault	Younger than 24–26 Ma; older than 20 Ma. Postdates Quechan volcanic rocks	Fault gouge, clay rich, red earthy hematite; incorporates clasts of rocks from Chocolate Mountains/Gatuna Fault	Low-angle normal fault subparallel to Gatuna Fault in Picacho Mine area. Reactivates Gatuna Fault in most places; locally lies within Quechan volcanic rocks.	Drobeck et al. (1986), Spencer et al. (1995) (age of Quechan volcanic rocks)
Younger faults	Post 20 Ma. Neogene. Post 13-Ma		N and NW-trending oblique slip faults. Strike-slip faults west of Picacho mine. Rotation around vertical axis	Drobeck et al. (1986) Richard (1992)S. Costello, unpublished data

Table 2 *Fault generations, alteration, and vein paragenesis in the Picacho deposit*

Fault rock/correlation	Location in mine	Characteristic textures/products	Veins/matrix[a]	Alteration minerals
Protomylonite *Chocolate Mountains Fault*	Deep portion of Dulcina Pit	Dynamically-recrystallized quartz, weak S–C fabric, developed in schist and granite. Dynamically-recrystallized, foliated calcite and chlorite "fish" incorporated into mylonitic fabric in places.	Calcite veins, deformed	Greenschist facies in schist: plagioclase is sericitized, biotite altered to chlorite c± titanite.
Cataclasite I *Gatuna Fault*	Throughout mine	Undulose extinction in quartz, deformation twinning of plagioclase; developed in granite and silicified schist in deep Dulcina pit.		Silicification
Cataclasite II *Gatuna Fault*	Throughout mine	Finer-grained than Cataclasite I; incorporates clasts of Cataclasite I; contains specular hematite as bladed aggregates that overgrow cataclastic fabric. Comprises several generations of cataclasite with associated mineralization—at least 3 in one sample (P63)	Qz ± py (one sample). Anomalous Au with pyrite in quartz vein (P94), but not ore-grade. Specular hematite ± cc; Specular hematite ± qz; Calcite in cataclasite groundmass in some samples; Cc-py-epidote (deep Dulcina)	Weak silicification in leucogranite
Cataclasite III *Gatuna Fault*	Throughout mine	Incorporates brecciated specular hematite in silicified cataclasite; crosscuts Cataclasite II; Cataclasites are developed in both granite and schist, but granite is more pervasively deformed	Calcite in groundmass in some samples	Calcite alteration of plagioclase in schist?
Green/gray sandy gouge *Gatuna Fault*	Low-angle normal faults bounding Jmw, deep Dulcina Pit	Silty/sandy gouge associated with cataclasite		
Breccia I *Picacho Fault*	Uncommon; shallow	Incorporates clasts of specular hematite-bearing cataclasite in calcite matrix	Calcite ± py	
Breccia II *Picacho Fault*	Throughout mine	Incorporates clasts of specular hematite and calcite-bearing cataclasite and breccia; groundmass contains red earthy hematite only at shallow levels, commonly contains euhedral pyrite, which has overgrown breccia. Synchronous with microbreccia zones that are overgrown by auriferous pyrite (P63, P212). Probably synchronous with Breccia I.	Disseminated Au-pyrite, chlorite, cc ± qz (deep Dulcina Pit, postdates cataclasite)	Calcite alteration of plagioclase in schist
Red gouge *Post-Picacho Fault*	Throughout mine in post-mineral faults, including the Picacho Fault. Red gouge-bearing faults are the only ones known to cut volcanic rocks	Red-hematite stained clay gouge; contains only brecciated clasts of opaque minerals	Calcite crosscutting Breccia II? No opaque mineralization	Red earthy hematite stain

Name of fault with which particular fault rocks in the mine are correlated is *italicized* in the first column.

[a]*Py* pyrite, *spec* specular hematite, *qz* quartz, *cc* calcite, *catacl.* Cataclasite

Table 3. *Trace element geochemical data*

Sample	Au (ppb)	Cu (ppm)	As (ppm)	Ag (ppm)	Sb (ppm)	Hg (ppm)	Bi (ppm)	Zn (ppm)	Pb (ppm)
Picacho Mine									
Leucogranite cataclasites/breccia									
P57	65	2.0	44.0	<0.5	1.7	110	<0.1	–	–
P58	<1	8.5	5.1	<0.5	0.3	14	<0.1	–	–
P63	2,600	7.5	27.0	0.5	1.8	14	<0.1	–	–
P65	25	5.0	27.0	<0.5	1.1	25	<0.1	–	–
P85	2,300	5.0	8.0	<0.5	1.9	160	<0.1	–	–
P91	400	11.0	110	<0.5	78.0	32	<0.1	–	–
P92	590	11.0	97	<0.5	140	72	<0.1	–	–
P94a	25	3.5	34.0	<0.5	5.1	36	<0.1	–	–
P96	62	5.5	16.0	<0.5	1.0	40	<0.1	–	–
P214	190	5.0	7.4	<0.5	1.8	14	<0.1	–	–
330W	1,330	28	250	<0.3	29	15	<3	65	6
C-D Ext	15,100	17	15	3.1	3	715	<3	86	23
C-Dul	4,474	129	12	4.9	<3	1600	<3	75	102
C-7	10,100	130	28	4.6	13	945	<3	11	73
Schist cataclasites/breccias									
P73	51	14.0	82.0	<0.5	4.7	40	<0.1	–	–
P76	<1	32.0	2.9	<0.5	0.4	43	<0.1	–	–
P94	180	13	14	<0.3	7	60	<3	64	<3
P170	2,400	22.0	65.0	<0.5	14.0	560	<0.1	–	–
P171	750	70.0	25.0	<0.5	1.8	>1000	<0.1	–	–
P172	3,500	61.0	53.0	1.0	4.0	280	<0.1	–	–
PIC7	3	0.9	6.7	<0.1	2.0	1040	–	139	9
PIC11	3	0.7	1.1	<0.1	0.7	1280	–	74.2	<2
PIC12a	3	18.6	1.6	<0.1	0.8	559	–	66.1	6
PIC13	2	11.0	2.0	<0.1	0.6	708	–	61.5	<2
PIC16	4	2.4	5.8	<0.1	13.0	352	–	57.3	7
PIC23	41	6.9	4.0	<0.1	0.9	152	–	42.8	<2
PIC24	16	7.2	4.8	<0.1	0.5	72	–	46.1	<2
C-8	4,092	39	23	2.5	7	65	<3	68	11
C-9	1,120	22	41	0.5	<3	10	<3	43	<3
Fanglomerate									
TB	5,640	30	107	0.5	23	120	<3	57	9
Little Picacho Wash, 1 km (1/2 mile) north of mine									
Silicified cataclasites (schist precursor)									
P3	220	17.0	4.4	1.5	0.7	14	0.2	–	–
P6	9	6.5	1.3	<0.5	0.9	25	0.2	–	–
P12	4	3.5	4.6	0.5	1.0	25	<0.1	–	–
P22	7	5.0	1.2	<0.5	0.5	11	<0.1	–	–
Schist									
P25	<1	10.0	1.7	<0.5	0.4	18	<0.1	–	–
Little Picacho Wash 3 km (2 miles) north of mine									
Schist									
P228	<1	30.0	2.7	<0.5	1.0	61	<0.1	–	–
Silicified schist									
P234	130	17.0	9.5	1.5	1.1	220	0.5	–	–
Gneiss									
P236	6	13.0	21.0	<0.5	0.6	83	0.2	–	–
Winterhaven Fm andesite									
P240	4	4.5	5.1	<0.5	0.4	43	<0.1	–	–
Copper Basin									
Winterhaven Fm andesite									
P47	9	24.0	0.6	<0.5	0.1	25	0.2	–	–
Gneiss									
P50	<1	7.5	2.7	<0.5	0.4	58	<0.1	–	–
P56	33	10.0	6.1	<0.5	0.2	18	<0.1	–	–
Sortan Wash–Leucogranite cataclasite									
PIC27	3	4.1	5.8	0.3	0.9	6	–	12.6	11

– not analyzed

Table 4 *Fluid inclusion data*

Sample	Host	Tmelt (°C)	Salinity (wt% NaCl equiv)*	Th (°C)
East of mine				
P104	Quartz+py vein	−1.0	1.74	199.6
				206.8
				214.7
				215.9
P110	Quartz-barren			170.5
				176.2
				188.7
Dulcina pit				
P129	Quartz			187.6
P130	Calcite			139.3
				155.6
				159.1
P215	Quartz	−0.4	0.71	206.1
P63	Quartz			170.9
				186.9
				187.9
				193.3
	Quartz-secondary			132.1
				144.2
				143.7
				197.1
P94	Quartz			204.3
		−0.9	1.57	207.4
		−1.3	2.24	210.4
94PIC13	Quartz	−1.1	1.91	197
		−1.2	2.07	214.5
		−0.8	1.40	217.7
		−0.9	1.57	192
		−0.5	0.88	200.1
		−0.8	1.40	191
		−0.4	0.71	183.6
				199
94PIC16	Calcite			138.9
				141.7
				131.4
				112.3
				130
				135.6
Apache Pit				
P212	Quartz			170
				173
				176
		−0.8	1.40	177
		−0.7	1.23	177
				182
		−1.0	1.74	190
				195
				196
				200
				210
		−0.7	1.23	215

*Salinities calculated according to Bodnar (1993). All inclusions are primary unless otherwise noted.

Table 5 *Stable isotope data*

Sample	Mineral	$\delta^{13}C$	$\delta^{18}O$	Associated minerals	Temperaturea	Fluid $\delta^{13}C$	Fluid $\delta^{18}O$
Picacho Mine—hypogene veins							
P57	Quartz		10.7, 11.1	Specular hematite	175–225		−2.6 to 0.9
P63	Quartz		13.3	Specular hematite	200–210 (FI)		1.7
P94	Quartz		10.2, 10.0	Pyrite (barren)	225 (FI)		−0.2 to 0.0
P212	Quartz		15.4	None	190–230		3.4 to 5.2
P170	Spec. Hematite		6.3	None			
P94a	Calcite	−2.4	9.5	None	140–160	−3.4	−2.4
P215	Calcite	−6.7	11.5	None	140–160 (FI)	−7.7	−0.4
94PIC11	Calcite	−9.5	14.5	Pyrite		−9.6	5.7
94PIC13	Calcite	−11.1	10.5	Specular hematite		−11.0	1.7
94PIC13	Calcite	−11.1	10.7	Pyrite + specular hematite, 1 cm from above	184–218 (FI)	−11.0	1.9
94PIC16	Calcite	−6.5	11.1	None	130–142 (FI)	−7.5	−1.9
		−5.5	9.9	Same vein		−6.5	−3.1
94PIC24	Calcite	−10.3	10.1	Pyrite + specular hematite		−10.2	1.3
Picacho Mine—mineral separates from schist and leucogranite							
P63	Feldspar		12.6				
P130	Chlorite		5.8				
	Quartz		10.2				
	Feldspar		8.5				
P135	Quartz		10.2				
	Chlorite		8.1				
P170	Quartz		9.5				
	Feldspar		7				
P172	Quartz		11.1				
P88	Biotite/chlorite		6.3				
P76	Feldspar		13.9				
	Chlorite		5				
94PIC7	Calcite	−9.6	14.7	Foliated cc in schist		−9.7	5.9
94PIC12a	Calcite	−13.3	11.9	Calcite alteration of plag		−13.2	3.1
	Calcite	−12.0	13	Calcite alteration of plag		−11.9	4.2
Little Picacho Wash—silicified breccia							
P3	Quartz		13.55	Pyrite	175–225		0.3 to 3.4
P16	Quartz		12.6	Pyrite	175–225		−0.7 to 2.4
P234	Quartz		9.2, 10.1	Pyrite	175–225		−4.1 to −0.1
Little Picacho Wash—quartz vein							
P236	Quartz		11.1	Pyrite	175–225		−0.9 to 1.1
Sortan Wash—calcite alteration							
94PIC27	Calcite	−8.1	13.9	Calcite alteration of plag		−8.1	5.1

Temperatures are from fluid inclusion homogenization temperatures (FI) or estimated ranges based on fluid inclusion data from other paragenetically similar samples. A pressure correction of 20°C would result in an increase in calculated fluid $\delta^{18}O$ of about 1.5‰ at 150°C. In the case of carbonates, the dominant carbon species in fluid is assumed to be H_2CO_3 (app). If bicarbonate dominates, calculated fluid $\delta^{13}C$ values are about 2‰ lower than shown.
[a] fluid inclusion (FI) homogenization temperature corrected for 2 km hydrostatic head.

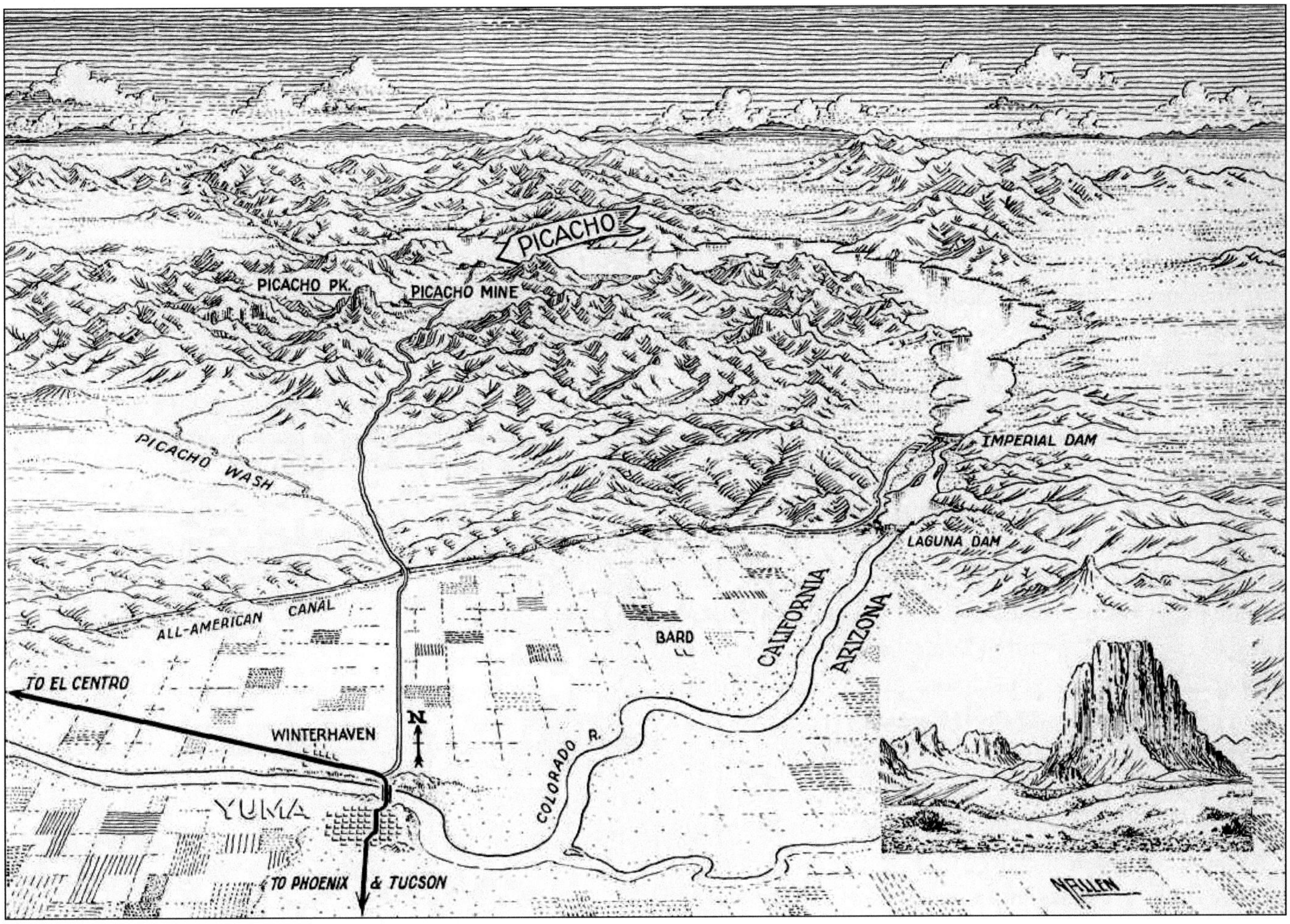

Saga of Old Picacho

BY J. WILSON MCKENNEY

Seventy-six years have passed since Jose Mendivil discovered gold near the Colorado river in the shadow of Picacho peak. The boom mining camp which mushroomed into existence following the original discovery was long ago abandoned. For nearly a quarter of a century Picacho has been just another ghost mining camp. But today there is new activity at the old ore shafts. Engineering science is preparing to reclaim gold which could not be mined profitably under the crude methods of earlier days. Here is the history of a mining property which may again become one of the most important producers in the Southwest.

AS THE MINING CAMP OF PICACHO sleeps, the brown waters of the Colorado creep up to drown the last flicker of its once robust vitality. Ancient arrastras and crumbling adobe walls will soon be under the surface of a lake formed by the recently completed Imperial dam.

The townsite of Picacho will disappear but there remain the open glory holes and dark shafts of the Picacho basin gold-mining district, five miles southwest of the river and 20 miles north of Yuma, Arizona. Abandoned for many years, the mines may soon be revitalized, reminiscent of bonanza days half a century gone, which a few veterans can still recall.

Mendivil is a magic name in the story of Picacho, for it was Jose Mendivil who discovered rich gold-bearing gravels there three-quarters of a century ago. Thus he

This article originally appeared in *Desert Magazine*, Volume 3, Number 5, March 1939, pp. 10–14 became the hero of a western saga rife with lust and chicanery, but brightened with the sounds and sights of Mexican cock-fights and *bailes*.

Ysidro, 66, eldest son of Jose, is one of the survivors of Picacho's turbulent life. From him and his son Michael; from Mrs. Clara Townsend, daughter of the first school-teacher and postmaster; from early state mining reports and county legal records, came fragments of information with which I pieced together this story of Picacho.

Jose Maria Mendivil was born in central Sonora, December 8, 1838, the son of Ramon, a Mexican geologist. His father was part Spaniard and part Oaxaca Indian, his mother half Spaniard and half Pima Indian. One morning in his seventh year he trudged off to school waving goodbye to his father and mother, not knowing it would be the last time he would ever see them. That day a band of roving Apaches captured him and took him away to the north. As an old man he told his grandson Michael the thrilling episodes of his life as a captive, wandering for eight years in the desert, always alert for an opportunity to escape.

The Mexican War was turning American eyes toward the Southwest, General Kearney's Army of the West had scored wheel tracks in the Arizona desert, and the first of the California gold-seekers were starting westward when Mendivil began his youthful adventures with Geronimo's fierce marauders.

He was about 15 years old when the band was pursued by American cavalry, sent by the U. S. government to check the Apache raids and guard the overland stage trail. By night he eluded the squaw appointed to guard him and crept into the camp of the Americans.

During the remainder of the army campaign on the border Jose served the Americans as guide, scout, and teamster. Granted an honorable discharge, he turned westward with the overland wagon trains. In his pocket he carried a paper bearing the name of Mendivil.

Miners Reach New Gold Field

At Fort Yuma Jose found feverish excitement as miners and deserting soldiers hurriedly left the frontier town for the newly discovered rich placers of La Paz, 60 miles north on the Colorado. The gold fever may have urged him northward.

Dates are uncertain in the Mendivil story, but it was about 1862 when Jose started north along the river from Fort Yuma. The Civil War was on and Jose had seen enough of army service, which was good reason for

Top: Mine buildings in Picacho basin about 1906. The monolith in the background is 1963-foot Picacho Peak, picturesque landmark of the famous mining region. All buildings in this picture are now removed. —Mendivil photo.

Center: The general store and post office at Picacho about 1900. On right is Mrs. Ferguson, postmaster, and next is her daughter Gertrude, who later became Billy Horn's wife. The small girl is Clara Ferguson, youngest daughter, who is now Mrs. Townsend of Yuma Valley. —Clara Townsend photo.

Bottom: This 450-ton stamp mill was built at Picacho about 1897 by Sen. Stephen Dorsey and associates. The long railroad trestle is the last lap of the five-mile run from the basin to the mill. —Clara Townsend photo.

him to avoid the traffic on the Arizona road. He crossed he river and started north alone on the California side. All day the bold outline of Picacho peak stood on the horizon before him.

In the gravel bed of a dry wash leading down to the river Jose found the Pachecos, a Sonoran family, searching for gold, He watched them curiously. They laid a blanket on the ground, shoveled a few pounds of gravel on it, then two men would grasp the corners of the blanket and flip it so that the breeze would carry away the dust. Then laboriously they picked the heavy gold particles from the nap of the wool.

Jose knew enough about prospecting to realize that the source of the gold was on the higher ground where the washes began. So instead of joining the Pachecos, he prospected the ledges several miles from the river in the shadow of Picacho peak.

From where he stood he looked to north and east, where sunlight glinted from the face of the sinuous Colorado. Like a great serpent, the river wound through hills splashed with myriad colors. And rising behind him was the majestic form of Picacho, perpendicular sides lifting 1500 feet up from the ground on which he stood. "La Reina del Desierto," he murmured to himself, as he tore himself from his dreaming to resume the search for gold.

He found rich ore at his feet, not in quartz veins but in fractured granite. He satisfied himself there was much gold, but decided he had not the energy to take it out. There were no nuggets lying around, the sun was too hot in midday for hard labor, and he had only just begun his journey. So he moved along, telling other prospectors about his find.

The nature of the low-grade ore made large milling investment necessary and mining capital was interested in richer fields at the time. For almost 20 years Picacho basin remained unvisited by mining promoters.

A miner named Dave (or James) Neahr built the first crude arrastra in 1873, crushed a small quantity of ore and loaded it on flatboats for transport down the river to Yuma.

First mention of Picacho appears in the state mineralogist's report for 1882 with the brief note that 6,000 tons of ore were raised during the census year, bringing an average of $21 a ton. Dr. DeWitt C. Jayne of Florida, N. Y., entered the field in 1879, built a 15-stamp mill on the bluff overlooking the river, and hauled a great deal of ore by mule-wagon from the basin. The mill building was constructed of rhyolite tuff quarried on the site, a substantial structure the walls of which still stand today. Dr. Jayne worked at Picacho for several years, suspended operations after being seriously injured by a runaway horse, and died in 1898. The Jayne claims covered Mendivil's discoveries and included the famous Goshen, Mars, St. George, and Venus mines.

In the meantime Jose Mendivil wandered north and got a job carrying the mail by pony express between Needles and Ehrenberg. Although he knew the Chemehuevis well and spoke their tongue he frequently rode into Fort Tison and Fort Mojave with reports of narrow escapes from unfriendly Indians.

Married at Yuma

About 1872 Jose married Jesus "Jessie" Romo, a Sonoran girl who lived at Fort Yuma. Their first child, Ysidro, was born at Fort Yuma the next year. As the years rolled by they became parents to five more boys and two girls, all born in or near Picacho. It is to be assumed that Jose worked in the district all the latter years of his life.

There are no available records that give Mendivil title to any of the larger gold mines, though he had an interest in and did assessment work on several properties. Imperial county (then San Diego county) records show that a homestead patent was issued to Joseph M. Mendivil for 155.63 acres of land in sections 19 and 24, T 13 S, R 22-23 E, SBM, on January 17, 1895. This piece of land is on the south bank of the river, a flat area around which the town of Picacho was to spring up overnight.

That same year the town of Picacho was born. Mendivil may have been negligent in securing his gold discoveries, but he showed rare diligence in protecting his home-site against the army of squatters. There had been prospectors' shacks in the mesquite brush for many years, but not until the mining czars cast their eyes with favor on the district did the town begin to boom.

Within a radius of six miles from Picacho, extending into the canyons and ridges of the Chocolate range, literally hundreds of claims were staked out. Some of the claims were known producers, like the Jayne group, but many remained undeveloped. When it became known that important capital was coming in, claim-holders rushed in to do needed assessment work. Prospect holes and small dumps soon pock-marked the faces of the hills like rabbit warrens. Every likely spot was attacked with pick and shovel except rugged Picacho peak, standing in solitary aloofness from these frantic proceedings.

Mining Company is Formed

Stephen A. Dorsey of Denver, one-time senator from Colorado, was a mining promoter of great ability. He saw the chaos, dreamed of the possibilities under large scale production, and proposed a consolidation of claimants in one operating company. His dream came to fruition in 1896 with the organization of the California Gold King mining company.

Dorsey cut out for himself a laborious task and he did not complete the consolidation until 1902. Thirty claims were included, with the four Jayne mines as the nucleus. Sixteen of the claims were patented, including the Apache, last of the great producers; the Dulcinea, the great glory hole near the basin mill-site; the Eastern California, Golden Casket, Extension, Mina Rica, Tierra Rica, and many lesser claims, most of them bearing Spanish names.

From 1904 to 1906 the stamps at the great river mill never stopped their thunderous roar. The little steam locomotives with their trains of swaying ore cars clattered back and forth on the crooked rails from the mines to the mill. Loaded, the trains went down-grade through high-walled washes and over dizzy wooden trestles, usually at breakneck speed. Dorsey's mill included a 450-ton unit, specially designed to treat large quantities of low-grade.

Great mining engineers of the day, including Seeley Mudd and John Hays Hammond, lent their genius to the development of the property.

Hopes ran high but apparently profits were low, for in 1906 the Colorado Iron Works company foreclosed after placing an attachment of $38,845 on the mill.

Then followed years of legal entanglement. The consolidation was reorganized as the Picacho Basin Mining company, and continued to operate the river mill two years longer.

Houses of adobe and arrowweed sprang up in the mesquite thicket within sound of the stamps. Saloons became the focal point of community life for the miners. In 1895 Mr. and Mrs. Walter Ferguson came to teach school and found awaiting them a neat white-washed adobe building which had been erected by public subscription. Mrs. Ferguson later became postmaster and operator of the only independent store. The mining company operated a cooperative mercantile establishment across the road and put up a long squat hotel building, topped by the largest roof in Picacho. Billy Horn, saloon keeper, mail contractor, freighter, deputy sheriff, and stage operator, now 81 years old and still living on his old stamping grounds, was a central figure in community life. The town grew lustily without government and the law was the knife and six-shooter — and Billy Horn.

Si Hamlund, old Picacho prospector, with his gold divining rod. He claimed it led him to rich deposits by pointing downward when he passed over gold. At right is Walter Ferguson, principal of the Picacho school, who had just returned from a prospecting trip with Hamlund when this picture was taken. —Clara Townsend photo.

New Gold Town Grows

At the peak of production about 1904 more than 700 men were employed at the mines and mill. The population of the town reached 2500.

Jose Mendivil built a substantial house on his homestead, planted grapes and date palms. His sons were growing up and finding work in the mines. Grandchildren climbed on Jose's knee during the long desert evenings. Jose was a poor trader in matters affecting his future. Life, he thought, should be an untroubled affair.

In 1901 he deeded 28 acres of his land to the California King for a mill-site, which was defined in the records with corners of mesquite trees and crosses marked on rocks. The next year he deeded the same company five acres for a pumping plant. Mendivil and his heirs subsisted for several years on the subdivision of the homestead.

Michael says that his grandfather might have become a rich man, "if he had not been so big-hearted." He was bored by frequent business trips to Los Angeles and San Francisco, much preferred to sit under his grape arbor at home. To the mining tycoons he would say, "Go ahead and take it if you want it. I got enough for me and my family." Michael relates the story of Jose refusing the beach-front of San Diego in exchange for the valuable Clip silver claim (Blaine mine) he had given away, simply because he did not want to be bothered with the worries of the wealthy.

In spite of the legal difficulties, production was pushed to fever heat for the four years following the reorganization of the company. Some ore assayed as high as $700 a ton, but toward the end of the boom it dropped to $3 a ton, finally becoming too poor to work profitably.

It became impractical to carry the ore to the river and in 1908 the mill was moved to the Diablo shaft, where most of the wooden framework still stands. About this time construction started on the Laguna dam between Picacho and Yuma and the steamers could no longer carry ore down the river. During the early years concentrates had been shipped by boat to San Francisco for smelting. With the change in mill locations Billy Horn's 12-mule freight wagons became the principal mode of transportation.

Son and grandson of Jose Mendivil, discoverer of gold in Picacho basin. Ysidro, left, has spent virtually all his 66 years in or near Picacho and was associated with his father in mining for 20 years. Michael, his son, speaks perfect English and knows the history and personalities of the Picacho region.

Zane Grey, famous western novelist, visited Picacho before the mill was shut down in 1910. Out of his sojourn in the camp came his "Wanderer of the Wastelands," published in 1922. His story gives a colorful, though romantically exaggerated picture of life among the Mexican miners. High point in his description is the *baile,* where gay music and dancing feet reflect the care-free character of the Sonorans. *Cascarones* burst in dark hair and the flashing eyes of the senoritas made a fitting background for the climactic struggle of his story.

When the mines shut down Jose Mendivil and his family moved to Yuma, where the 78-year-old man died in 1916. With him passed all that was left of Picacho's gay Sonoran life. Little remained of the Mendivil homestead. As the government took claim to the Imperial dam flood basin this year, title companies found great difficulty in untangling the involved claims and counterclaims.

Changes in corporate control settled the mining properties on Jacob Ridgeway of Philadelphia, who died in 1908, leaving the estate in the hands of several heirs. Title lay in a Quaker City bank for 20 years, forgotten by all except a few promoters who knew the value of rich ore still lying in the shadow of Picacho peak.

Five years ago an energetic mining promoter, William C. Doak, bought the titles and proceeded to lay plans for the rehabilitation of the Picacho mines. Death overtook him in 1935, leaving the unfinished work in the hands of the Doak estate and E. H. Page and wife. This combination in turn conveyed options to the Picacho Gold Mining company which began operations in 1936.

During the past two years the company has been exploring and blocking out ore in the basin. Within recent weeks the Picacho Gold Mining company has pooled its interest in the property with the Nipissing Mining company and the consolidated group is operating as the Picacho Mining company.

A 2000-ton mill recently purchased in Nevada has been transported to the Picacho site and is to be erected as soon as a five-mile pipe line can be completed from the Colorado river to supply water for mixing cement for the foundations.

Since the completion of Imperial dam and improvement of roads in the Yuma Indian reservation, many motorists have found the trip to Picacho an interesting adventure. Turning north off U. S. 80 highway at a point a mile east of Winterhaven, California, the Picacho road dips under a railroad trestle and continues due north across the reservation five miles, where it crosses the new All-American canal. The road continues north up Picacho wash, climbing to an elevation of about 900 feet as it passes through a rugged range. Always ahead is the profile of Picacho peak, shaped like a huge organ console. At a point 17 miles from the highway may be seen a group of mining buildings: this is the center of the basin mining district.

Picacho did not find fame comparable to the great boom camps of Nevada: Tonopah, Goldfield, Bullfrog, Rhyolite, and Rawhide, probably because its span of life was shorter and its gold production was much lower. Statistics are meager, but Ysidro Mendivil says Wells-Fargo handled $13,000,000 in Picacho gold. Picacho had romantic color other boom towns did not have. It was set in a background of turbulent river and mesquite, many-hued hills and picturesque Picacho peak. It was the battle-ground of mining tycoons, the field of honor for knife-and-pistol fights between hot-tempered Sonorans, encounters which usually ended in rows of white crosses on Cemetery hill.

Picacho basin may again become the scene of mining activity on a grand scale. But it will be a coldly modern and efficient system. The Fergusons, the Horns, and the Mendivils will not be there; they look on from a distance, remembering when Picacho was the home of care-free miners and rough adventurers.

PICACHO MINES.

Senator Dorsey Expects to Develop Them Extensively.

[ASSOCIATED PRESS NIGHT REPORT.]

SAN FRANCISCO, Nov. 1.—Stephen W. Dorsey, former United States Senator from Arkansas and now of Denver, arrived here tonight from Colorado. About a year ago Dorsey and William Teller, brother of Senator Teller, secured control of a large group of low-grade gold-bearing claims at Picacho, about thirty miles from Yuma, in San Diego county. Last winter they interested English capital in their properties and began development. Senator Dorsey was unwilling to divulge his plans for the future, but he admitted that he hoped soon to have a large force of miners at work in the Picacho district. His plans also include the erection of a large stamp mill on the Colorado River, not far from the mines.

From Los Angeles Times, November 2, 1897.

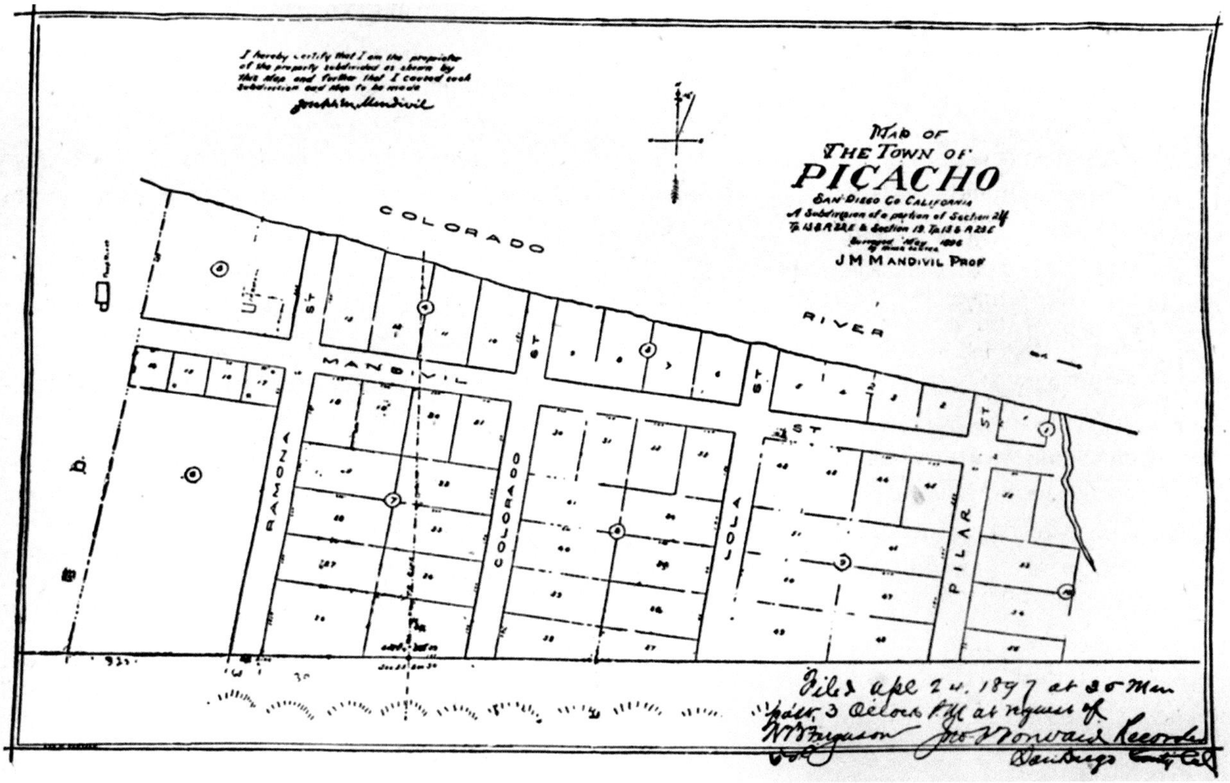

Original plat of Picacho townsite, now submerged, which Mendivil subdivided from part of his homestead.

Gold, Guns, and Fiestas in Old Picacho

MIKE A. MENDIVIL

Mike Mendivil, born at Picacho, is the grandson of Jose Maria Mendivil, fabulous figure in the camp's history. He grew up in the big Mendivil home in Picacho, saw the famous week-long fiestas and bailes held there. Now living in Winterhaven, he and his sons carry on the tradition that all Mendivils are musicians by playing in their own Spanish orchestra.

Mike A. Mendivil

THERE WAS SO MUCH GOLD IN PICACHO in the early days that men were careless about it. My grandfather, Jose Maria Mendivil, told me about seeing the Pacheco boys throwing away pans that had $5 to $10 worth of gold in them, while they were prospecting down the wash.

"No," they would say, "that's not worth placering!" Yet that much in a pan meant the gravels there were worth every bit of $500 a yard. And they wouldn't be

This article originally appeared in the *Calico Print*, a mid-20th century Western history periodical edited by Harold and Lucile Weight (Vol. VII, No. 2, February 1952, p. 4–6).

bothered with that much. They'd go on until they found a place where they could get $20 a pan.

And when my grandfather located the first ledges up in Picacho Basin, he saw the gold lying right on the surface of the washes, concentrated by the rains on the bedrock. That is actually how he made the discovery.

My grandfather had been kidnapped, down in Sonora, by the Apache Indians when he was seven and he was their captive eight years before he escaped. Then, because he knew the Apache language and ways so well, he served with the army as a scout. After that he carried mail — I think it was for the Territorial government of Arizona — from Yuma way north up the Colorado River.

It was then he found gold at Picacho. There were people in and out there, prospecting and hunting, and he came onto parties he thought he had known down in Sonora. He asked them questions, looked around and saw lots of mineral signs there. The next trip down, he did some prospecting. His father, in Sonora, had been a geologist and he knew that the best part would be up above. He went up a wash until he really ran into it rich! He got onto the main lodes and took some samples and showed them to the men placering down below. He told them he had gotten them up the wash — but there were quite a few washes! And he and the Pacheco family — old Francisco Pacheco — were the only ones interested in real mining.

Picacho—the Sharp Peak—also known as Belltower and Chimney Peak, for centuries guided travelers and adventurers in the Colorado Desert, while at its feet, shining in the sandy washes, lay millions in placer gold.

He sent those samples to San Francisco and found they had great values. That set everybody wild — but grandfather located five good claims. The Pachecos mixed up with him and around him, but they were more for placer mining than quartz. My grandfather didn't really want to bother with either one, but it was too rich to pass up — so he held onto what he could in the lode claims.

The Pachecos all got busy placer mining, from daybreak to way dark. In those days they worked 12 hours at least — great husky hombres, you know. Using either pans or dry washers, they would make two or three hundred dollars per day. It wasn't long until their families moved up there — and they were all large families. I got to know quite a few of them after they were old. They were all miners and all big, well-built men. I think you could still see some of their work up there — they'd dig those great six-foot tunnels. How they struck up that high, I don't know. But they used nothing but double-jacks to strike the drill that we use jackhammers for. And that's the way they carried on the first rough part of their mining.

After the strike my grandfather married and moved up to Picacho. Families from all over the country moved up. The Romos — that's the family he married into — the Pachecos, two Martinez' not related, quite a few Lopez', the Hodges. The first groups were all Mexican, then some of the Indians went to mining and it wasn't long before there were a few Chinese. They'd leave their kitchens, their jobs, because there was gold all over.

Prospecting out from Picacho, down the river they found Providencia and Oven washes — that was the nugget placer. It was nothing to find $70 nuggets around there — big slugs of gold. Up the river and down the river, there were a lot of camps started by the discovery at Picacho.

After the Picacho Basin mines were operating and the big mill going down by the river, things changed a little. But I've heard there was lots of highgrading. They tried to stop it by using inspectors, but it wasn't practical. They'd put the stuff in their shoes and socks — and they'd bring it out in their sandwiches. Yes — golden sandwiches. At Picacho they had these tacos, they call them nowadays. When the inspector asked to see what they had, they'd say: "You see, that's part of my lunch I didn't eat."

Mike Mendivil in the foreground with climbing partner on top of Picacho Peak. For geologists, an exclusive look at the Bear Canyon conglomerate atop the peak. Photo courtesy of Ernie Mendivil.

I don't know why he didn't heft it. Of course those tortillas and beans would be pretty heavy naturally — but never metallically! If they just brought a normal load out in their sandwiches or tacos twice a day, it would accumulate to quite a bit. They didn't care what wages they were paid. And the gold was accepted anywhere — no questions asked. The first place they would dispose of it, naturally, was the saloon. They wouldn't even have it weighed — there was plenty of gold. They'd just shove $4 or $5 across for one shot of whiskey.

If they didn't highgrade, it was still easy to get gold. They didn't work Sundays — so they'd just go down the wash themselves, and in an hour they would have enough to carry on to pay day. There was gold all over the place — and that easy money meant there were lots of drunks too. That was one of the main entertainments tor us youngsters in Picacho-watching the drunks up and down the street.

My grandfather sold part of his interest in his claims and took stock for most of the rest. That, the biggest portion, he was just frozen out or, when the company brought assessments on the stock so high and values so low. He just dropped it, or sold it to meet assessments. In other words, he gave it away. He was a strange man in the way of making deals. He gave lots of things away — just didn't want to be bothered with them. "I've got enough on my hands, now," he would say. "I've got enough for the rest of my days."

I don't know, I guess he figured on the Picacho mine carrying him through. He was a busy man, then, with many activities. He had his homestead at Picacho, and he sold part of it to the mining company for a mill site, and laid out Picacho townsite on some, and sold lots there. He built that big fine house by the river, all of redwood. It had a room in each corner and hallways through and the big hardwood dance floor in the middle. Every kind of fruit tree he could get hold of, he tried. He had date palms, pomegranates, figs, apricots, grapes — I think he tried everything but bananas.

And he got the most expensive buggy he could — with rubber tires, lights all over the thing, fine top. Of course he had some beautiful horses. And he had some interest in the river steamer lines — the captains all used to stop to see him. After the mines closed down, in 1910, he continued to live in the big house at Picacho until 1916. Then he was sickly, and came down into Yuma for a little medical care. That year the Colorado River was up pretty high, and the Gila happened to come down at the same time — there were no dams on the Gila then. And both rivers hit at the same time and went right through Yuma. My grandfather was quite an invalid and he had to get up to save some of his property and help others. He caught a terrible cold and it settled in his kidneys. He didn't last long.

I was born in Picacho in September 1900, and I lived there until 1917. All the education I got, I got at Picacho schools — but they had quite a time with me at first. From the time I could walk around, I was with my dad — Ysidro Mendivil — everywhere he went. They brought a lot of freight from Yuma to Picacho by burro then, and my dad had the biggest packtrain of them all. He would have 20 or more burros in his train, and he loaded every one of them himself — putting on big mattresses and hundred pound sacks of potatoes, and many and many a barrel of whiskey. Whiskey was just as important those days as beans and potatoes.

For many years he kept that up. He probably catered to the poor people around them, in little camps along the route, between Yuma and the Landing. One of the main places he would unload would be at the mines, where they were doing the actual placering. There was just a burro trail there then.

I went with him on many and many a trip. I would just lay up there on a burro — he'd tie me on — and sleep away. But I was important. I had to be along with him on the pack trips to supervise the freightage. I had to be!

As for school — that just wasn't in the picture as I saw it. The teacher, Mr. J. E. Carr, was nice, but when I was turned over to him and my father lit out, I was a very lonely kid. Then I got a little careless and had the misfortune of dumping two or three inkwells over the floor, and that scared me. They had to take me to school by hand — I wouldn't go. And I knew my father's route very well, and just about where he would be at any time.

Young Mike Mendivil at Picacho, circa 1912. Photo courtesy of Ernie Mendivil.

I'd leave school — somewhere I would leak away — and set out to find him. They'd have all the kids in school looking for me.

Finally the teacher got tired and started getting rough — and that didn't do any good. I'd have a whipping every day and still pull out the next day. I remember once I broke away from him while he had me by the hand, going up a trail. He thought I was doing all right and got a little careless — and I just darted off left-handed — and boy, I took off. A little bit of kid, but I jumped over an eight-foot bluff and made him jump after me before he caught me again. He was mad! Then he kept me tied to his desk — just like breaking a little wild animal.

For punishment they had some nice, clean straight switches and some from the catsclaw bushes — and the last operation was untrimmed, unprocessed switches — I remember that just as well as if it was yesterday. He gave me those so hard I had the blood streaming out of my bare legs. I tell you, my mother wept over my legs.

The next day Mr. Carr asked me if my legs were hurting and would I — just once — try school one day. I said: "Yes — but are you going to hurt me?"

He said, "Why, absolutely not. I whip you because you leave."

It was as simple as that! And I, kept watching and waiting all day, but he didn't whip me. It seemed there was something wrong. And from then on I went to school every day. I made scholarship medals and emblems and prizes. Mr. Carr said he never saw a boy take to school so rapidly, and I got to liking that man so much. He threw my mind wide open to schooling, even though he had to start with a catsclaw switch.

As soon as I learned a little, my grandfather had me writing letters for him. He could dictate letters like nobody's business, and all I had to do was write them down. Then I'd read them over and over to him. He knew very little English, but what he had was very business-like.

We had three powerful celebrations up there at Picacho. May 5, which celebrates the victory of the Mexican people over the French soldiers of Maximilian in 1862, and September 16, Mexican Independence Day, were the biggest times. But as I remember, the Fourth of July — the only American celebration there — was just about as important. They decorated the whole works — buildings had papers of all colors. They had balloons, filled with hot air, that tried to go up to Mars, and sometimes they would burn before they got off the ground. Fourth of July they celebrated with regular dynamite and there were quite a few one-armed men — consequences of that.

When those big days came — and sometimes for weddings and christenings — my father would hold dances and barbecues that might last a week — right on through. You'd see piles of pumpkins and jerky and fresh meat — in those days $2 would buy a whole steer. And musicians would take turns playing — day and night — while the dancing went on.

Sometimes there would be two or three hundred at one of those affairs. About everybody in town would attend — and guests would come by river steamer or horseback or stage. Since mail was slow, invitations were sent out two or three weeks in advance. Those invited would come over — move in — for a month or two of visiting. Strangers were welcomed too.

They were polite in those days. Only a drunkard would cause a little disorder. Even the drunk ones would go politely and ask for a dance. Chances are he'd approach the husband or the mother of the girl before he would speak to her — and perhaps he would get turned down. They were all dancers — and even when they were drunk they would be sociable, usually.

I was only a kid, supposed to be sleeping, but now and then I could hear awfully harsh language outside, and then here would come a boot or a bottle to hit the side of the building, and everybody would scatter out. When it got rough, the musicians would play harder than ever, to detract from it and keep the dance going.

Besides the special days, we had dances at the house almost every Saturday and Sunday — and had good modern music for those days: accordion, guitar, violins, kettles. My grandfather spent a lot of money on my Uncle Tito, bringing in professors from way down in Mexico — university men — to teach him music.

At those dances we'd have two or three regular numbers, and then waltzes, and the *jarabe tapatia*, *chiapanecas*, a few *jotas* — and square dances. They were prominent — and plenty long. That's when the families would sit watching the callers, and they knew their stuff. Mike Garcia was one of them, I remember. But the musicians dreaded the square dances — they were so long. Of course, in those days if musicians got two or three dollars for a night, it was a well-paid affair. They'd play two or three days for $10 or $15. And two or three men, the owner of the house, the master of ceremonies or the man who was doing the baptismal side of it — would pay for the whole thing.

When Picacho was booming, there was a lot of shooting but not much killing. But it got pretty hot some paydays and the town jail — which was just a tunnel cut back into a white tufa hill with an iron gate across it — was full. Billy Horan was San Diego County deputy sheriff at the time, and he had three or four special deputies under him.

One of the big events in town was the shooting of one of Billy's deputies — Pete Bohorkus. I did not see them kill Pete — it was about 1905, I guess, and I was only a little boy then. But I heard about it from those who saw it and who took part. The way I have heard it — though some people will deny it — Pete was one of the kind who would kill just for nothing. He'd killed several peaceful men. He listed them as his "prize victims," and it was said that he would kill, then throw in a knife or another gun he carried and would say: "I had to do it in self defense!" He also had a list — quite a list — of those he was "going to get." He said: "They'll all go." One of my uncles was on that list.

Finally, there was so much feeling when he shot one prominent citizen, they decided to arraign him for it. But they were going to take him way over to San Diego for his hearing. The people in town thought he would just be released over there, and they had enough of that.

Billy and another deputy already had Pete loaded up on the stage when these men came up. Most of them were on Pete's list. They all had been drinking, but they weren't drunk. One of them — Manuel Lopez, brother of the man Pete had just killed — told Pete he didn't like the idea of him shooting his brother like that, and that he was going to kill him. Superintendent Ridgeway of the mine company — he was a special deputy — pulled a gun and put it on Lopez and said that it was the law — Pete was going to be taken to San Diego.

Another fellow — we'll call him Rafael — went over to Ridgeway and told him to leave Lopez alone. Ridgeway said: "I'm a special deputy, and I'm going to take care of this" He swung his gun on Rafael, and Rafael beat him to it and shot him in the hip. Ridgeway swung around and dropped — and threw his gun away.

Billy Horan, who had been hoping things would smooth over, got hot and said: "Cut it out!" and he shot

Billy Horan in later life.

Rafael and clipped him in the side. Things were out of control by then, and Rafael swung right around and turned loose at Billy. He didn't hit him, but Billy and the other deputy left.

By that time, Pete knew it was getting mighty hot, and he wouldn't sit up there any more. He jumped from the stage and started running across country. There was another fellow — they called him Old Indio (he was a very respectable character). He was way off to one side and out of shooting distance to Pete, but he yelled to Rafael to shoot Pete before he got too far.

Rafael said: "I don't want to shoot him in the back!" So he yelled at Pete, again, to stop. Naturally, Pete wouldn't stop.

Then Old Indio called: "Rafael, you shoot him, or I'll shoot you. I don't want to have to go run old Pete down."

Rafael had to do something and turned loose with his gun — just one shot — and he split Pete's skull open from behind. Today, up on one of those little porphyry hills just below the placer pit where the ore bin is, you can still find a little cross on the rocks where old Pete fell.

Old Indio went to Mexico, and so did Rafael. But later Rafael came back and gave himself up and they gave him five years for shooting this deputy, Ridgeway. He was released after serving three.

There wasn't any morgue in those days — they just loaded Pete on a wagon, bouncing all the way to the river. They buried him there — and they never held much of a wake over him. That night they celebrated in Picacho.

Identified from left to right, Clyde Stewart, Ed Rochester, Jeff Richards, Tony Mendivil (son of Joseph), Earl Kerr, and George Converse. Courtesy of the Arizona Historical Society, Yuma

Blame the Gold or the Climate— We Can't Get Away from Picacho

ED ROCHESTER

Ed Rochester, born in a covered wagon near Ehrenberg, was operating a gold dry washer when he was seven or eight. His life has been spent prospecting, trapping and exploring Western deserts. The past 12 years he has been dividing his time between Picacho and Winterhaven, and has gained the unofficial title of Mayor of Picacho.

WHEN I FIRST VISITED PICACHO, I didn't know I'd keep coming back here the rest of my life. That was in 1911 and another young fellow and I were making a horseback trip from Yuma to Blythe, up along the river. The Picacho mine — up in the Basin, had only recently shut down, and there still was a store at the mine. But they were selling out the supplies cheap, and we bought a few things, such as bacon and coffee. Just about everyone had left the diggings.

We rode on five miles down to the river. Quite a lot of the old town was left there, though it had grown up pretty well with mesquite trees. And there were a good many houses up at Tortilla Flat — I imagine 25 or 30.

We visited with Old Man Riley — Tom Riley — who had the store and post office at Tortilla Flat. Of course he

This article originally appeared in the *Calico Print*, a mid-20th century western history periodical edited by Harold and Lucile Weight (Vol. VII, No. 2, February 1952, p. 4, 7).

wasn't really old then, but when a fellow's young, anybody much older than he is, is an old man. He showed us some of the highgrade from the Picacho mine and told us a lot about it. He had built his store — which is the building we live in, in Picacho now — about 1900, as I heard the story. Then he went up to Alaska, but he came back. This building was the store here until some time after 1922, and also the last official post office Picacho had.

It's the only post office Picacho has today — and probably the only clothespin post office in the world. We've got incoming and outgoing clothespins. Whatever mail anyone brings in from the box at Winterhaven is hung on the incoming clothespin. Whoever has mail to go out puts it in the outgoing clip, and the first one making the 24 mile trip to the post office takes the mail.

The second time I remember stopping in Picacho was in 1914, when I came down the river by boat on a trapping expedition. I bought supplies at Riley's store, and spent a few days around there. From then on I was in and out quite a bit. I guess the gold was the real interesting thing about the district. But, it was pretty country, too, with nice winter climate, a lot of canyons where you could get out of the wind, a lot of firewood — and always somebody to visit with.

During these stops, I met all the prospectors and miners. Ysidro Mendivil and his brother, Tony, were the active miners of the Mendivil family at that time, and they were at Picacho mining. Tony especially was the miner of the bunch. There were five sons and two daughters in the family. Ysidro was the oldest, and Ramona the youngest.

Tony could neither read nor write, but he had a wonderful gift of memory. He lived at Picacho until he was 70 years old and the farthest he got away was over toward the coast somewhere. But he didn't like the fog so he came back to Picacho. He stayed around this country and died a few years ago.

There are many conflicting stories about the strike at Picacho. Tony told me that when his father first came here, the Pachecos were placering up at the Old Camp, and that his father went on up the wash and found the vein up there. Others have told me that the Pachecos were working right on the vein when Joe Mendivil first came into the country. Anyway, I believe there were quite a few miners in the country when Mendivil arrived, but he went up and filed a claim on what he thought was the best part of the vein. It turned out to be pretty good — the Apache and two or three other claims. But he got mixed up with the high-rolling boys and was beaten out of the mine.

The 200-ton mill built at the Diablo shaft in the basin in 1908 has not ground ore since the mine closed in September, 1910. Photo by Harold Weight.

The mill as it looked in 2012, 60 years later. Note the modern mine tailings that surround the mill. Photo by Todd Wirths.

The very first settlement, before there was a Picacho, was established by a Martinez, as a sort of trading post in the rincon we call Windmill Valley, where there were a lot of Indian farms. I suppose Picacho itself was really at its best — oh, from '85 to '95. That's when I think most of the people were there. That's when the most gold was being taken out by small miners from placers.

Of course there might have been a bigger industry when the mine first started to work in a big way. It took an awful lot of wood, you know, to run those big boilers. There were a great many more woodchoppers than there were men in the mines. So it would be pretty hard to say just when the boom was.

I don't know just when the big 450-ton mill down on the river started operation, but it was prior to 1900.

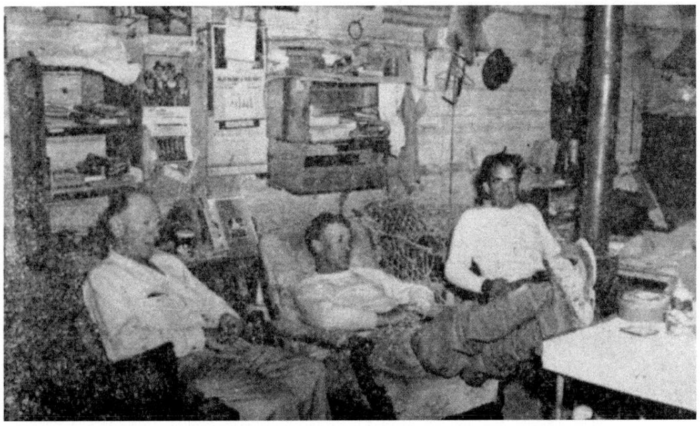

Picacho is not dead. Some fishermen, vacationers or residents are there throughout the year. Here (from left to right) Ed Rochester, Earl Kerr and Charlie Curtis, the "Picacho City Council," hold a session in their home, which is the old Tom Riley store and Picacho's last post office, built about 1900. [*Editor's note: Both Ed and Earl are buried at Picacho Cemetery. Photo by Harold Weight.*]

At first they hauled the ore down to the mill in wagons and then they built a railroad. A Mr. Gardner told me he had a contract for building the railroad down the wash in 1900. There is a story that when the railroad reached Red Mountain, the workers hit a rich placer spot in the wash, and work on the railroad stopped while they placered.

When the railroad was built, it happened to be a wet cycle, and the road washed out so often they finally abandoned it. Then they set up the mill up at the Basin.

The rich ore in the Picacho mine was spotty. John Nevills, an old desert miner who lived around the river here for 50 years, told me he had worked in the stope, that there was some very high grade ore, and that there was some highgrading being done. The average dropped to about $2.50 a ton when the mill at the mine closed in 1910. I don't believe there has been any production from the mine since then.

When Earl Kerr and I came down to Picacho from the Palo Verde country, in the spring of 1940, the Nipissing Company, a Canadian outfit, was prospecting at the mine. They drilled the country extensively, and bought the last of the Mendivil holdings from Ysidro — which consisted of six claims to the east of old Picacho mine. They hauled in the old Belmont mill from Tonopah in 1939, but before they got it erected, war in Europe started. Being a British company, they didn't put the mill up, and in 1941 they hauled it back out. But they still have the ground and keep a watchman, and no doubt they intend to work it some time.

Some placering has continued most of the time since the 1860s. Some areas have been completely worked over many times. The gold is found in old channels here. We just have to picture mountains lots higher than they are now. As they eroded, the gold was concentrated in channels not so steep as the present ones, and the concentration extended well down toward the old channel of the Colorado river, which was about a mile from the present channel. After a while the country may have sunk. At any rate, those old channels were filled with debris.

In the course of time the country was elevated again and other channels washed out that followed pretty well the courses of the previous ones. The rich deposits of placer were created where the most recent channels cut the old ones, making a second or perhaps a third concentration. There is evidence of three periods of erosion, and the last concentration, of course, was the richest. Those were the deposits first found, the ones that occurred in the present drainage system.

As these were exhausted, the miners began to work the older channels, working underground in small tunnels and shafts. That condition of deposit prevailed all along the river here, for a distance of about 15 miles. The peculiar thing about Picacho is that it is all small gold. A two-pennyweight piece would be a very big piece, while down the river three or four miles — in Oven, Providencia, Mayflower and Sortoyomo canyons, pieces up to ¾ of an ounce are not unusual.

We came here 11 years ago, just to work out a little of that placer we had located. We had no intention of staying on here. But somehow or other, we just never got away.

IMPERIAL MINE DEAL REPORTED

Group of Gold Claims in Picacho Basin Area Sold for $135,000

EL CENTRO, Feb. 2. (Exclusive) Details of one of the largest mining deals in the history of Imperial county were made public today when an option was recorded, involving the sale of twenty-six claims in the Picacho basin to Richard C. Willis, who is said to represent the Treadwell-Yukon Company. The sale price was $135,000.

William C. Doak of Los Angeles is owner of the property, which is located along the Colorado River on the California side, forty miles north of Yuma. The claims named in the option include three groups operated extensively in 1887-89 when Picacho had a population of 1500 persons.

Gold was first discovered in this district by Joe Mendivil in 1860. Work was discontinued on a majority of the properties twenty years ago and the old town now is practically deserted. The option includes the townsite of Picacho and a well and pump.

From Los Angeles Times, *February 6, 1933.*

SECTION 3

THE CARGO MUCHACHO MOUNTAINS and ENVIRONS

MINES AND MINING.

THE DESERT MINES OF SAN DIEGO COUNTY.

Active Development Work in the Perris Mining Section—Nevada County's Gold Record—The Klondike — Antimony — Concerning English Syndicates.

Encouraging news has been received from the mines out on the desert during the past week. Development is especially active in the mines located along the Colorado Desert, in the eastern part of San Diego county.

CARGO MUCHACHO.

The Signal Hill group of twenty-six claims is situated in the Cargo Muchacho district of San Diego county, adjoining the famous old property of that name owned by H. W. Blaisdell of Yuma, where the rich tailings produced by the Cargo Muchacho during the years the mine was being operated, are being worked. Los Angeles parties—Richard Garvey, J. A. Stevenson and others—are the owners, and as soon as sufficient stock can be placed, it is proposed to begin extensive development work. The Yuma Sun says that eastern parties are now on their way to examine the property, and if satisfactory, will invest in a large amount of stock.

From Los Angeles Times, *February 4, 1898.*

Cargo Muchacho Mines
A Photo Essay

WILLIAM J. ELLIOTT
Consulting Engineering Geologist
P. O. Box 541, Solana Beach, California 92075

At the Other End of the Jump Rope....

MY INTRODUCTION TO COLLEGE LEVEL geology occurred during a weekend camping trip to the Cargo Muchacho Mountains. Professor Baylor Brooks took a small group of first year San Diego State College (now University) physical geology students out for a weekend of exploring rocks, dirt, and old gold mines during the fall semester of 1960.

Fifty-two years later I am revisiting where it all began. I was happily suckered into a lifetime of geo-adventures while sitting around a crackling campfire with a spaghetti dinner and a couple of "brewskis." Lifelong dreams and aspirations were freely shared with cohorts and our geo-leader, "Skip" Brooks. Not much has changed. We all still enjoy the great out-of-doors pretty much the same way as we did back then!

Location and Elevation. The Cargo Muchacho mines are located in the Cargo Muchacho Mountains of southeastern Imperial County, California. They can be found on the Ogilby, California, 15′ and 7½′ topographic quadrangle maps (U. S. Geol. Survey, 1963 and 1997) in the south-center portion of Section 20 (projected), T15S, R21E, San Bernardino Base and Meridian. Google Earth global positioning coordinates are: 32° 50.08′N, 114° 46.76′W. Mine openings and associated milling operations are found at about 600 to 725 feet above mean sea level.

Geologic Maps. Regional geologic mapping of southeastern Imperial County is shown on maps published by Brown (1923, Plate III), Morton (1977, Plate 1), Olmsted and others (1973, Plate 3), and Strand (1962). A detailed look at the Cargo Muchacho mining district was made by Henshaw (1942) for his thesis during the 1937–1939 field seasons.

Mining Activity. The following is abstracted, messaged, and reconstituted from Henshaw (1942), Morton (1977), and Shumway and others (1980). As with most reports of early mining activity, details vary from author to author. The intent herein is to provide a brief sketch of the main attractions, the details of which may never be known for sure.

"The Yuma region was first visited by the Spaniards in 1540" (Henshaw, 1942, p. 151). Cargo Muchacho district mining activity dates to the early 1780s, when placer deposits in Jackson Gulch were briefly exploited immediately south and east of the future location of the hardrock Cargo Muchacho mine. Later, in about 1823, shortly after establishment of the Mexican Republic, mining in the district resumed using a donkey walking around in endless circles pulling a large heavy rock over ore to crush it and free the gold. This primitive device was known as an *arrastra*.

The Cargo Muchacho name is reported to have originated when two young boys (*muchachos*), wanting to imitate their father's daily work at the mines, returned to camp one evening with their shirts loaded (*cargado*) with rich gold ore. Hence, the shortened Spanish *cargo muchacho*, or loaded boy!

With completion of the Southern Pacific Railroad between Yuma, Arizona, and Los Angeles, California, in 1877, mining activities again flourished until about 1900, producing about $5.1 million in gold. This would be about $13.2 million in 2010 dollars (U.S. Dept. of Labor, 2011).

Hard-rock underground mining at the Cargo Muchacho mines began in 1877 and continued until 1882. They were surveyed for patent in 1892. Other periods of major activity were in 1890–94, 1936–42, and again in 1949–52(?). Quartz veins in a 2- to 8-foot wide silicified shear zone were followed downdip within a quartz diorite host rock, striking N4°E, and dipping 48°E. "Tailings of [Cargo Muchacho] ore milled in the 1890s were cyanided(?) in about 1940" (Morton, 1977, p. 49).

No evidence was found of recent mining activity at the Cargo Muchacho mine in 2011. The closest recent mining activity has been at the American Girl mine, located approximately 1½ miles north-northwest of the Cargo Muchacho mine (U. S. Geol. Survey, 1963, 1997).

A Close Look at Tailings Pond Sediments. If an unsuspecting rational geologist were parachuted into this location blindfolded, with no prior knowledge of local or regional geology, it could lead to a heretofore unheard of urge to … reach for a brew. For all of the world, some of these mine tailing sediments look like they came right out of a full scale pond experiment designed for students of sedimentology. Thinly- to massively-bedded silts and sands with some scattered small gravel combined with primary structures, soft sediment deformation, and apparent cross-bedding could drive one to the nearest bar … and I don't mean sand bar. For example, see Photo 1.

To provide additional insult to already festering confusion, well rounded pebble-sized Colorado River "B" gravels (Metzger and others, 1973, pp. G22–G23) are present near the southeastern end of this deposit (Photo 2). Did the Colorado River venture this way and leave these gravels behind? Or were these imported to be used in some part of the mining operation?

Photographs. Three antique black and white photographs of the Cargo Muchacho mines are filed at the Arizona Historical Society Library in Yuma, Arizona. These are reproduced here (Photos 3, 5, and 7), along with modern photographs of the same views taken on September 1, 2011, by the author (Photos 4, 6, and 8). The photographs tell the story and captions help provide historical insight.

Infrastructure, head-frames, hoists, boilers, the mill building, tailings dam, housing and out-buildings are all gone. About all that remains today are mine openings (Photo 9), two cyanide(?) vats and/or water cisterns(?) (Photo 10), mine tailings (Photos 4, 6, and 8), a few scattered scraps of weathered wood and rusting iron, and rapidly deteriorating primitive access roads and trails.

In stark contrast with darkly varnished granitic and metamorphic country rocks, light colored patches in the 2011 photographs (Photos 4, 6 and 8) represent all that remains of the old tailings pond sediments (Photos 5 and 7). Photo 1 is a close up view of eroded and exposed sediments along the eastern edge of the tailings pond.

A look into the mine itself is provided by Photos 11, 12, 13, and 14. These photos were taken by Robert Ian Farber on January 28, 2012, at about the 500-foot elevation level. Photo captions add context to these images of one hundred plus year old underground mine workings.

ACKNOWLEDGEMENTS

Todd Wirths found the three old black and white photos of the Cargo Muchacho hard-rock mining area at the Arizona Historical Society in Yuma, Arizona. Carole Ziegler allowed extra time for photos to arrive and made constructive suggestions to improve overall flow and content. Carol Brooks at the Arizona Historical Society assisted with making arrangements to purchase copies of each image. Thanks to a chance meeting with Dr. Monte Marshall, Robert Ian Farber gave permission to use four of his exceptional images taken on January 28, 2012 deep down inside the main Cargo Muchacho mine shaft. Thank you, Robert!

And a special thanks to my wife, Wendy, who only fussed a little when I spent the night solo at the Cargo Muchacho mines site in 103°F. heat to be able to match the three old photographs with the same views in soft early morning light.

SELECTED REFERENCES

Brown, J. S., 1923, The Salton Sea Region, California: U. S. Geological Survey Water-Supply Paper 497, 292p.

Crane, Walter R., 1917, Ore Mining Methods, Comprising Descriptions of Methods of Support in Extraction of Ore, Detailed Description of Methods of Development of Mines, of Stoping and Mining in Narrow and Wide Veins and Bedding and Massive Deposits Including Stull and Square-Set Mining, Filling and Caving Methods, Open Cut Work and a Discussion of Costs of Mining: John Wiley and Sons, New York, 305p.

Henshaw, P. C., 1942, Geology and Mineral Deposits of the Cargo Muchacho Mountains, Imperial County, California: California Journal of Mines and Geology, v. 38, n. 2, p. 147–196.

Metzger, D. G., Loeltz, O. J., and Irelna, B., 1973, Geohydrology of the Parker-Blythe-Cibola Area, Arizona and California: U. S. Geological Survey Professional Paper 486-G, 130p.

Morton, P. K., 1977, Geology and Mineral Resources of Imperial County, California: California Division of Mines and Geology, County Report 7, 104 p., Pl. 1; Geological Map, scale 1:250,000, contour interval 200′.

Olmsted, F. H., Loeltz, O. J., and Irelan, B., 1973, Geohydrology of the Yuma Area, Arizona and California — Water Resources of Lower Colorado River–Salton Sea Area: U. S. Geological Survey Professional Paper 486-H, 227p.

Strand, R. G., compiler, 1962, Geologic Map of California — San Diego-El Centro Sheet: California Division of Mines and Geology Map Sheet, scale 1:250,000, contour interval 200′.

U. S. Dept. of Labor, 2011, "Consumer Price Index, Inflation Calculator," Bureau of Labor Statistics. http://www.bls.gov/data/inflation_calculator.htm.

U. S. Geological Survey, 1963, Ogilby, California, 15′ topographic map, scale 1:62,500, contour interval 40′.

U. S. Geological Survey, 1997 (provisional), Ogilby, California, 7½′ topographic map, scale 1:24,000, contour interval 40′.

Shumway, G. L., Vredenburgh, L., and Hartill, R., 1980, Desert Fever — An overview of Mining in the California Desert Conservation Area: Bureau of Land Management, U. S. Department of the Interior, 3610 Central Avenue, Suite 402, Riverside, California, Contract No. CA-060-CT7-2776, February 1980, 141p.

ILLUSTRATIONS

Photo 1. *Close-up view of fine-grained tailings pond sediments showing primary structures and evidence of soft sediment deformation. Photo by W. J. Elliott, 2011.*

Photo 2. *Birds-eye view of Colorado River "B" gravel (Metzger and others, 1973, p. G22–G23) resting on tailings pond sediments near eastern side of tailings dam. Photo by W. J. Elliott, 2011.*

Photo 3. *Cargo Muchacho mines, circa 1890(?). Northerly view of mining camp in foreground, tailings dam (white elongated horizontal rectangle) in middle distance, and underground mines (on right) at far end of short winding hillside dirt road. Horse-drawn wagon in shadow at bottom center. Arizona Historical Society, 2011, used with permission.*

Photo 4. *Northerly view, same as Photo 3. Photo by W. J. Elliott, 9-1-2011.*

Photo 5. *Cargo Muchacho mines, circa 1890(?). Southerly birds-eye view of tailings pond, two-story/two smoke-stacks mill building on left, and mining camp beyond. Small buildings and mine openings (shafts) are above and left of mill building. Mining camp is downstream from tailings pond. Arizona Historical Society, 2011, used with permission.*

Photo 6. *Southerly view, same as Photo 5. Photo by W. J. Elliott, 9-1-2011.*

Photo 7. *Cargo Muchacho mines, circa 1890(?). Easterly view across tailings pond to mill building. Small building and head-frame at upper left mark location of mine shafts. Note wagon-horses on left of mill building. Arizona Historical Society, 2011, used with permission.*

Photo 8. *Looking east, same view as Photo 7. Photo by W. J. Elliott, 9-1-2011.*

Photo 9. *Looking down (about 48° below horizontal) into the main Cargo Muchacho mine shaft opening. Photo by W. J. Elliott, 9-1-2011.*

Photo 10. *One of two cyanide(?) vats and/or water cisterns(?) (near Photo 9). Photo by W. J. Elliott, 9-1-2011.*

Photo 11. *Stacy James taking a break in the main Cargo Muchacho mine shaft at about the 500 foot level (below the entrance). Note the wooden ladder on his left (viewer's right), adjacent to steel rails upon which ore carts traveled in and out of the mine. The rails and pipe in the middle are the same ones visible in Photo 9 at the mine entrance. Stressed (failed) vertical timbers are visible at the right side of the image. Photo by Robert Ian Farber, January 28, 2012.*

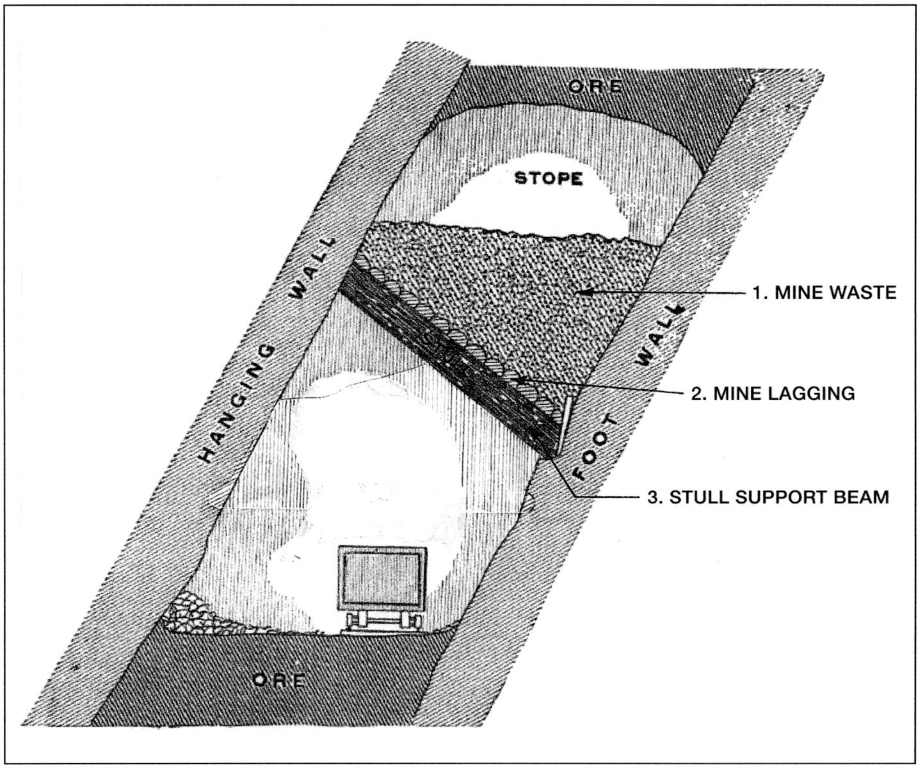

Figure 1. *Example of stull support used in tunnel mining to store mine waste underground rather than hauling it to the surface (Crane, 1917, Figure 1). See Photo 12. Note that the stall beam is longer than the perpendicular width of the stope walls. Therefore, it has to be installed, as shown, "out of square." This allows the natural downward gravitational (normal) g force of overlying mine waste to wedge the timber even more securely into the hanging and foot walls. (1) Mine Waste. (2) Mine Lagging. (3) Stull Support Beam.*

Photo 12. *Horizontal mine drift roofed by a stull (Figure 1), above which mine waste rock is stored rather than being hauled up and out of the mine. Note that the large diameter main cross members are longer than the width of the opening, thus allowing gravity to wedge them firmly into the wall rock. Mine waste is stored on top of the smaller diameter longitudinal lagging "roof" members. Photo by Robert Ian Farber, January 28, 2012.*

Photo 13. *Inverted "U", timber-set mine supports in a horizontal mine drift. Note how the first two sets are bent and cracked, indicating failure over time. Photo by Robert Ian Farber, January 28, 2012.*

Photo 14. *Artifacts of underground mining. An empty cardboard box that held Trojan Explosives and Chemicals and a wooden box in which bearings were shipped by The Timken Roller Bearing Company from Mount Vernon, Ohio, sit forlornly on a disheveled array of dusty broken mine timbers. Photo by Robert Ian Farber, January 28, 2012.*

Rotation of crustal blocks around vertical axes as estimated by paleomagnetic studies of Miocene volcanic rocks in southern California and southwestern Arizona:

A review of previous studies and some unpublished data from the southern Colorado river region

MONTE MARSHALL, STEVEN COSTELLO, JOSEPH BUTTERWORTH, AND MICHAEL VESETH
Geology Department, San Diego State University, San Diego, California

INTRODUCTION

PALEOMAGNETISM, ALONG WITH EARTH-quake seismology, made possible the discovery of plate tectonics in the 1960s. Reversals in the polarity/direction of the geomagnetic field every few hundred thousand years cause the linear marine magnetic anomalies that proved the sea floor was spreading from oceanic ridges and rises and the continents were indeed drifting. But these reversals, as phenomenal and important as they are, make up only part of paleomagnetism's contribution to geotectonics.

The purpose of this paper is to review what the paleomagnetic study of Miocene rocks has revealed about the last 20 million years of microtectonics of southern California, and to publish 'new' data from southeast California and southwest Arizona. We start with a short summary of the behavior of the geomagnetic field that makes such studies possible, then review some of the most outstanding results, and end with a puzzle about the tectonics that may have affected the most southeastern and southwestern parts of California and Arizona, respectively. The time span that we will be dealing with essentially begins with the first, chaotic contact of the Pacific Plate with the North American Plate at about 30 million years ago. The state of stress is one that involves northwest-trending dextral shear and northeast-trending sinistral shear — or, in terms of its resulting strain, a N-S compression and E-W extension. This is a time period when the docking of allochthonous terranes, like possibly the western Peninsular Ranges, along the western margin of North America has ended and the western margin of the continent is modified or deformed a bit less dramatically and more locally by the San Andreas system of faults.

A SHORT REVIEW OF THE EARTH'S MAGNETIC FIELD AND HOW IT IS USED IN TECTONIC STUDIES

Careful study, during most of the last century, of the direction of the remanent magnetism frozen in many rocks when they form has revealed much about the behavior of the geomagnetic field. Over most of at least Phanerozoic time, the geomagnetic field has had, to a first approximation, the shape of a simple dipole like the field around a bar magnet. And, even though the magnetic dipole axis wobbles somewhat around the earth's rotation axis, its magnetic poles, on average, coincide with the geographic poles. This property of our earth's magnetic field has profound paleographic and tectonic implications! The inclination, I, of the flux lines at the earth's surface of a co-axial, geocentric, dipole field has a simple trigonometric relation to geographic latitude, L: $\tan I = 2\tan L$. The field is horizontal at the earth's equator, is inclined about 60 degrees to the horizontal at the latitude of San Diego, and is vertical at the poles. This simple equation is the main way of calculating paleolatitudes, which, along with matching rock types and ages, allows the reconstruction of former supercontinents — especially for times prior to 200 million years ago, when the oceanic crustal record of continental movements has all been subducted.

But the inclination of a remanent magnetization frozen in rocks when they form is not the only useful component of the remanence direction. Again, in the case of an ideal co-axial, geocentric, dipole field, any departure of the azimuth, or the direction of the horizontal component, of the magnetization from north suggests that the rock has been rotated. Shortly after the discovery of plate tectonics and long before the closing up of the Mediterranean Ocean and the interaction of the African and European plates was well understood, European paleomagnetists found that Iberia (Spain and Portugal), Corsico-Sardinia, and northern Italy had not only been translated but also rotated counterclockwise around vertical axes, relative to northern Europe.

Before proceeding into paleomagnetism's contribution to our understanding of the tectonics of southern California, we need to understand how paleomagnetic studies try to get a data set that yields at least an approximation of the direction of a co-axial, geocentric, dipole field at a given location. The dipole field that makes up most of the geomagnetic field not only wobbles, but it is combined with secondary magnetic fields generated by eddy currents in the outer core that move, grow and decay. All of these time-variant departures from an ideal magnetic dipole field cause 5, 10, 20, or even larger, angular deviations of the compass needle from True North. It is called secular variation by paleomagnetists, and is well known to geologists and mariners as the Declination of Magnetic North. Since these variations usually seem to be random, paleomagnetic studies try to average out the secular variation at a given locality by sampling the geomagnetic field direction at many points in time over a time interval of at least several tens of thousands of years. When dealing with sedimentary rocks, for example, oriented cores are taken from many adjacent layers. When working with volcanic rocks, an ideal outcrop to sample would have many flows intercalated with sedimentary units that would evidence some passage of time. Finally, and counter-intuitively, magnetization directions of the individual units (bed or flow) should be sufficiently scattered — to prove the passage of time and provide a mean direction that sufficiently samples the range of secular variation. Now you can understand, in part, why some earth scientists (being jealous!) call paleomagnetists, "paleomagicians"!

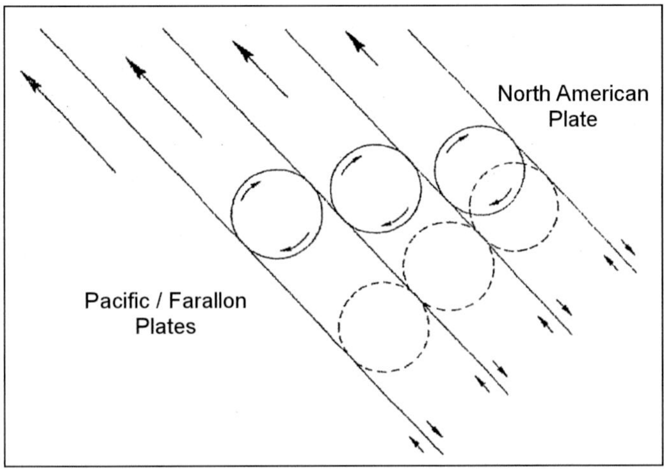

Figure 1. *Cartoon showing northwestward translation and clockwise rotation of equant crustal blocks. Their sharp edges are worn off in a zone of distributed, right-lateral shear.* (*Modified from Beck, 1980*)

A SHORT HISTORY OF THE DISCOVERY OF THE VERTICAL AXIS ROTATION OF CRUSTAL BLOCKS IN SOUTHERN CALIFORNIA

Myrl Beck, of Western Washington University, was one of the first paleomagnetists to work along the western edge of North America. And he was the first to point out that most of the paleomagnetic studies in the western Cordillera found shallow paleomagnetic inclinations, indicating northerly translation, and declinations that suggested the rocks studied had undergone vertical axis clockwise rotations (Beck, 1980). In light of Tanya Atwater's revolutionary paper that showed how the history of subduction of the Pacific sea floor beneath the west coast of North America could explain much of west coast Cenozoic tectonics (Atwater, 1970), he attributed the translations and rotations to oblique subduction and, especially after 5 million years, to distributed shear associated with the San Andreas Fault System. His first model for the mechanics of these block translations and rotations is shown in Figure 1, and the blocks, once their four edges were rounded, were popularly known as "Beck's Ball Bearings"!

The true mechanism is obviously more complicated and the vertical axis rotation of a block of the earth's crust that has a prismatic shape in map view involves at least two mechanical considerations. First, either all the sharp edges must be rounded so that the prism becomes essentially a vertical cylinder, or the adjacent crust must splinter into fault-bounded slices that slide past one another in such a way as to make room for the rotating block. A good example of this will be seen shortly in the discussion of the rotation of the Transverse Ranges. In this second case, spaces having the shape of vertical triangular prisms will be generated, creating a line of geologic basins. The second mechanical requirement is a horizontal zone of detachment or decoupling at some depth that allows the block to rotate.

In the late 1970s and early 1980s, Bruce Luyendyk and his students at University of California Santa Barbara began paleomagnetic studies of the Channel Islands and Transverse Ranges in southern California and found large vertical axis clockwise rotations of much of this area (Fig. 2; Luyendyk et al., 1985). Probably the geologically most important and one of the largest rotations found by the University of California Santa Barbara group was a 90 degree clockwise rotation of the western Transverse Range Block. The Transverse Ranges, as their name implies, are anomalous in that they lie at a large angle to most of the mountain ranges in California and trend east-west instead of north-south or northwest-southeasterly. Prior to about 20 million years ago, they apparently trended N-S and were attached to the rocks of the Santa Ana Mountains. Since that time the crustal block that contains these mountains as well as the northern Channel Islands has rotated some 90 degrees. An early, and apparently still largely valid, model for the rotated Transverse Ranges Block and the strike-slip faults to the north and south whose slip accommodated this rotation is shown in Figure 3. In a more recent article, Luyendyk (1991) has summarized the estimated amount and timing of these rotations.

A more recent study has focused on the basins created in the Los Angeles Area by the large block rotation of the western Transverse Ranges as they pivoted around a point south of the San Gabriel Mountains (Ingersoll and Rumelhart, 1999; Ingersoll, 2008). The transrotation and transtension centered on the Los Angeles Area between about 20 and 6 million years ago is not only tectonically dramatic but financially significant (Fig. 4). The largest of the basins, the Los Angeles Basin, is one of the most petroliferous per-unit-volume basins in the world (Biddle, 1991). With an estimated ultimate recovery of 9 billion barrels, it approaches the size of Prudhoe Bay, Alaska. The basement on the western side of the basin is Catalina Schist and on the eastern side granitic and metamorphic rocks. The maximum depth of the central trough, located just 8 miles southeast of downtown Los Angeles,

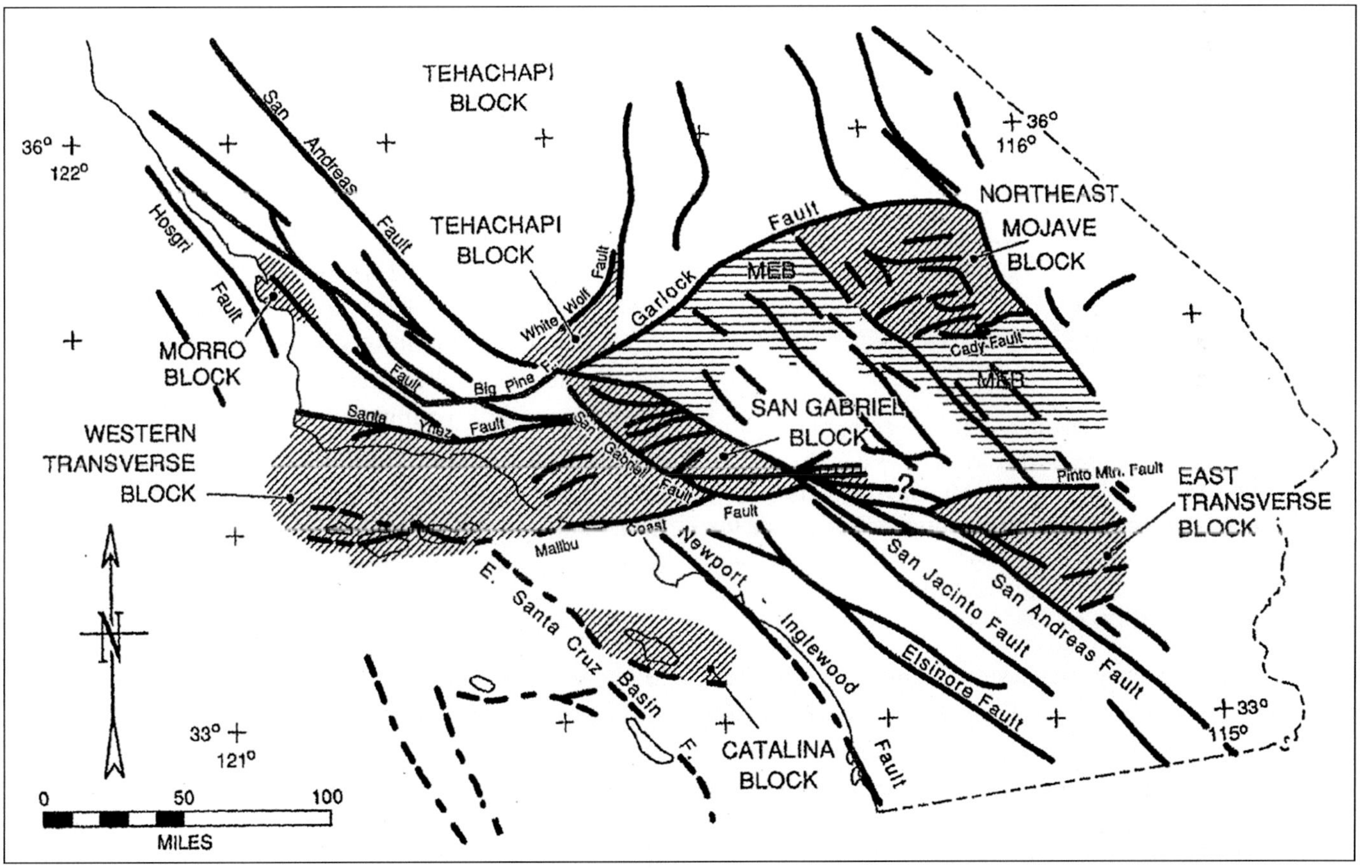

Figure 2. *Areas in central and southern California that, based on anomalous paleomagnetic declinations, have undergone vertical axis block rotation during Neogene time. Diagonal lines are blocks that have rotated since about 20 Ma in the zone of deformation between the Pacific and North American plates. The horizontal lines show the Mojave Extensional Block (MEB) and contains areas that have even earlier rotations. (Modified from Luyendyk, 1991)*

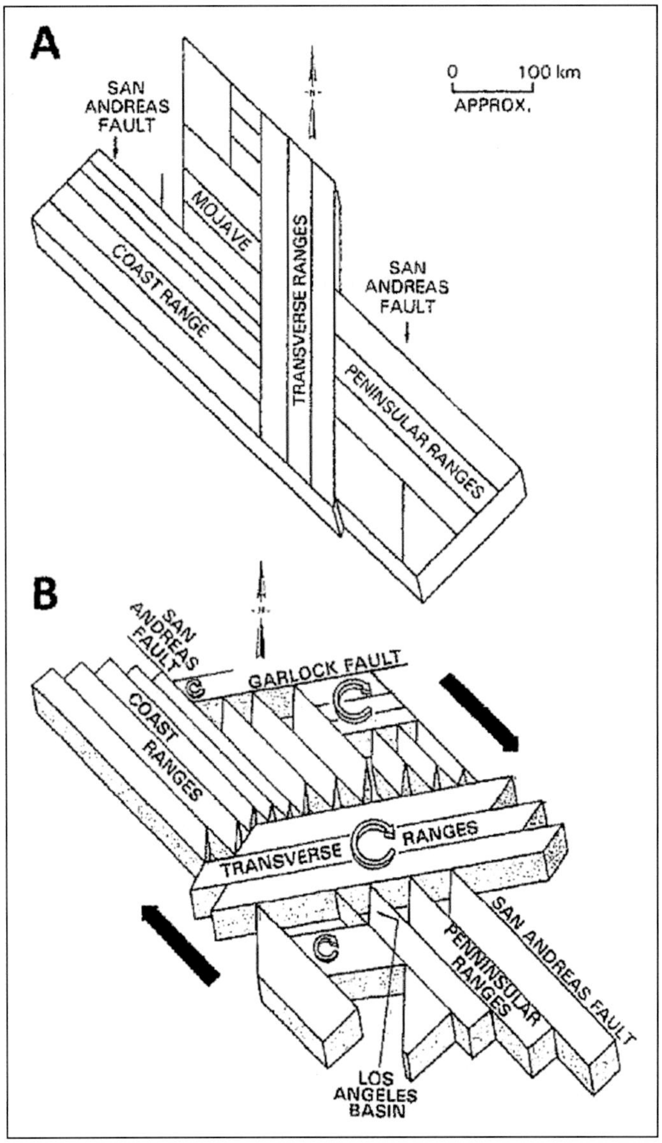

Figure 3A. *Model showing the N-S trend of the Transverse Ranges at 20 million years ago. To the northwest and southeast (at that time) are lines showing known and inferred, NW-SE trending faults whose progressive right slip allowed (caused?) the Transverse Ranges to rotate 90 degrees clockwise.*

Figure 3B. *Current Geometry. The right lateral faults east of the San Andreas, between the eastern Transverse Ranges and the Garlock fault form the Eastern California Shear Zone. Note the rows of mostly triangular-shaped basins, such as the hypothetical Los Angeles Basin (Luyendyk et al., 1980, as modified by Biddle, 1999).*

is estimated by gravity surveys to be about 30,000 feet. Wells have encountered 14 million year old, deep marine sedimentary rocks at 20,000 feet. The bottom literally fell out when the western Transverse Range Block rotated and tore away from the mainland!

The next obvious questions are why did this large, 200 by 100 Km, rectangular-shaped sliver of the earth's crust tear away from the mainland and rotate 90 degrees? And how deep is the detachment plane? In most of the literature one reads that the cause is "dextral shear", but rarely do the authors discuss a detachment plane at depth. A more precise explanation ironically comes from the careful study of the linear marine magnetic anomalies on the sea floor off the continental borderland/shelf of southern and central California. In her first great contribution to western North America tectonics, Tanya Atwater (1970) showed that the linear marine magnetic anomalies off our coast revealed that the subduction of the Farallon Plate (FP) became discontinuous when the spreading center between it and the Pacific Plate (PP) had migrated east to the point where it came into contact with the North American Plate (NAP). This happened at about the latitude of Los Angeles and at about 30 million years ago. From this time, the PP — FP Spreading Center migrated south as a triple junction until it reached the tip of Baja California at about 6 million years ago. To the north of the contact point, the PP — FP Boundary, which in this case was a transform fault, migrated up the coast of California to Cape Mendocino as part of a different kind of triple junction.

Geological processes are usually not as simple as we would like, and the complexities of a subduction zone — especially when it swallows plate boundaries, were less well understood in 1970 than now. Tanya Atwater's 1970 article raised the question of whether the Farallon Plate was completely subducted in the zone between the two ever-more-distant triple junctions, or even if some of the leading edge of the Pacific Plate might have followed the last remnants of the Farallon Plate down into the mantle. Currently, most plate tectonicists think that the main force driving subduction is the negative buoyancy of the down-going oceanic lithospheric slab — it simply sinks into the aesthenosphere because it is denser. But, because it is new and hot, young oceanic lithosphere is probably less dense than the aesthenosphere and is buoyed up. As spreading centers near oceanic trenches this change in density contrast probably leads to a tendency for the upper part of the down-going slab to subduct ever more slowly — until it stops being subducted and remains as part of the sea floor outboard of the trench. In this case, the lower or deeper part of the slab will continue to subduct and a gap or "window" will open in the slab, allowing the hot aesthenosphere to rise up and come into contact with the upper/continental lithospheric plate.

In their detailed studies of the linear marine magnetic anomalies offshore of central California, Tanya and her student, Jeff Severinghaus, found several fragments, or microplates, of the Farallon Plate that form the oceanic crust outboard of the now defunct subduction zone (Atwater

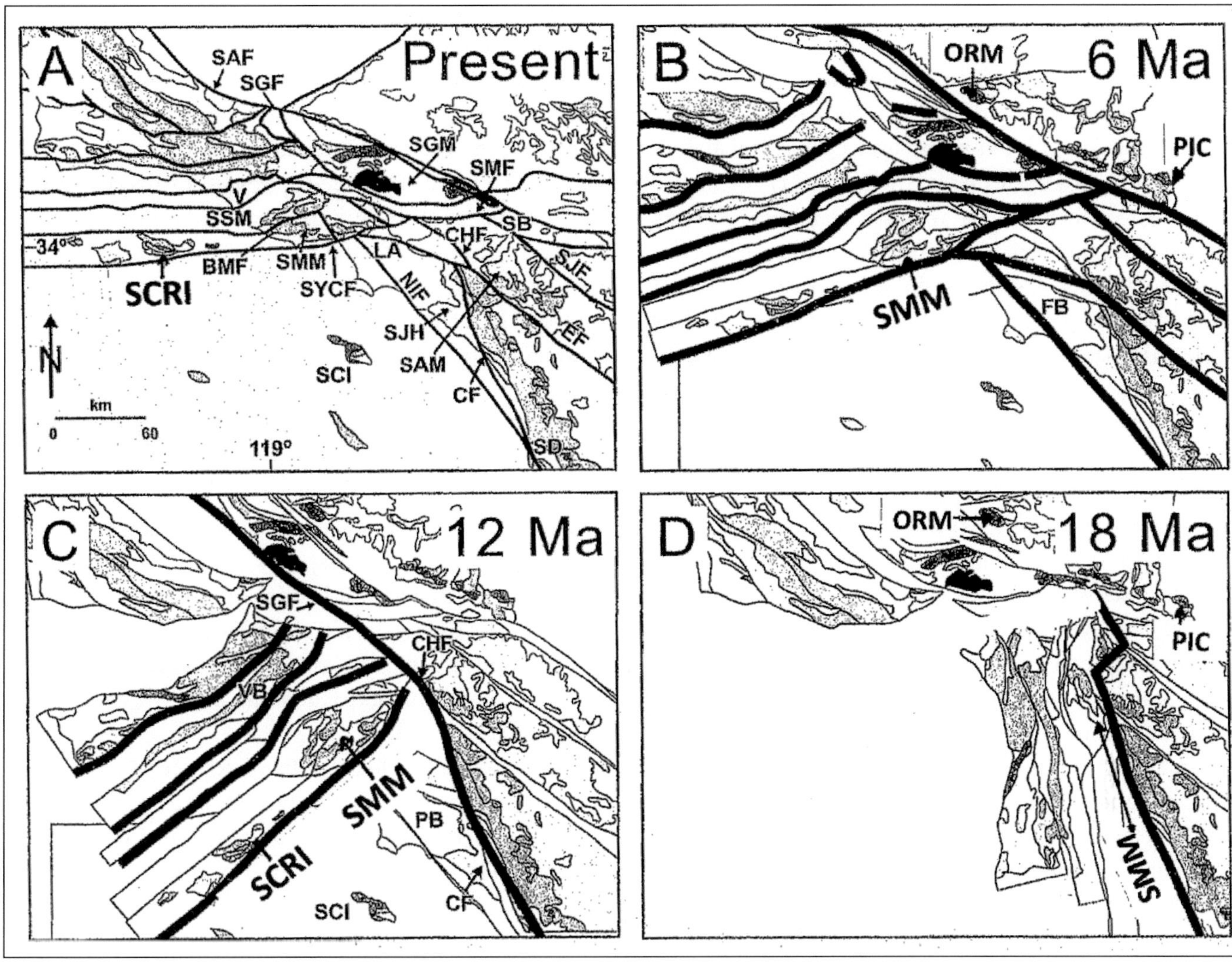

Figure 4. *Present and palinspastic geologic maps of the Los Angeles area. Heavy and light lines are known or suspected strike-slip faults involved in the rotation of the western Transverse Range Block. Some of the more important faults are: SAF, San Andreas fault; SGF, San Gabriel fault; SJF, San Jacinto fault; EF, Elsinore fault; CF, Cristianitos fault; CHF, Chino Hills fault; SCRI, Santa Cruz Island fault. Some of the more important places are: LA, Los Angeles; SMM, Santa Monica Mountains; V, Ventura; SGM, San Gabriel Mountains; SB, San Bernardino; SAM, Santa Ana Mountains; SD, San Diego; SCI, Santa Catalina Island; ORM, Orocopia Mountains; PIC, Picacho Peak Area.* (*Modified after Ingersoll and Rumelhart, 1999*)

and Severinghaus, 1989; Atwater, 1989). Called the Monterey and Arguello plates, their easterly extent under the western edge of the North American Plate and to which plate, the North American Plate or the Pacific Plate, they were more strongly coupled with remained uncertain. A more recent off-shore seismic survey suggests that these two microplates extend under the continental shelf at least 100 Km inland from the paleotrench (Nicholson et al., 1994). In their final stage of subduction, the angle of subduction became ever more oblique, i.e., more northerly, and caused right lateral, strike slip faults to form in the overlying North American Plate. The authors suggest that by 20 million years ago the old spreading center between the Pacific Plate and the Farallon Plate annealed, and that the microplates, even though they were buried under the North American Plate, became part of the Pacific Plate and shared in its northwesterly motion. To the south two other microplates of the partially subducted Farallon Plate became part of the Pacific Plate at 17 and 12 million years ago. This formed a zone of transtension that extended from central California to the tip of Baja California and explains much of southern California coastal tectonics in the last 20 million years. An important feature of this model is that the PP — NAP Boundary in the area of discussion involved a gently-dipping slip surface that evolved into a family of vertical strike-slip faults.

A related important corollary of this model is that the shallow subduction zone with its cap of marine sediments, including turbidites from the shelf, and pillow basalt probably extends at least as far east as the San Andreas Fault.

After being subducted several tens of kilometers, all these sediments and mafic igneous rocks have been metamorphosed to schist — green, blue, or black, depending on the pressure, temperature and length of time to which they were exposed! And it is this schist layer, called the Catalina Schist in the Los Angeles Area, that probably forms the detachment surface that the western Transverse Ranges rotated on and the deep water Miocene sediments on the west side of the Los Angeles Basin were first deposited on! Note that in the palinspastic reconstructions of the Los Angeles Area (Fig. 4) for 6 million years and older, where several hundred kilometers of dextral offset on the San Andreas have been removed, the Pelona-Orocopia Schist (ORM) is only 100 kilometers to the northeast of the basin. Although the Catalina and Orocopia Schist may differ somewhat in age, lithology, and metamorphic facies, there seems to be so much variability in these three parameters within each schist, the Rand-Pelona-Orocopia Schist may well be the landward extension of the Catalina Schist. All seem to have been emplaced beneath North American continental crust by very low angle/shallow subduction from Late Cretaceous to Mid Eocene time. And, finally, one can ask if this collection of metamorphosed marine sediments and fragments of oceanic igneous crust are essentially an extension to the south of the famous Franciscan Formation of Northern California?!

PALEOMAGNETIC STUDIES IN SOUTHEASTERN CALIFORNIA AND SOUTHWESTERN ARIZONA

The best documented vertical axis, crustal block rotation east of the San Andreas in southern California is that found in a paleomagnetic study by Carter et al. (1987) in the eastern Transverse Ranges, or Little San Bernardino Mountains (the East Transverse Block in Fig. 2). Based on magnetic measurements of cores taken from twenty seven, 10 to 15 million year old basalt flows between the E-W trending, left-lateral Pinto Mountain and Chiriaco faults, this crustal block has been rotated 40 degrees clockwise (Fig. 5). They considered that the mechanics were essentially the same as for the 90 degree clockwise rotation in the western Transverse Ranges, i.e., a crustal block bounded by NE-SW trending left-lateral faults (probably conjugate faults to the San Andreas), that was able to rotate 40 degrees clockwise because of the right-lateral offsets on a series of NW trending faults to the north, called the Eastern California Shear Zone (the faults in the MEB region of Fig. 2).

The careful reader at this point should then ask the question: "If the Eastern Transverse Range Block is at the northern end of the block labeled "Transverse Ranges" in Fig. 3a and at the eastern end in Fig. 3b, wouldn't that imply that the eastern block has rotated 90 degrees like the western block?!" Since the rocks studied by Carter et al. are only about 10 million years old and the western Transverse Ranges started rotating at about 20 million years ago, do these basalts record only the last half of the rotation? Did the eastern Transverse Ranges originally trend more northerly, then they started to rotate clockwise at about 20 Ma, and at about 5 million years ago got cut off from their western counterpart as it was displaced about 200 kilometers to the north by the San Andreas? Another consequence of this model, as has been mentioned earlier, is that if the crustal block is fairly rigid, then more or less triangular gaps or spaces must have opened on BOTH the north and south sides of the block, as shown in Figure 3B. The Eastern California Shear Zone should have its counterpart/rough mirror image set of right-lateral, NW trending faults to the south of the Eastern Transverse Ranges.

That such large blocks of the earth's crust could have rotated about a vertical axis by such large amounts and not left more obvious deformation, and thus went unnoticed for so long by geologists, is amazing. As with plate tectonics, the effects of an invisible force opened our eyes to the world of block tectonics! However, questions about the magnitude of the rotation of the Transverse Ranges have remained. In his efforts to palinspastically restore the offsets on the San Andreas system of faults, Bob Powell needed only half as much rotation as that found by the University of California Santa Barbara group—45 and 20 degrees of clockwise rotation for the western and eastern

Figure 5. *Geologic sketch map of southeastern California and southwestern Arizona. Circles show the location and results of the paleomagnetic studies discussed in this article. The arrows show the average paleomagnetic declination and the number is the number of flows used in calculating the site mean. Any departure of the arrows from due north is called the declination anomaly and suggests vertical axis rotation of the site area. The three circles in the northwest part of this map are from Carter et al., 1986. These are: ETR, eastern Transverse Ranges, regional result; DB is the two sites in the Diligencia Basin. The two circles on the east side are the regional results of Michael Veseth, 1985, MV; and Joseph Butterworth, 1984, JB. The cluster of 11 circles along the southern Colorado River is the site results of Steven Costello, 1985. Because they are so geographically, and perhaps tectonically, scattered, each of Costello's sites are named. PP is the Picacho Peak area. The mountain ranges important to this study are named.*

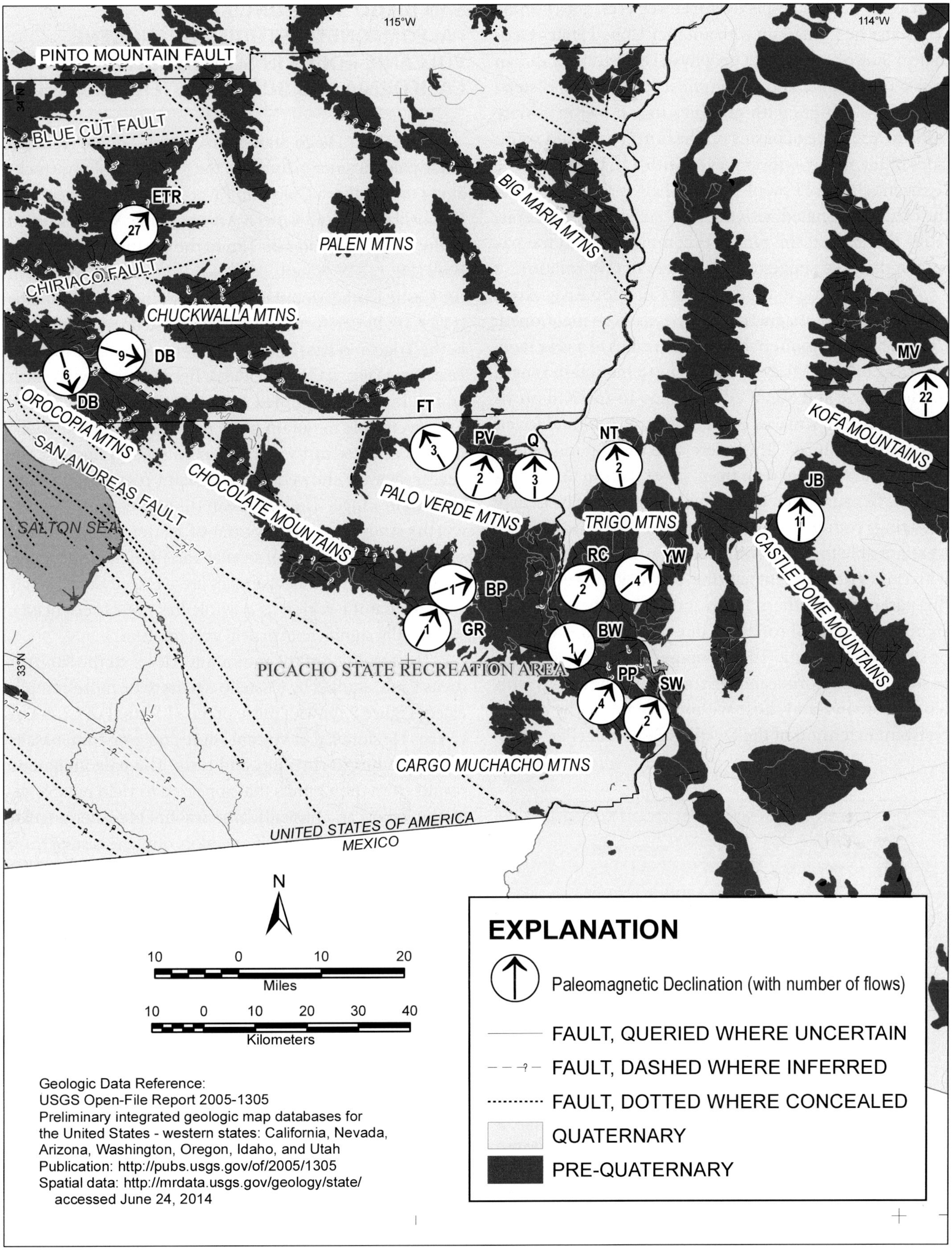

115°W
114°W
34°N
33°N
PINTO MOUNTAIN FAULT
BLUE CUT FAULT
ETR
27
CHIRIACO FAULT
CHUCKWALLA MTNS.
PALEN MTNS
BIG MARIA MTNS
DB
9
6
DB
OROCOPIA MTNS
SAN ANDREAS FAULT
SALTON SEA
CHOCOLATE MOUNTAINS
FT
3
PV
2
Q
3
PALO VERDE MTNS
NT
2
TRIGO MTNS
MV
22
KOFA MOUNTAINS
JB
11
CASTLE DOME MOUNTAINS
RC
2
YW
4
BP
1
GR
1
BW
1
PICACHO STATE RECREATION AREA
PP
4
SW
2
CARGO MUCHACHO MTNS
UNITED STATES OF AMERICA
MEXICO
N
10 0 10 20
Miles
10 0 10 20 30 40
Kilometers
Geologic Data Reference:
USGS Open-File Report 2005-1305
Preliminary integrated geologic map databases for
the United States - western states: California, Nevada,
Arizona, Washington, Oregon, Idaho, and Utah
Publication: http://pubs.usgs.gov/of/2005/1305
Spatial data: http://mrdata.usgs.gov/geology/state/
accessed June 24, 2014
EXPLANATION
Paleomagnetic Declination (with number of flows)
FAULT, QUERIED WHERE UNCERTAIN
FAULT, DASHED WHERE INFERRED
FAULT, DOTTED WHERE CONCEALED
QUATERNARY
PRE-QUATERNARY

Transverse Ranges, respectively (Powell et al., 1993). Some years later he teamed up with another United States Geological Survey scientist (a geophysicist!) and they did an excellent geologic and geophysical study of the eastern Transverse Ranges with graphics that show by gravity lows the presence of basins on the north, east, and south sides of the range — several even with a triangular shape (Langenheim and Powell, 2009). Their determination of the offsets associated with the E-W sinistral (left-lateral) faults that dissect the ranges is compatible with the paleomagnetically suggested 40 degrees of CW rotation.

Before turning to the southern Colorado River Area, one further paleomagnetic study needs to be mentioned. Calderone et al. (1990) published the results of a very large sampling of mid-Miocene basalts in 16 mountain ranges in the Mojave and Sonora deserts, 120 to 300 Km north of and 50 to 300 Km east of the southern Colorado River, respectively (Fig. 6). They were able to get good paleomagnetic directional data from 179 of the 190 flows they drilled. Because they deliberately chose flat, or nearly flat-lying, young (12–20 million year old) basalts, they were not able to say anything about rotations associated with earlier detachment faulting, but they showed that in this part of the southern Basin and Ranges there were no significant regional rotations due to high angle dip slip or strike slip faulting. They proposed that the boundary in southern California and Arizona between regions with block rotations, and those without, might be the southeastward extension of the Death Valley Fault (Fig. 6).

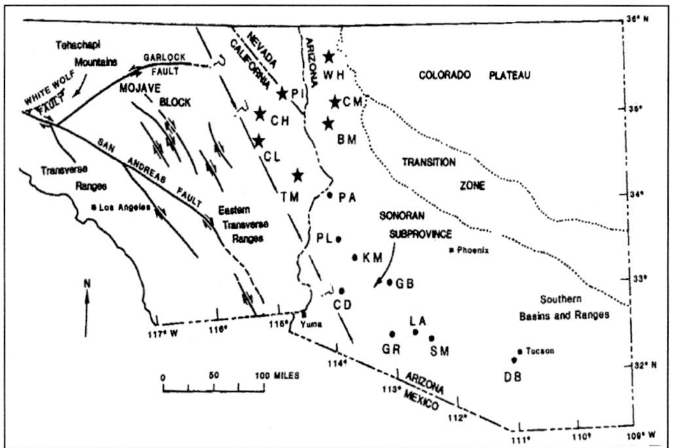

Figure 6. *Map of southwest Arizona and southern California. It shows the location of the 16 mountain ranges, indicated by two capital letters, where Calderone et al. took oriented cores from 190 mostly basalt flows of mid-Miocene age. Stars indicate ranges used in the northern subregion. The dashed line with question marks is the extension of the Death Valley fault. (Modified from Calderone et al., 1990)*

SAN DIEGO STATE UNIVERSITY PALEOMAGNETIC STUDIES OF MIOCENE VOLCANIC ROCKS IN SOUTHEASTERN CALIFORNIA AND SOUTHWESTERN ARIZONA

Three San Diego State University graduate students made paleomagnetic studies of the Miocene volcanic rocks along the southern Colorado River region in southeastern California and southwestern Arizona. Mike Veseth worked in the Kofa Mountains 90 Km to the northeast of Picacho Peak (Fig. 5, MV; Veseth, 1985). Joe Butterworth worked in the Castle Dome Mountains and western Kofa Mountains (Fig. 5, JB; Butterworth, 1984). Steve Costello sampled flows in the Trigo and Palo Verde Mountains and in the Picacho Peak Area (Fig. 5; Costello, 1985). Between the two, Veseth and Butterworth sampled 67 basalt and tuff flows/units. They were able to obtain good paleomagnetic directions from 51 of these units and as can be seen in Figure 5 the vector average showed no significant rotation of either mountain range. The site mean directions from Butterworth's study show the amount of scatter expected from sampling the full range of secular variation of the geomagnetic field (Fig. 7). And when they are averaged they strongly suggest that the region as a whole has not experienced a statistically significant vertical axis rotation.

The tectonics of the mountains closer to the San Andreas Fault, studied by Costello, seems to be more complicated. Figure 8 shows a more precise location of Costello's 11 sites. He sampled 32 volcanic units, ranging from basalts and andesites to rhyolites and tuffs. The paleomagnetic results from the 24 units that appeared to yield reliable directions show an unusually large amount of scatter/angular dispersion (Fig. 9A). But if one plots only the results from his 7 southern sites, the angular dispersion is much more reasonable and the resultant vector of these normally and reversely magnetized units has a declination of N36E — which is 40 degrees clockwise from the expected mid-Tertiary geomagnetic direction (Fig. 9B). Such a limited number of sites and flows doesn't prove conclusively that this area has been rotated clockwise about 40 degrees sometime during the last 25 million years (the average age of his rocks), but his data do suggest it. The dotted traces of the Sand Hills and Algodones faults, possible former extensions of the San Andreas Fault, can be seen on the southwestern border of the southern Chocolate Mountains (Fig. 5). The clockwise rotation of the Picacho Area is possibly an 'extension' of the even larger block rotations suggested by the paleomagnetic study of Carter and others (1987) in the Diligencia Basin 110 Km to the northwest (DB, Fig. 5). Like a string of pearls,

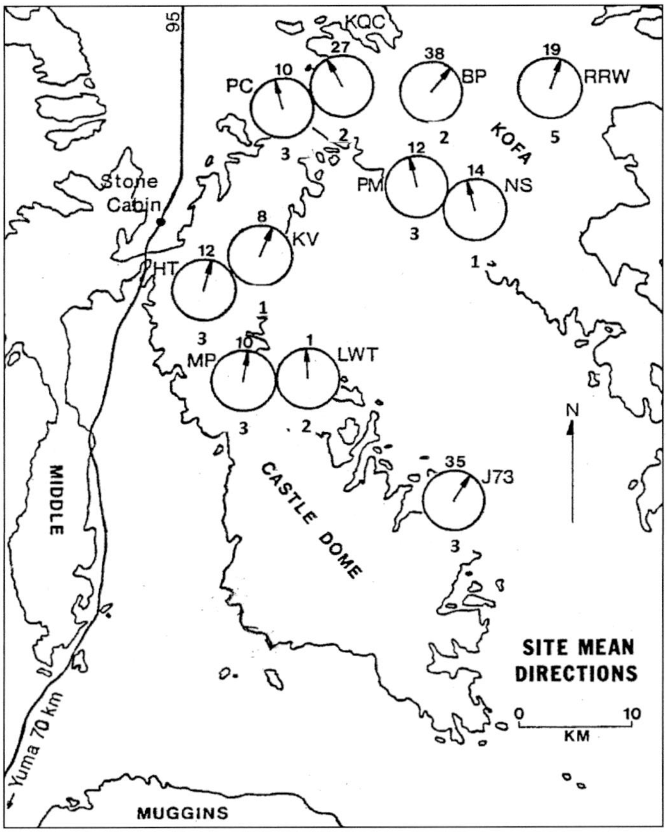

Figure 7. *Site mean paleomagnetic declinations from the Castle Dome and southern Kofa Mountains in southeast Arizona about 50 km east of the Colorado River (Figure 5). Number below each circle is the number of flows, mostly welded tuffs, used from the site. Number above each circle is the degrees of apparent rotation and the arrow shows sense of apparent rotation. (Modified from Butterworth, 1984)*

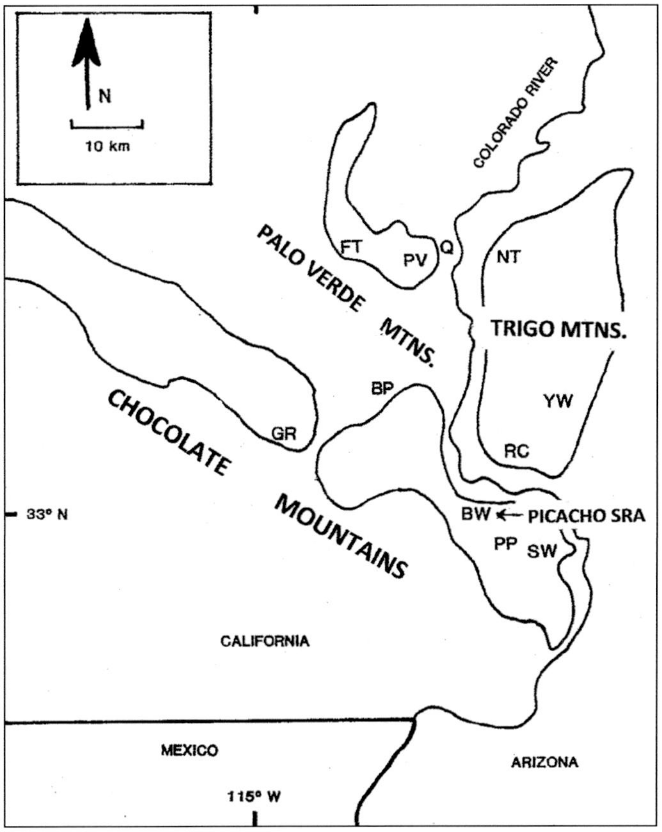

Figure 8. *Map showing the location of Steve Costello's eleven paleomagnetic sites. The sites are shown with double upper case letters which are abbreviations of the place names: FT, Flat Tops; PV, eastern Palo Verde Mountains ; Q, quarry; NT, north Trigo Mountains; YW, Yuma Wash; RC, Red Cloud Wash; SW, Senator Wash; PP, Picacho Peak; BW, Bear Canyon; BP, Buzzard Peak; GR, Glamis Road.*

the Mecca Hills and Orocopia and Chocolate mountains could consist of a series of small blocks that have been rotated clockwise by dextral shear in the San Andreas Fault Zone. A model for this deformation would be much closer to Beck's Ball Bearings (Fig. 1) than Luyendyk's rigid block that somehow remains not cut by the right-slip faults that border and rotate it (Fig. 3).

SUMMARY AND CONCLUSIONS

This paper began with a review of what is currently known about the behavior of the geomagnetic field and how the record of past field directions as recorded in volcanic rocks as they cool can tell us not only if they have been translated, but even rotated. It has focused on paleomagnetic studies of volcanic rocks of Miocene age first in the Los Angeles Area and then in southeastern California and southwestern Arizona. Rocks of this age were chosen because of their abundance and because we hoped to avoid the effects of earlier terrane accretion events. This region of the United States Southwest was chosen for study because it lies in the broad shear zone of the PP-AP, currently centered on the San Andreas Fault System. This review of previous work of others and the first publication of work done by SDSU was prompted by the choice of the Picacho Peak Area to be the location for the 2011 San Diego Association of Geologist's annual field trip!

The largest vertical axis rotation of a crustal block in southern California, both in terms of the size of the crustal block and the amount of rotation, is the rotation of the western Transverse Ranges. The 90 degree clockwise rotation of the western Transverse Ranges during the past 20 million years is one of the most significant tectonic events in southern California. The cause of this rotation is the right shear couple associated with the collision of the Pacific and North American plates, but not in the simple way usually envisioned. The Farallon Plate

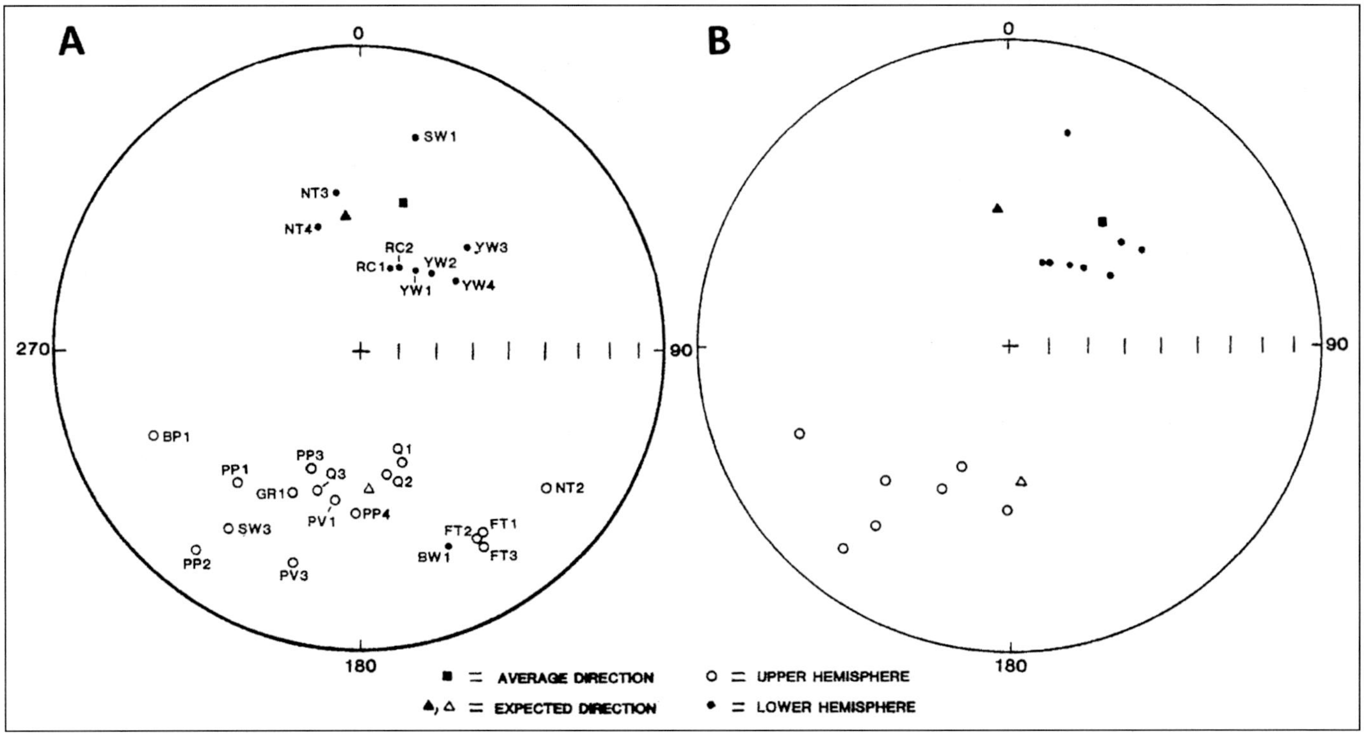

Figure 9A. Stereographic projection of the cleaned and tectonically corrected magnetization directions of flows from Costello's entire study area. The sites are shown by double capital letters and their location is shown on Figures 5 and 8. Of the 32 volcanic units, ranging from basalts to rhyolites to tuffs, only 24 yielded 'cleaned' directions—9 with normal and 15 with reversed polarities. The declination of the vector sum of these 24 units is N17E.

Figure 9B. When the directions from the four northerly sites, FT, PV, Q, and NT, are removed, the scatter is considerably reduced and the vector average of the southern sites has a declination of N36E (Costello, 1985).

was not entirely subducted and its upper/trailing edge apparently became fused to the Pacific Plate and shared its northwesterly motion. The part of the North American Plate that was immediately overlying it was torn away from North America — rotating the western Transverse Ranges 90 degrees and leaving a chasm, the Los Angeles Basin that extends down to mid-crustal depths.

The second largest crustal block rotation is the 40 degree clockwise rotation of the eastern Transverse Ranges. The mechanical model for this rotation appears to be less dramatic. The space for this block to rotate as well as the force to make it rotate is due to the dextral offset along members of the Eastern California Shear Zone and equivalent faults to the south of the range. These latter faults are not so well mapped/understood, but gravity surveys show the presence of basins on the south side of the Little San Bernardinos as well as on the north. The rotation of the range was accompanied by a set of currently E-W, left-slip faults. One of the interesting questions about this model is why the San Andreas Fault is the only one of northwest-trending faults that has cut and offset the Transverse Ranges.

Paleomagnetic studies of California south of the eastern Transverse Ranges in the southern Basin and Range province of southeastern California and southwestern Arizona have failed to find any equally large crustal blocks that have rotated since mid-Tertiary time. However, on the west side of the Salton Trough there is an area along the southern end of the Elsinore Fault that has undergone complicated rotations (Marshall et al., 2006). And, on the eastern side of the trough, there are two blocks with paleomagnetically suggested rotations — one at the northern and the other at the southern end of the Chocolate Mountains. If more paleomagnetic sampling proves these rotations to be real, that would open the possibility of additional rotated blocks along the entire length of the Mecca Hills and Orocopia and Chocolate mountains, i.e., along the length bounded by the current San Andreas Fault and its possible earlier extension to the southeast. Such a series of rotated blocks would be like a string of pearls, or, less poetic but historically more accurate, a string of Beck's Ball Bearings!

ACKNOWLEDGEMENT

I thank Dave Bloom for applying his mastery of computer graphics in simplifying my Figure 5 — making it much clearer and more professional.

REFERENCES CITED

Atwater, T., 1989, Plate tectonic history of the northeast Pacific and western North America, in Winterer, E. L., Hussong, D. M., and Decker, R. W., eds, The Eastern Pacific Ocean and Hawaii: Boulder, Colorado, Geological Society of America, The geology of North America, v. N.

Atwater, T., and Severinghaus, J., 1989, Tectonic maps of the northeast Pacific, in Winterer, E. L., Hussong, D. M., and Decker, R. W., eds, The Eastern Pacific Ocean and Hawaii: Boulder, Colorado, Geological Society of America, The geology of North America, v. N.

Beck, M.E., 1980, Paleomagnetic record of plate-margin tectonic processes along the western edge of North America: Journal of Geophysical Research, v. 85, p. 7115–7131.

Butterworth, J.E., 1984, Paleomagnetic investigation of mid-Tertiary volcanic rocks in the Castle Dome and southern Kofa Mountains, Yuma County, southwestern Arizona: M.S. thesis, San Diego State University, San Diego, CA.

Calderone, G., and Butler, R.F., 1984, Paleomagnetism of Miocene volcanic rocks from southwestern Arizona: Tectonic implications: Geology, v. 12, p. 627–630.

Calderone, G.J., Butler, R.F., and Acton, G.D., 1990, Paleomagnetism of Middle Miocene volcanic rocks in the Mojave-Sonora Desert Region of western Arizona and southeastern California: Journal of Geophysical Research, v. 95, p. 625–647.

Carter, J. N., Luyendyk, B.P., and Terres, R.R., 1987, Neogene clockwise tectonic rotation of the eastern Transverse Ranges, California, suggested by paleomagnetic vectors: Geological Society of America Bulletin, v. 98, p. 199–206.

Costello, S.C., 1985, A paleomagnetic investigation of mid-Tertiary volcanic rocks in the lower Colorado River area, Arizona and California: M.S. thesis, San Diego State University, San Diego, CA.

Ingersoll, R. V. and Rumelhart, P. E., 1999, Three-stage evolution of the Los Angeles Basin, southern California: Geology, v. 27, p. 593–596.

Ingersoll, R. V., 2008, Reconstructing southern California, in Spencer, J. E., and Titley, S. R., eds., Ores and orogenesis:Circum-Pacific tectonics, geological evolution, and ore deposits: Arizona Geological Society Digest 22, p. 409–417.

Langenheim, V.E. and Powell, R.E., 2009, Basin geometry and cumulative offsets in the Eastern Transverse Ranges, southern California: Implications for transrotational deformation along the San Andreas Fault System: Geosphere, v. 5, p. 1–22.

Luyendyk, B.P., Kamerling, M.J., and Terres, R., 1980, Geometric model for Neogene crustal rotations in southern California: Geological Society of America Bulletin, v. 91, p. 211–217.

Luyendyk, B.P., Kamerling, M.J. , Terres,R.R., and Hornafius, J.S., 1985, Simple shear of southern California during Neogene time suggested by paleomagnetic declinations: Journal of Geophysical Research, v.90, p. 12, 454–12, 466.

Luyendyk, B.P., 1991, A model for Neogene crustal rotations, transtension, and transpression in southern California: Geological Society of America Bulletin, v. 103, p. 1528–1536.

Marshall, M., Mace N., Erskine, B., and Jansen, E., 2006, What regional and local paleomagnetic data tell us about the tectonics of the Coyote Mountains, southwest Imperial Valley, southern California, in Bloom, D. M., and Farquharson, P. T. eds. Geology and history of southeastern San Diego County, California: San Diego Association of Geologists, p. 55–62.

Nicholson, C., Sorlien, C. C., Atwater, T., Crowell, J.C., and Luyendyk, B.P., 1994, Microplate capture, rotation of the western Transverse Ranges, and initiation of the San Andreas transform as a low-angle fault system: Geology, v. 22, p. 491–495.

Powell, R.E., 1993, Balanced palinspastic reconstruction of pre-late Cenozoic paleogeology, southern California: Geologic and kinematic constraints on evolution of the San Andrea Fault System, in Powell, R.E., Weldon, R.J.,II, and Matti, J.C., eds., The San Andreas Fault System: Displacement, Palinspastic Reconstruction, and Geologic evolution: Boulder, Colorado, Geological Society of America Memoir 178.

Veseth, M.K., 1985, Paleomagnetics of mid-Tertiary and younger volcanic rocks of the Kofa Mountains, southwestern Arizona: M.S. thesis, San Diego State University, San Diego, CA.

OPEN RARE MINERAL FIND

Vitrefrax Company Acquires Vast Deposit of Fire-Resisting Material at Ogilby

Recent acquisition of a big Kaynite mine at Ogilby, Cal., by the Vitrefrax Company, 5100 Pacific Boulevard, manufacturers of fire brick and other fire resisting material has increased the assets of the corporation more than $600,000, according to an announcement yesterday by by General Manager A. E. Small.

The mine covers an area of 200 acres seventeen miles from Yuma, Ariz. Many years ago it was the scene of intense activity when miners and others from all over the country went there in search of gold. For a long time all operations on the property were abandoned.

Mr. Small estimates the deposits in the mine of more than 1,500,000 tons of a rare mineral which has not been discovered in any form elsewhere. From this mineral the Vitrefrax Company produces the basic product known as Vitrox.

Discussing the product Mr. Small said yesterday:

"This material has a combination of extraordinary qualities. It has the greatest resistance to heat of any commercial product. Four or 5000 degrees of heat have no effect upon it. It has no expansion or contraction under any conditions ever met with in industry. It is one of the hardest substances known. It has extraordinary toughness and resistance. A small percentage added to dishware, for example, makes dishes practically unbreakable.

"Vitrox is an absolute nonconductor of electricity and is becoming a tremendous factor in the entire electrical industry. More than 15,000,000 spark plugs now in use are composed, in part, of Vitrox. It conducts heat with exceptional rapidity and is being employed in all sorts of electrical heating elements.

"It is highly abrasive and is coming into use in all sorts of grinding mills and for other grinding and abrasive purposes. It is absolutely insoluble in sulphuric and other acids, and has scores of applications in chemical industries, in mining operations, in brass and other foundries, in steel mills, etc. Used as a coating for fire brick, or as an integral part of the brick itself, and for cementing the bricks together, it imparts qualities which have never before been considered possible to the linings of smelters, kilns, ovens and furnaces in every important type of industry in the world.

"These are a few of the uses of this remarkable material, and the Vitrefrax Company not merely owns the only known source of supply, but also controls the patents and secret processes by which alone it can be produced as a commercial product, and has developed a marvelous plant for its manufacture, and has opened markets for it throughout the world.

"While developing Vitrox, the Company also has been making many other heat-resistant products which are in use throughout the industries of the country, and it has grown from a $40,000 company, five years ago to one with $1,500,000 of capital and $600,000 surplus."

From Los Angeles Times, *July 19, 1925*

Photo 1. Araz outcrop adjacent to Southern Pacific Railroad tracks. View to northeast. Photo by W.J. Elliott, 2011

Araz Outcrop

WILLIAM J. ELLIOTT
Consulting Engineering Geologist
P. O. Box 541, Solana Beach, California 92075

INTRODUCTION

ON APRIL 11, 2011, A RECONNAISSANCE along both sides of the Interstate 8 corridor from about 3 to 7 miles west of Winterhaven, California, was made to search for the elusive Bouse Formation. This effort was prompted by a letter from Dr. Anna Buising to Mr. Joe Corones, dated January 31, 1994, wherein she mentions the presence of "… suspiciously Bouse-like trough cross-bedded pink siltstone on both sides of Interstate 8 several miles west of Winterhaven" (Buising, 1994).

It was a warm spring day, and my curiosity was piqued. What are thought to be older Colorado River alluviums defined by Metzger and others (1973) and by Olmsted and others (1973) were to be found in just about every dry wash and road cut.

Becoming weary of digging my "Q-stick" (à la Dr. Roy Shlemon*) into these geologically younger abandoned fluvial terrace deposits, I headed back south toward Imperial County Highway S-24 (old US-80). Just before driving through a Southern Pacific Railroad underpass to reach pavement, an unusual looking outcrop caught my eye. Could this be an errant exposure of Bouse Formation? The jury is still out, and the decision will be yours. Read on.

This outcrop has not been mapped or studied in any kind of rigorous detail. The objective here is to identify its location and provide a brief description upon which future inquiries may be based. It is hoped that questions posed herein will stimulate careful investigation, lively discussion, and finally, defendable answers.

* Dr. Roy Shlemon is a world renowned Quaternary geologist and soil scientist. His endowment to the Geological Society of America provides the foundation for the Roy J. Shlemon Mentor Program in Applied Geoscience. This outreach is designed to acquaint undergraduate and graduate students with careers in applied geoscience.

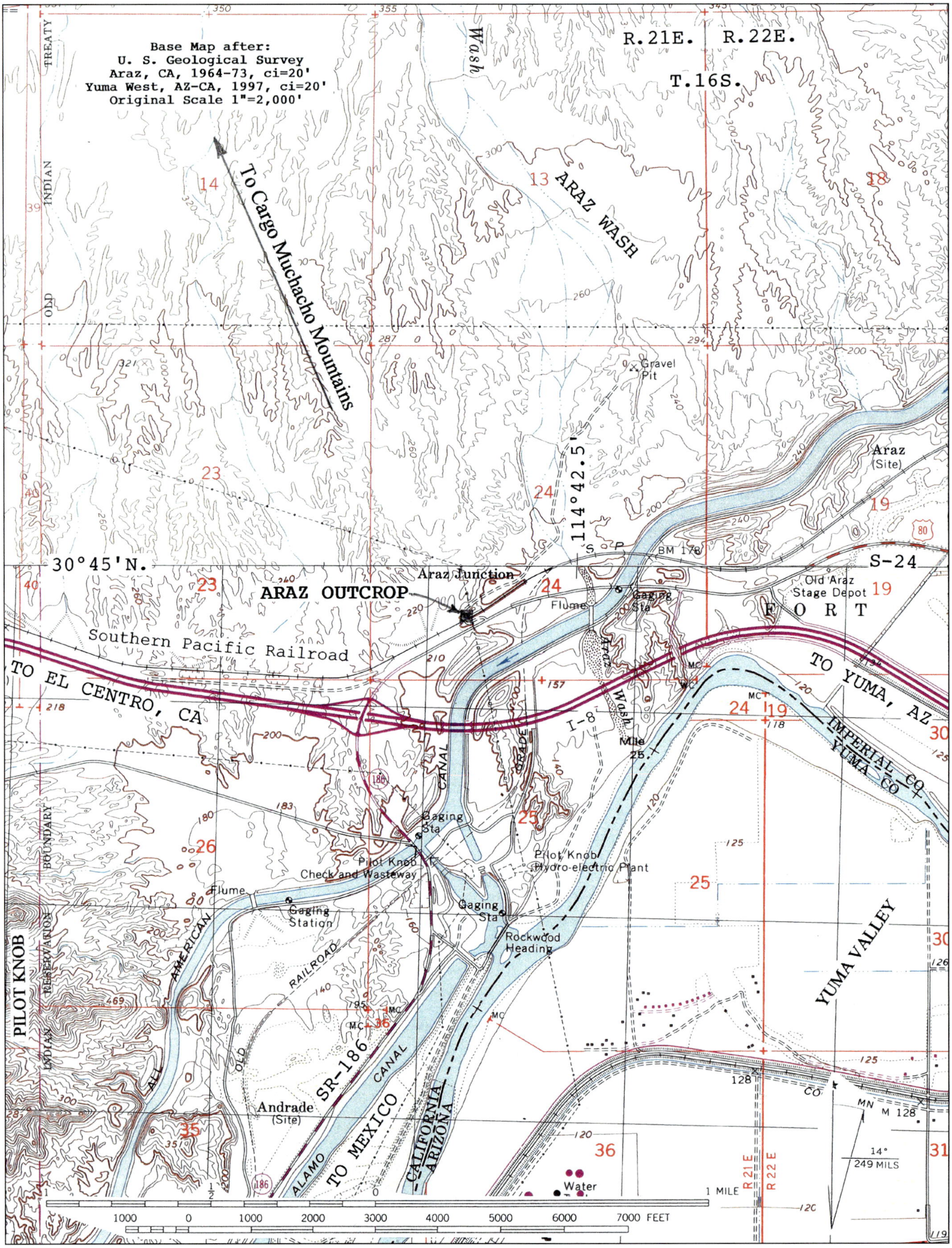

Map 1. *Portion of U. S. Geological Survey 7½' topographic maps, Araz, CA, and Yuma West, AZ-CA. Scale 1"=2,000', contour interval 20'.*

LOCATION AND ELEVATION

The Araz outcrop is located on the east side of an unnamed dry wash, immediately north of the railroad underpass (Photo 1). It is approximately 4.7 miles west of Winterhaven, California, and about 0.3 mile north of Interstate 8 (Map 1). It is situated in the NW¼, SE¼, SW¼, Section 24, T16S, R21E, San Bernardino Base and Meridian. Google Earth GPS coordinates are: 32°44.88′N, 114°42.89′W. On Google Earth, the elevation is about 203 feet above mean sea level at road level.

Photo 2. *The Yuma crossing. Southern Pacific Railroad and Old U.S. Highway 80 steel truss bridges span the Colorado River (lower left). Modern reinforced concrete truss bridge beyond, carries four lanes of Interstate-8 traffic across the Arizona/California state line. View to southwest. Photo by W.J. Elliott, 9-21-11.*

EARLY HISTORY AT AND NEAR THE CONFLUENCE OF THE COLORADO AND GILA RIVERS—THE YUMA CROSSING

Standing at the Araz outcrop, one is surrounded by more than 200 years of western Europe's adventures in the American Southwest. Much of this early history pivots around crossing the Colorado River at its confluence with the Gila River, near present-day Yuma, Arizona. This historical narrows, where hard, dense, middle Proterozoic gneissic rock (Marshall, 2012) is exposed on each side of the river, is commonly referred to as *The Yuma Crossing* (Photo 2).

The following snippet stories begin in the late 1700s, with Spain's desire to colonize Alta California, and ends with a reminder of the "Greatest Generation's" determination to win a war (Brokaw, 1998). General Patton's desert training area, which extended 190 miles, from Pilot Knob (just west of Yuma) to Searchlight, Nevada, was used to prepare young American troops for combat against German ground forces in North Africa's harsh desert environment. A wounded, but resolute America simultaneously won wars on two fronts, Europe and the South Pacific, in only 3¾ years: December 7, 1941, through August 15, 1945.

Juan Bautista de Anza Expedition

In 1769, the Spanish began efforts to colonize Alta California. Prevailing winds and an adverse California Current made, at best, a journey by sea long and difficult. Hence, a land route was explored by Juan Bautista de Anza. He left at what is now Tubac, Arizona, on January 8, 1774, and crossed the Colorado River at its confluence with the Gila River and continued on to Monterey, the capital of Alta California, arriving on April 19, 1774. His expedition included 3 padres, 20 soldiers, 11 servants, 35 mules, 65 head of cattle, and 140 horses (Guerrero, 2006; Trafzer, 1980, pp. 15–18; Wikipedia, 2011 "Juan Bautista de Anza"). My, my, how customs, political alliances, and cultural norms have changed, not to mention how much more efficiently and quickly we can now travel this same route—in a day!

Butterfield Overland Mail and Stage

The Butterfield Overland Mail and Stage carried United States mail between Saint Louis, Missouri, and San Francisco, California, from September 1857 through March 1861. The route maintained numerous outposts and stations along the way, including the still-standing, but barely, Araz stage depot. Passengers could join in on this 22-day grueling journey for $200 one way! The "oxbow" southern route through Yuma, Arizona, and across the Colorado River, avoided winter weather to the north but was plagued by conflicts with Native Americans. This service ended abruptly with the onset of the American Civil War in April, 1861 (Brown, 1923, p. 6; Dowd, 1960, p. 8; Ormsby, 1942; Trafzer, 1980, pp. 97–111; Wikipedia, 2011, "Butterfield Overland Mail").

The remains of the old Araz stage depot adobe are located approximately 1.1 miles east of the Araz outcrop, on the south side of Imperial County Highway S-24 (old U.S. Highway 80; Map 1, Photos 3, 4 and 5). Other more well-known Butterfield stage stops are located in San Diego County, along San Diego County Highway S-2, at Vallecito Stage Station County Park, at Warner Springs (in Henshaw Valley), and at Oak Grove along State Route 79, southeast of Aguanga.

Photo 3. *Old Araz stage depot, circa 1946(?). View to southwest. University of California Davis Special Collections, 2011. Reprinted with permission (from Dowd, 1960, p. 8.).*

Photo 4. *Old Araz stage depot, constructed in 1856, circa 1950s(?).*

Photo 5. *Melting remains of the old adobe Araz stage depot. View to southwest. Photo by W.J. Elliott, 8-31-2011.*

Southern Pacific Railroad

Railroad tracks were laid and completed in 1877 between Los Angeles, California, and Yuma, Arizona, crossing the Colorado River at Yuma (Brown, 1923, p. 7; Morton, 1977, p. 29; Nelson, 2006, pp. 32–34; Trafzer, 1974, p. 28; Trafzer, 1980, pp. 112–121). The railroad's Araz Junction is located about 0.2 mile northeast of the Araz outcrop (Map 1). At this point, a former spur track split off the main stem, going south-southwesterly toward Mexico (old railroad grade on Map 1). All that remains at this location is a main-line change from single- to double-track. Lacking tracks and ties, the old railroad spur grade is badly eroded and mostly obscured by desert brush.

Journeying approximately 1¼ miles further northeast along the Southern Pacific tracks, one comes to the site of a former railroad outpost named Araz (Map 1). In 2011, a few tamarisk trees, some hardy desert brush, and deteriorating portland cement concrete foundations are all that remain to mark the location of this former way-station.

Old Plank Road

Built in 1915, the old one-lane plank road provided a relatively safe, though bumpy, way to cross the Algodones sand dunes when traveling in motorized vehicles between San Diego, California, and Phoenix, Arizona via Yuma, Arizona. This approximately 7 mile long collection of boards (Brown, 1923, p. 145; Nelson, 2006, pp. 35–38; Trafzer, 1974, p. 30; Rube, 1996) was used because at the time it was thought impractical to lay either portland cement concrete or asphaltic concrete on a foundation of loose, shifting sand [fine-grained, mostly equidimensional (rounded) and frosted quartz sand].

By 1926, a 20-foot wide, two-lane paved road replaced the difficult to maintain, and sometimes hard to find, plank road (Wikipedia, 2011, Old Plank Road). The 1926 road was replaced a short time later by U.S. Highway 80, a "modern" two-lane paved road. And now, since the mid 1960's, one can travel (at warp speed) between San Diego and the Phoenix area (Casa Grande) on a four-lane divided highway, Interstate 8.

Site of Camp Pilot Knob

"Camp Pilot Knob was a unit of the Desert Training Center established by General George S. Patton, Jr., to prepare American troops for battle during World War II. It was the largest military training ground ever to exist. At the peak of activity here at Pilot Knob, June-December 1943, the 85th Infantry Division, and the 36th and 44th Reconnaissance Squadrons of the 11th (Mechanized) Calvary trained for roles in the liberation of Europe, 1944–45." Quote from: California Registered Historical Landmark No. 985. The plaque was placed by the California State Department of Parks and Recreation in Cooperation with the Bureau of Land Management and Squibob Chapter, E. Clampus Vitus, November 10, 1990. It is located on the west side of Sidewinder Road, approximately 0.3 mile north of Interstate 8.

THE ARAZ OUTCROP

This approximately 6- to 8-foot thick, 200-foot wide wrap-around outcrop is overlain by nearly flat-lying, poorly- to moderately well-consolidated claystones, siltstones, and sandstones which are presumed to be members of the older Colorado River alluviums of Metzger and others (1973) and Olmsted and others (1973). The stratigraphic base of the Araz outcrop was not found. A detailed description of overlying, and geologically younger, Colorado River deposits is beyond the scope of this effort.

Lithology

This well-sorted (poorly-graded), well-rounded dry outcrop (Photo 6) is described, using a 10× triplet hand lens in bright sunlight, as falling in the vicinity of a very coarse silt to very, very fine sand. Somewhere in the neighborhood of at least half of a disaggregated sample is easily blown away in a slight breeze, suggesting a high fines (silt and clay-sized particles) content.

Composition appears to be largely clear to milky, equidimensional rounded quartz grains, ±3% biotite (flakes), ±1% vari-colored (red, green, black) grains, ±3% clay and clay-sized fragments, and very few heavy minerals. No lithic fragments were noted. Less than about 0.1% of the sample was attracted to a powerful magnetite magnet. A quick bubbly reaction was observed with 10% hydrochloric acid.

Color varies in the range of pale yellow-ochre to pale grayish-tan, with approximately 10% to 30% streaks of discontinuous rusty orange-brown weathering fronts along bedding. It is very thinly bedded, poorly indurated (not cemented), and easily friable between the thumb and fingers. In water, it disintegrates rapidly into a slushy mush.

Structure

On average, bedding was observed to strike N15°E, and dip down 20° in an east-southeasterly direction (Photo 7).

Photo 6. *Close-up of Araz outcrop showing thinly bedded coarse silt to very very fine-grained sand. Bedding dips away from the camera. View to northeast. Photo by W.J. Elliott, 2011.*

Photo 7. *Base of overhang illustrates approximately 20° east-southeasterly dipping Araz Outcrop bedding. Thinly bedded Araz sediments can be seen behind the 5 ft. long Jacob staff. View to northeast. Photo by W.J. Elliott, 2011.*

Age

At this point, about all that can be said about the age of the Araz outcrop is that it is older than overlying deposits of clay, silt, sand and gravel. In a broad regional context, these coarse silts to very very fine sands appear to be geologically relatively young, certainly no older than Neogene (<23 million years).

Genesis of the Araz outcrop

Tectonic Tilting(?): At *first blush*, viewed from the distance of the Southern Pacific Railroad underpass, the Araz outcrop looks for all the world like tectonically tilted beds. Flat-lying younger deposits of clay, silt, sand, and gravel overlie these beds with an apparent angular discordance of about 20° (Photos 1 and 7).

If this is the case, these relatively uniformly thinly bedded coarse silt to very very fine sands may well have originated as an over-bank deposit on a broad floodplain or in the central portion of a quiet-water basin or trough. Can you think of some other environments of deposition where these fine-grained sediments might have accumulated?

Subsequent to this relatively low-energy sedimentation / accumulation cycle, tectonic (and maybe even gravitational) forces tilted these beds to their present inclination of approximately N15° E, 20° ESE. Erosion then planed off a relatively smooth surface upon which younger sediments were deposited.

Cross-Bedding(?): On the other hand, at *second blush*, viewed up close and personal along the entire length and height of the outcrop, one is lead to wonder if the Araz outcrop is perhaps an exposure of sedimentary bedding inclined relative to the principal surface of accumulation, suggesting foresets or cross-bedding (Potter and Pettijohn, 1963, p. 69). If this is the case, these uniformly thinly bedded coarse silt to very very fine sands may well have originated in a fluvial setting as current bedding, for example, as a point bar or within a deltaic environment. Can you think of some other environments of deposition where this fine-grained sediment might have accumulated as inclined bedding dipping approximately 20° in an east-southeasterly direction, indicating a generally easterly direction of water flow?

Subsequent to this relatively high-energy sedimentation and accumulation cycle, erosion planed and smoothed the top of these cross-beds, followed by yet another cycle of sedimentation and accumulation of clay, silt, sand, and gravel deposited on the Araz outcrop surface, with an angular discordance of approximately 20°. At this juncture it is presumed the overlying clay, silt, sand, and gravel beds are to be correlated with the older Colorado River alluviums of Metzger and others (1973) and Olmsted and others (1973).

General Comments: Broad regional uplift and active erosion of all these now inactive depocenters has exposed the outcrop as we see it today. As you explore the outcrop, see what geologic evidence you can find to support either tilted bedding or cross-bedding, or perhaps some variation thereof, keeping in mind the principal of multiple working hypotheses (Chamberlain, 1890). Beyond the distinctive, relatively uniformly thinly bedded coarse silts to very very fine sands, can you find any primary or secondary sedimentary structures, or fossil or trace-fossil remains, that might provide some clues to the Araz outcrop genesis?

DISCUSSION

Bouse Formation

The "Pliocene" Bouse Formation, named and described by Metzger (1968, p. D135) was at that time believed to be a record of sedimentation in an embayment of the proto-Gulf of California that extended at least as far north as about Needles, California along the Colorado River Corridor. It was further noted that "... a correlation between the Bouse Formation and some of the marine sediments of the Imperial Valley undoubtedly exists" (Metzger, 1968, p. D133). Recent work by Poulson and John (2003), and Roskowski and others (2010), suggest an ~5 million year old (latest Miocene to early Pliocene) age for the Bouse Formation.

As described by Metzger (1968, p. D132), "The silt and fine-sand layers [of the Bouse Formation] are commonly grayish orange, or very light gray to very light pink. Most of the sand is only weakly compacted or cemented; however, some thin sandstone layers are well cemented with calcium carbonate. Some of the light-gray sand layers contain numerous black minerals that result in a 'salt and pepper' appearance."

Smith (1970, p. 1417) concluded that, "Faunal evidence from a large area along the lower Colorado River indicates that the area, during the Pliocene, was connected with the sea." Though somewhat confusing and ambiguous, a sparse fauna of foraminifera, mollusks, barnacles and ostracodes collected from the Bouse Formation suggest a marine to brackish water environment of deposition. Additionally, in comparing the nearby Salton Trough Basin with Bouse basins and embayments, Smith (1970, p. 1417) noted that: "Paleoecologically the two areas are quite different, the Imperial Formation assemblage indicating deeper relatively open conditions, while the Bouse Formation assemblages indicate a shallow, more restricted environment."

Later work by Roskowski and others (2010, p. 1625) has shown, by comparison of strontium, carbon, and oxygen isotopic ratios, that the Colorado River corridor's Bouse Formation is, in fact, of fresh water origin. Their "results reinforce the conclusion that the Bouse Formation was deposited in a descending series of basins connected by overflow of Colorado River water."

That said, little compelling evidence remains to connect the southern reaches of the string of fresh water Bouse Formation lakes with the clearly open-ocean, ~6.3/5.5 million to ~4.2 million year old marine Imperial Formation, exposed extensively on the western side of the Salton Trough (Deméré, 2006; Dorsey, 2006).

Based on descriptions in the literature and my field observations (including the type reference section of Metzger (1968, p. D130), finding a clear correlation between the Araz outcrop and the Bouse Formation seems elusive at best. You be the judge.

Chemehuevi Formation

Mappable surface exposures (units D and E of Metzger and others, 1973, p. G29) were collectively referred to by earlier workers, including Longwell (1936, p. 1444), as Chemehuevi Formation.

Following work in the late 1960s and early 1970s by Metzger, Olmsted, and others (see references), recent workers (Agenbroad and others, 1992; Malmon and Howard, 2007; Malmon, 2009; Malmon and others, 2011) have for the most part, dropped references to units D and E and have returned to the use of the Chemehuevi Formation as one unified mappable unit (Dunbar and Rodgers, 1957. p. 259).

Longwell. Chester Longwell (1936, p. 1443–1445) notes that the slightly consolidated Chemehuevi Formation consists chiefly of cream-colored to pale yellow sand, silt, and clay. In places it is buff colored, and exceptional layers are light-brown, or even reddish. Eroding to a badlands topography, these horizontally bedded fine-grained sediments appear to extend continuously along the Colorado River for hundreds of miles.

A cream-colored upper member forms steep slopes, and consists chiefly of prominently cross-bedded coarse sand with a few scattered layers of hard clay. This is likely unit E of Metzger and others (1973). Underlying this member, a buff-yellow unit stands as cliffs where clay and silt predominate over fine sand. This is likely unit D of Metzger and others (1973).

Metzger and others. Metzger and others (1973, p. G20-G31, Figure 11, Plate 1) divide post-Bouse Formation Colorado River deposits into younger and older alluviums (p. G20). Younger alluvium, representing the latest aggradation of the Colorado River, is identified as Holocene (p. G28).

Older alluviums are divided, from oldest to youngest, into units A through E (p. G21-G25). This alphabetic nomenclature was apparently a convenient way to separate and identify the various Colorado River deposits, especially in drill logs. Based on a sparse vertebrate fauna, units D and E are thought to be middle to late Pleistocene, while units A and B are probably Pliocene and Pleistocene. Unit C, the piedmont gravels, lie somewhere in between (Metzger and others, 1973, p. G28). Units A, B, and C are not addressed herein.

Unit D is composed of two facies (Metzger and others, 1973, p. G24–25):

> Facies 1. A locally-derived *basal gravel* composed of pebble- to cobble-sized clasts that vary from sub-angular to sub-rounded, as well as rounded to well-rounded clasts imported by the Colorado River, overlain by
>
> Facies 2. *Interbedded sand, silt, and clay* that typically exhibits a tan color and is often times tinted with a slight pinkish to reddish cast. Darker shades of brown define some clay beds.

Save for initial depositional dip, flat or nearly flat-lying Colorado River unit D beds suggest an absence of measurable structural deformation since deposition (Metzger and others, 1973, p. G25).

Unit E is composed of two facies:

> Facies 1. Tan, unconsolidated, fairly well sorted, medium-grained sand with scattered rounded to well-rounded pebbles deposited by the Colorado River.
>
> Facies 2. Gravel deposited by local tributaries.

Malmon and Howard. Malmon and Howard (2007, p. 57) note that "A distinctive set of fine-grained deposits [Chemehuevi Formation] occurs throughout the lower Colorado River Valley, extending from just below the mouth of the Grand Canyon to well into the river delta below Yuma, Arizona.... Below Parker, they occur in isolated outcrops but also underlie large continuous terraces such as... Yuma Mesa. These deposits form the Chemehuevi Formation of Longwell (1936).... The deposits usually consist of a lower, cliff-forming unit composed mostly of fine sand, silt, and clay (unit D, in the terminology of Metzger and others, 1973), and an overlying sub unit of loose, poorly bedded, medium to coarse sand that forms gentle slopes (unit E) [of Metzger and others, 1973]."

Olmsted and others. Olmsted and others (1973, p. H45–H53, Plate 3) identified terrace deposits underlying the western borders of Yuma Mesa as older alluvial deposits of the Colorado and Gila rivers. In an effort to lay eyes and fingers on an outcrop of Chemehuevi Formation, a sample of what is probably unit E (of Metzger and others, 1973) was taken from the western edge of Yuma Mesa. Located approximately seven miles south of Yuma, and about a mile southwest of the U. S. Marine Corps Air Station, it was found to be a thinly-bedded, well-sorted, poorly graded equidimensional, rounded, fine-grained quartz sand. One would be hard pressed to equate these sediments with the Araz outcrop, but should have no trouble seeing the similarity with overlying deposits.

Comment: Based on field observations and descriptions in the literature, finding a clear correlation between the Araz outcrop and the Chemehuevi Formation seems elusive at best. You be the judge.

The Araz outcrop — Bouse Formation or Chemehuevi Formation?

Spencer and others (2010, p. 3) consider the outcrop mapped as Bouse Formation at the United States Army Proving Grounds (Map 2; Olmsted, 1972), located approximately 16 miles northeast of Yuma, Arizona, to be Chemehuevi Formation (Colorado River units D and E of Metzger and others, 1973, p. G29). Equally, Spencer and others consider drilled and logged subsurface sediments identified as Bouse Formation in the Yuma Valley and Yuma Mesa Sedimentary Basin (Olmsted and others, 1973, Figure 15, p. H43, Plate 3) to be Chemehuevi Formation (Colorado River units D and E of Metzger and others, 1973, p. G29).

Based on a review of a few Araz outcrop photographs, Dr. Robert Reynolds, of LSA Associates, Inc. in Riverside, California, commented that: "… the Araz sediments look a little too 'pink' to be Bouse — more the color of the Chemehuevi beds near Parker [Arizona]" (e-mail from Dr. Robert Reynolds, dated June 6, 2011.) If this is the case, Araz outcrop sediments described herein may well be correlated with Chemehuevi Formation (Colorado River units D and E).

On the other hand, if the Araz outcrop is correlative with the Bouse Formation, it would push Olmsted and others' (1973, Figure 15, p. H43) inferred sub-crop margin of Bouse Formation several miles northwest (Map 2). If not, nothing changes.

Either way, this interesting and errant outcrop is a puzzle with answers awaiting further investigation. So, the decision is up to you. Is it Bouse Formation or is it Chemehuevi Formation? To go way, way out on a limb, the Salton Trough's Imperial Formation could even be added to one's list of multiple working hypotheses (Chamberlain, 1890). Or, perhaps, is it an artifact of something none of us has ever seen before — a new, and heretofore undescribed, unit or sub-unit?

And, as a parting note, perhaps there are other as yet undiscovered outcrops of this nature in southeastern Imperial County and southwestern Yuma County. You may be the lucky prospector.

FURTHER PONDERINGS REGARDING THE ARAZ OUTCROP

Some of the following text is intended to create a bit of uncomfortable cerebral pressure for the novice or wanna-be geologist. If you are an old-timer, answers to some questions will be easy. But wait, others may keep you awake at night. Have fun — isn't that why most of us stumbled into geology in the first place?

Whether you have decided the Araz outcrop is of tectonic origin (tilting), or simply an artifact of running water (cross-bedding), the following questions will be equally applicable.

1. Where did the sediments come from (Dunbar and Rodgers, 1957, pp. 3–4)? What was their provenance? Without sophisticated laboratory analyses of the sediments and possible target sources, only a guess seems possible at this point.

Whether from igneous, sedimentary or metamorphic origins or some combination thereof, it would appear the source(s) was likely to be quartz-rich, with biotite, garnet(?) and sphene(?) as accessory minerals. If present at the source(s), lithic fragments, feldspars and hornblende did not survive the journey — either through chemical weathering to clay minerals and/or mechanical abrasion of these cleavable minerals.

Originating from igneous terrane, one might expect light-colored acidic rock types to include any combination of granite, syenite, rhyolite porphyry, or trachyte porphyry. From sedimentary terrane, one would most likely expect something akin to a quartz-rich sandstone. And, if the source were metamorphic terrane, quartz-rich gneiss or schist, with or without quartzite might be expected. Can you expand on this?

2. By what means were these sediments transported from their source or sources to the site of deposition? Possibilities would seem to include one or more variations of the following (Dunbar and Rodgers, 1957, pp. 4–27): a. Aqueous transport in solution, suspension, or by bottom traction (rolling and/or saltation), including turbidity currents? b. Wind? c. Ice? d. Gravity? Can you add some ideas of your own?

3. And here are a few more questions to agitate the gray matter.

a. What and where were the sources of water to the basin?

b. If not a landlocked basin, where did the water discharge?

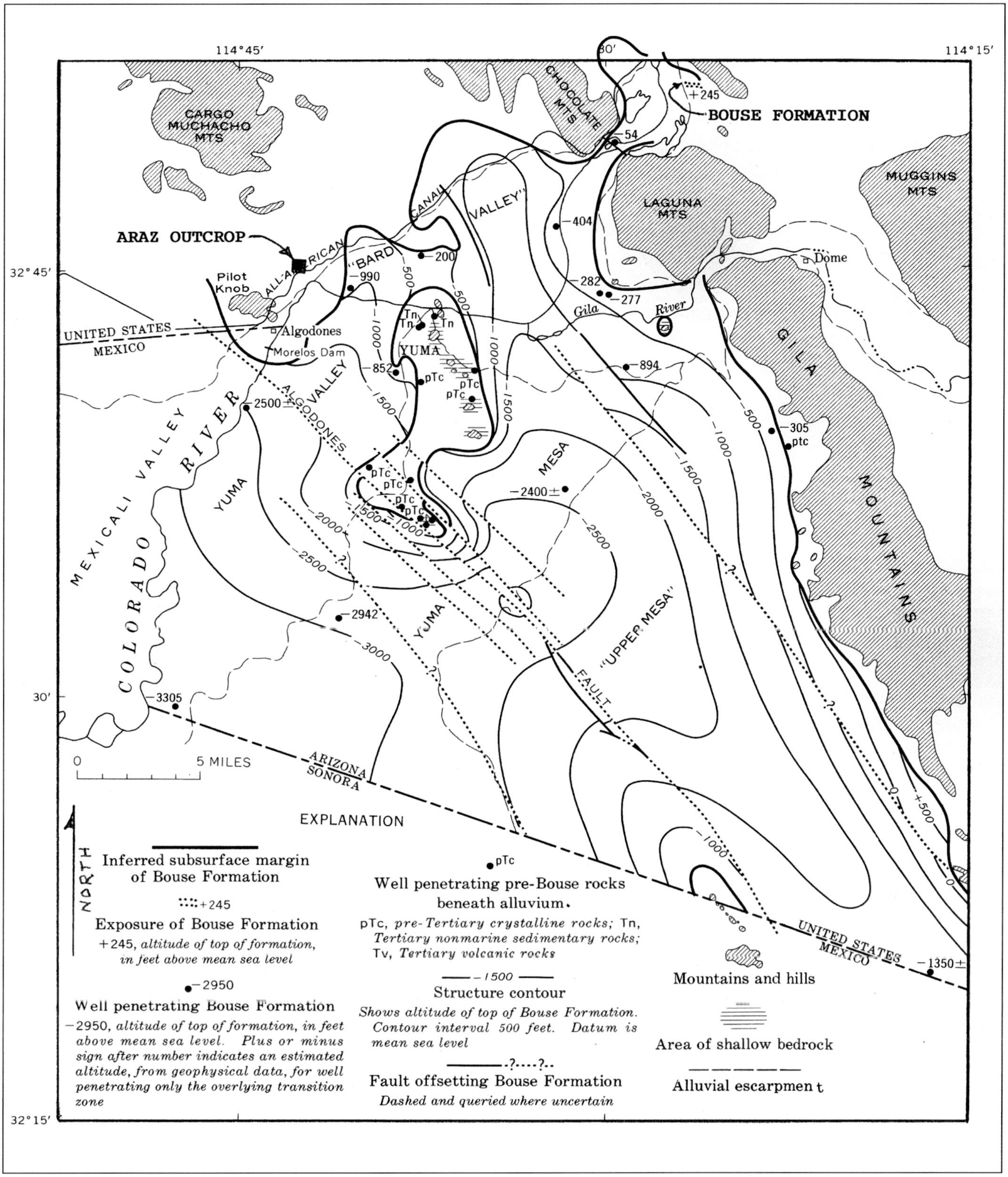

Map 2. *Figure 15 from Olmsted and others (1973) showing Bouse Formation outcrop north of the Laguna Mountains, and inferred extent of Bouse Formation sub-crop. If the Araz outcrop is correlated with Bouse Formation, then the sub-crop boundary would need to be moved about 2 miles northwesterly.*

c. Since the Araz outcrop is a small piece of a once active depositional environment, what happened to change the geologic environment from active deposition to active erosion?

d. What did the depositional basin look like?

e. Where were the geographical basin boundaries?

f. What was the composition of the basin borders?

g. Where are the basin boundaries now? Were they eroded away? Were they faulted away during the past 5 million years as the Gulf of California opened (Dillon and Ehlig, 1997; McKibben, M. A., 2006)? Do artifacts of those boundaries still exist, perhaps buried under modern sediment, and/or are they distant mountains/hills that currently bear little or no resemblance to their architecture at the time of Araz deposition?

h. How much geologic time is represented by the apparent angular discordance between the easterly dipping Araz outcrop and overlying, nearly flat-lying, presumed undifferentiated older Colorado River sediments?

i. With uplift and exposure to sub-aerial atmospheric conditions, how long did it take oxygenated terrestrial waters circulating through Araz outcrop sediments to chemically decompose iron-rich biotite (and/or other iron-bearing minerals) to produce orange or brown staining along bedding planes?

j. What relatively recent natural and/or anthropogenic processes led to the present exposure of Araz outcrop sediments? Clearly, as the adjacent desert arroyo widened its channel with each gulley-washer, more and more obscuring debris has been washed away.

What about construction of the Southern Pacific Railroad grade and bridge over the wash? Could construction of the low portland cement concrete wing-wall, between the train bridge and the Araz outcrop have hastened day-lighting the exposure? You will have to look hard for the wing-wall, it is almost hidden behind brush, and is partly covered with slope wash sediments.

And finally, pondering all of the above, it is curious to wonder how many people have passed by this exposure without observing its distinctive, out-of-place character in an otherwise barren desert landscape of rock-hard clay beds, cross-bedded medium to coarse sands, and abandoned gravel-surfaced (desert pavement) outwash fans and terraces.

ACKNOWLEDGEMENTS

A special thanks goes to Todd Wirths for encouraging me to pursue this effort. Carole Ziegler's encouragement and careful, critical manuscript review resulted in a much improved presentation. Joe Corones provided me with the 1994 letter from Dr. Anna Buising which sparked this challenge. Charlie Lough and Lowell Lindsay steered me toward helpful references. Dr. Michael Conway, Dr. William Presch, Dr. Robert Reynolds, and Mr. Mark Roeder were all very helpful with telephone conversations, e-mail correspondence, and assistance with obtaining hard-to-locate and out-of-print publications. John Glover kindly provided access to the Bouse Formation Outcrop of Olmsted (1972) and Olmsted and others (1973) on the United States Army Proving Grounds, north of Yuma, Arizona. JR and George Morgan kept me scrambling with obscure references, thought provoking discussions, and warm summertime adventures to the Colorado Desert. Marion Champion, Imperial Irrigation District, granted permission to reprint the Old Araz Stage Depot image on page 8 of Dowd, 1960; Photo 3. Marion Champion and Tony DeZego, Imperial Irrigation District led me to John Skarstad at U. C. Davis, Special Collections, and the black and white photo of the old Araz Stage Depot (Photo 2). John Skarstad, U. C. Davis Special Collections, assisted me with obtaining a scanned copy of the old Araz stage depot (Photo 3).

Following our mid-August 2011 SDAG meeting, Dr. Monte Marshall and I drove to, and camped at, the top of Indian Pass in the Chocolate Mountains of Picacho, arriving at 1 am. The purpose of this insane journey was to beat the summer heat on a just-after-sunrise jaunt to look for Bouse Formation sediments near the mouth of Gavilan Wash. Stimulating discussions kept us both awake and made the less-than-24-hour trip almost bearable.

Additionally, I would be remiss if I failed to thank my most helpful and understanding wife, Wendy. Without her computer tech support, patience, and frequent offerings of something to eat and drink, this effort would not have been such a fun and interesting adventure!

REFERENCES

Agenbroad, L. D., Mead, J. I., and Reynolds, R. E., 1992, Mammoths in the Colorado River Corridor: San Bernardino County Museum, Spec. Pub. 92-2, pp. 104–106.

Brokaw, T., 2004, The Greatest Generation: Random House, New York, 412 p.

Brown, J. S., 1923, The Salton Sea Region, California: U. S. Geol. Survey Water-Supply Paper 497, 292 p.

Buising, A., 1994, Personal correspondence to Mr. Joe Corones, San Diego, CA, regarding the Bouse Formation, letter dated January 31, 1994, 1 p.

Chamberlin, T. C., 1890, The Method of Multiple Working Hypotheses: Science (old series) v. 15, p. 92–96; reprinted 1965, v. 148, pp. 754–759.

Deméré, T. A., 2006, The Imperial Sea — Marine Geology and Paleontology, in Jefferson, G. T., and Lindsay, L., eds., Fossil Treasurers of the Anza-Borrego Desert — California State Parks, Colorado Desert District Anza-Borrego Desert State Park, The Anza-Borrego Foundation and Institute: Sunbelt Publications, San Diego, California, pp. 30–41.

Dillon, J. T., and Ehlig, P. L., 1997, Displacement on the Southern San Andreas fault, in Baldwin, J., Lewis, L., Payne, M., and Roquemore, G., eds., Southern San Andreas Fault — Whitewater to Bombay Beach, Salton Trough, California: South Coast Geological Society, Santa Ana, California, Field Trip Guidebook No. 25, November, 1997, pp. 188–205.

Dorsey, R., 2006, Stratigraphy, Tectonics, and Basin Evolution in the Anza-Borrego Desert Region, in Jefferson, G. T., and Lindsay, L., eds., Fossil Treasures of the Anza-Borrego Desert — California State Parks, Colorado Desert District Anza-Borrego Desert State Park, The Anza-Borrego Foundation and Institute: Sunbelt Publications, San Diego, California, pp. 90–104.

Dowd, M. J., 1960, Historic Salton Sea: Imperial Irrigation District, El Centro, California, 32 p.

Dunbar, C. O., and Rodgers, J., 1957, Principles of Stratigraphy: John Wiley and Sons, Inc., New York, 356 p.

Guerrero, V., 2006, The Anza Trail and the Settling of California: Santa Clara University, Santa Clara, California, Heyday Books, Berkley, California, 223 p.

Longwell, C. R., 1936, Geology of the Boulder Reservoir Floor, Arizona-Nevada: Geol. Soc. Bull., v. 47, pp. 1393–1476.

Malmon, D. V., 2009, The Chemehuevi Formation — A Geologic Example of Extraordinary Sediment Loading in the Colorado River During the Late Pleistocene [abs.]: Geol. Soc. America, Annual Meeting, Portland, Oregon, 1 p.

Malmon, D. V., and Howard, K. A., 2007, Overview — The Chemehuevi Formation Along the Lower Colorado River: San Bernardino County Museum, 2007 Desert Symposium, pp. 57–60.

Malmon, D.V., Howard, K. A., House, P. K., Lundstrom, S. C., Pearthree, P. A. , Saran-Wojcicki, A. M., wan, E., and Wahl, D. B., 2011, Stratigraphy and Depositional Environments of the Upper Pleistocene Chemehuevi Formation Along the Lower Colorado River: United States Geological Survey Professional Paper 1786, 95 pp.

Marshall, M., 2012, The Unappreciated Role that the Amalgamation and Break-Up of Supercontinents has Played in the Geologic Evolution of the Western Margin of North America, in Snyder, S., Miller-Hicks, B., and Miller, J., eds., Palms to Pines — Geological and Historical Excursions through the Palm Springs Region, Riverside County, California: San Diego Association of Geologists and South Coast Geological Society, distributed by Sunbelt Publications, San Diego, California, pp. 49–72.

McKibben, M. A., 2008, Introduction to the Salton Trough Rift, in Reynolds, R. E., ed., Trough to Trough — The Colorado River and the Salton Sea: Calif. State Univ., Desert Studies Consortium and LSA Assoc., Inc., Riverside, CA., Field Trip Guide and 2008 Desert Symposium Proceedings, pp. 91–95.

Metzger, D. G., 1968, The Bouse Formation (Pliocene) of the Parker-Blythe-Cibola Area, Arizona and California: U. S. Geol. Survey Prof. Paper 600-D, pp. D126–D136.

Metzger, D. G., Loeltz, O. J., and Irelna, B., 1973, Geohydrology of the Parker-Blythe-Cibola Area, Arizona and California: U. S. Geol. Survey Prof. Paper 486-G, 130 p.

Morton, P. K., 1977, Geology and Mineral Resources of Imperial County, California: Calif. Div. Mines and Geology, County Report 7, 104 p., Pl. 1 — Geol. Map, scale 1:250,000, contour interval 200′.

Nelson, R., 2006, Images of America, Early Yuma: Arcadia Publishing, San Francisco, 127 p.

Olmsted, F. H., 1972, Geologic Map of the Laguna Dam 7.5-Minute Quadrangle, Arizona and California: U. S. Geol. Survey Map GQ-1014, scale 1″=2,000′, contour interval 20′.

Olmsted, F. H., Loeltz, O. J., Irelna, B., 1973, Geohydrology of the Yuma Area, Arizona and California; U. S. Geol. Survey Prof. Paper 486-H, 227 p.

Ormsby, W. L., 1942, The Butterfield Overland Mail: The Huntington Library, San Marino, California, Wright, L. H., and Bynum, J. M. eds., 179 p.

Potter, P. E., and Pettijohn, F. J., 1963, Paleocurrents and Basin Analysis: Springer-Verlag, Academic Press, Inc., Berlin, 296 p.

Poulson, S. R., and John, B. E., 2003, Stable Isotope and Trace Element Geochemistry of the Basal Bouse Formation Carbonate, Southwestern United States: Implications for the Pliocene Uplift History of the Colorado Plateau: Geol. Soc. America Bull., v. 115, no. 4, pp. 434–444.

Roskowski, J. A., Patchett, P. J., Spencer, J. E., Pearthree, P. A., Dettman, D. L., Faulds, J. E., and Reynolds, A. C., 2010, A Late Miocene-Early Pliocene Chain of Lakes Fed by the Colorado River: Evidence from Sr, C, and O Isotopes of the Bouse Formation and Related Units Between Grand Canyon and the Gulf of California: Geol. Soc. America Bull., v. 122, no. 9/10, pp. 1625–1636.

Rube, B. J., 1996, A Wooden Road Through the Hollow of God's Hand — Taming the American Sahara: Yuma Historical Society, Yuma, Arizona, 69 p.

Smith, P. B., 1970, New Evidence for Pliocene Marine Embayment along the Lower Colorado River Area, California and Arizona: Geol. Soc. America Bull., v. 81, pp. 1411–1420.

Spencer, J. E., Patchett, J. P., Roskowski, J. A., Pearthree, P. A., Faulds, J. E., and House, P. K., 2010, A Brief Review of Sr Isotopic

Evidence for the Setting and Evolution of the Miocene-Pliocene Hualapai-Bouse Lake System: U. S. Geol. Survey Workshop, Flagstaff, Arizona, Summer, 2010, 10 p.

Trafzer, C. E., 1974, The Yuma Crossing — A Short History: Yuma County Historical Society, Yuma, Arizona, 41 p.

Trafzer, C. E., 1980, Yuma, Frontier Crossing of the Far Southwest: Western Heritage Books, Inc., Wichita, Kansas, 151 p.

University of California, Davis, 2011, Special Collections, Eastman's Studio Originals Photographic Collection, Old Araz Stage Depot, near Yuma, Arizona, black and white image No. B-4076, circa 1946(?). Photo 3.

United States Department of the Interior, 1974, Earth Manual, A Water Resources Technical Publication: United States Department of the Interior, Bureau of Reclamation, second edition, 810 p.

United States Geological Survey, 1964–73, Araz, Calif., 7½′ topographic quadrangle map, scale 1″=2,000′, contour interval 20′.

United States Geological Survey, 1997, Yuma West, AZ-CA, 7½′ topographic quadrangle map, scale 1″=2,000′, contour interval 20′.

Photo 1. Colorado *River "B" gravel outcrop, approximately one mile south of I-8, on the west side of SR-186. (Photo by W. J. Elliott 2-21-2012)*

In search of Colorado River "B" gravel, northeastern Imperial County, California

With notes on Early Tertiary red beds

WILLIAM J. ELLIOTT

Consulting Engineering Geologist

P. O. Box 541, Solana Beach, California 92075

DR. MONTE MARSHALL

235 Quince Street, San Diego, California 92103

INTRODUCTION

OUR INTEREST IN AN ANCIENT COLORADO River course southwest of Cibola, Arizona, was piqued by a figure shown in a Journal of Geophysical Research paper by Sherrod and Tosdal (1991). The ancient river path is shown exiting the current Colorado River corridor at the mouth of Milpitas Wash between the Palo Verde and Midway mountains, and then approximately following present-day State Route 78 in a southwesterly direction toward Glamis and the Salton Trough (Figure 1). Other than optimistically searching for candidate valleys and water gaps for this hypothesized potential route on Google Earth images and a folder full of United States Geological Survey topographic maps, we wondered about the possibility of finding a few sheltered sedimentary deposits left behind by the river.

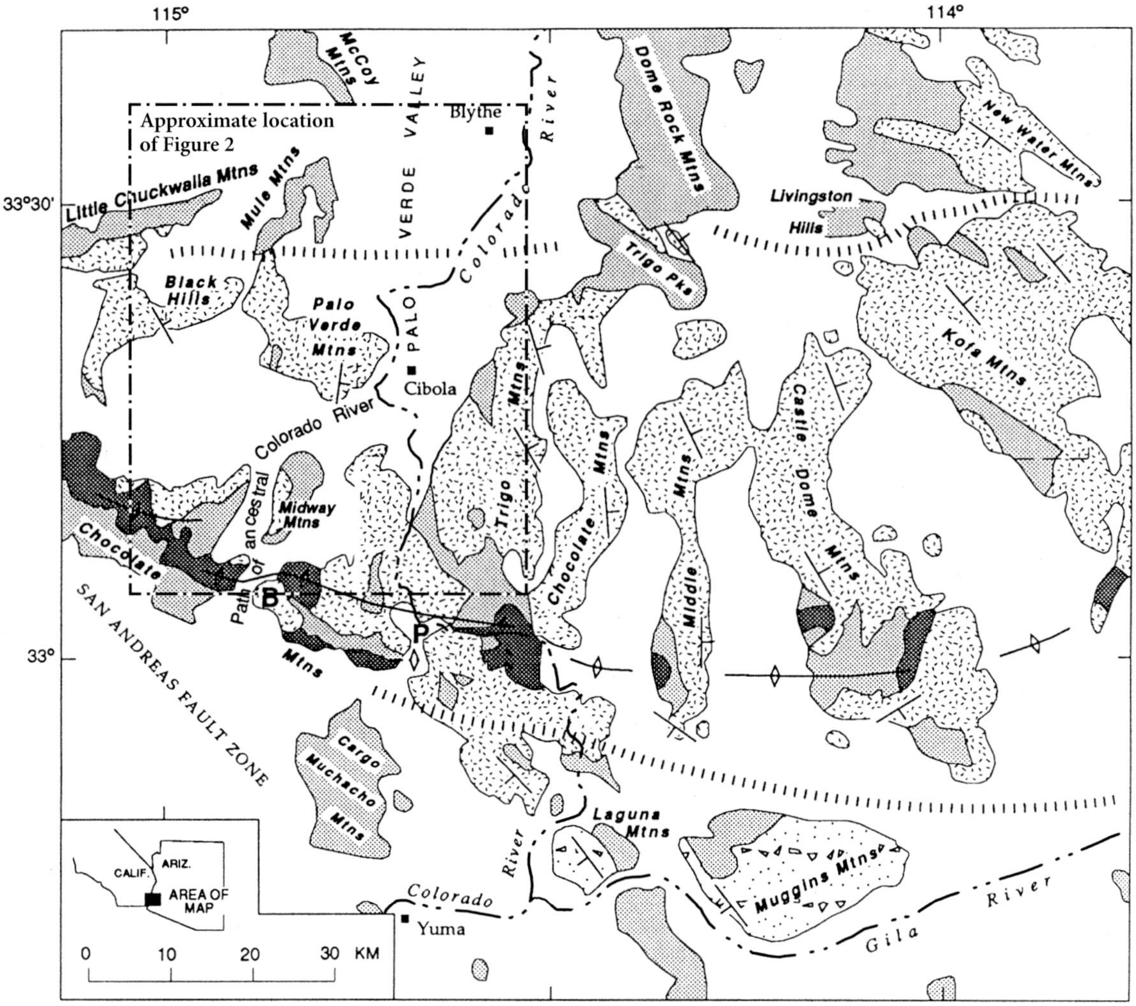

EXPLANATION

- Sedimentary rocks and deposits (Holocene to Miocene)—Includes middle Miocene fanglomerate deposited during extensional deformation
- Tertiary volcanic rocks (Miocene and Oligocene)—Mainly 32-18 Ma. Includes minor Tertiary granite
- Tertiary sedimentary and volcanic rocks (Miocene and Oligocene)—In Muggins and Laguna Mountains
- Orocopia Schist (Mesozoic)—Includes some large granitic intrusive masses of Mesozoic and Tertiary age in western Chocolate Mountains
- Meta-igneous and metasedimentary rocks (Mesozoic)—Includes Proterozoic metaplutonic rocks found mainly south of exposures of the Orocopia Schist
- Chocolate Mountains anticlinorium
- Northern and southern boundaries of structurally extended volcanic terrane discussed in text
- Generalized orientation of bedding in volcanic and sedimentary rocks
- **P** Picacho State Park
- **B** Black Mountain

Fig. 1. Location map showing generalized geology and geographic setting of the study area in southwestern Arizona and southeastern California. Data from *Jennings* [1977] and *Reynolds* [1988]. Chocolate Mountains anticlinorium from *Haxel et al.* [1985].

Figure 1. *Postulated path of Ancestral Colorado River (south of Palo Verde Mountains and west of Midway Mountains). Modified after Sherrod and Tosdal, 1991.*

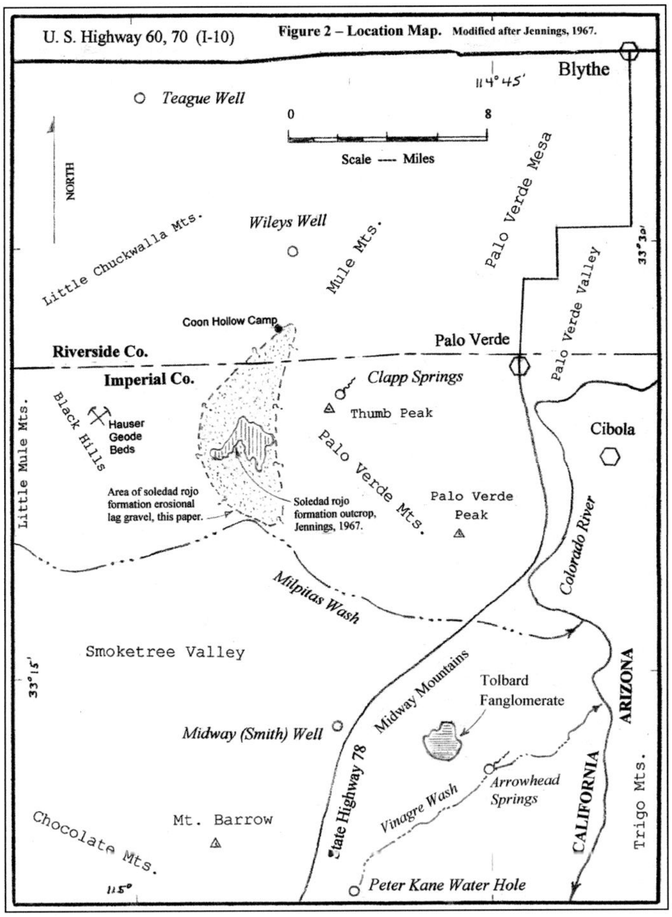

Figure 2. *Location Map. Approximate areal extent of gravel that appears very similar to Colorado River "B" gravel is shown with a stipple pattern and enclosed with a queried dashed line. A solid boundary surrounds a vertical striped pattern that delineates early Tertiary red beds (Jennings, 1967; Morton, 1967); here these strata are informally named Soledad Rojo formation. Modified after Jennings, 1967.*

Scattered remnant deposits of the approximately five million year old Bouse Formation (lake deposits) are present in the general area of the proposed ancient river route (Metzger and others, 1973, Plate 1; Spencer and others, 2008, Figure 2), so why not some leftover Colorado River deposits (Photo 1) to verify the presence of an ancient river path?

Our best bet, we thought, would be to look for far-traveled well-rounded chert and quartzite clasts, so common in Plio-Pleistocene Colorado River "B" gravel deposits (Metzger and others, 1973, p. G22-G23). Lag deposits of these exotic clasts should give us a good clue that the river had to have been present, or at least somewhere nearby. The idea of finding distinctly identifiable finer-grained deposits of sand, silt or clay seemed remote, at best, considering the significant amount of time available for erosion to have removed these relatively fragile sediments.

DISCUSSION

Let's take a look

A cursory reconnaissance along State Route 78 between Milpitas Wash and the intersection of State Route 78 and Ogilby Road (County Route S-34) provided some teasing encouragement, but nothing upon which we could hang our hats. So, we thought, why not do some exploring around Smoketree Valley to see if the river might have taken a circuitous detour through this broad, nearly flat, and nearly featureless valley (Figure 2).

We began our search with a late afternoon, dusty and bumpy drive across the middle of Smoketree Valley. All we found was a lot of sand and relatively small angular gravels which were most likely derived from nearby hills and mountains. After spending the night at Coon Hollow Campground (United States Geological Survey, 1983) we drove back south along Wiley's Well Road–Milpitas Wash Road (BLM Road Number 704) to resume our search along the edges of Smoketree Valley (United States Geological Survey, 1988a).

Our first stop was on a dark gray, abandoned fan surface near the junction of BLM Road Numbers 704 and 901 (Photo 2). This spot can be easily reached by driving approximately 10¼ miles northwest along Milpitas Wash Road (BLM Road Number 704) from State Route 78. Here we found what looked like the typically rounded quartzite clasts (Photo 3) encountered so frequently in sedimentary deposits and as lag gravel on abandoned fan surfaces, within the present-day Colorado River corridor (Metzger and others, 1973, p. G22–G23).

Eureka!

We've found it — or so we thought. A few clasts were cracked open, and sure enough, we observed common ordinary quartzite (Photo 4). Checking on a few more nearby fan surfaces, we found similar or the same type of scattered clasts.

Continuing our search and driving southeast along the well-traveled, gravel-surfaced Milpitas Wash Road, we ran out of the rounded gravel and into fans that were covered only with locally derived angular gravel. Thinking that we'd found an isolated exposure of Colorado River "B" gravel, plans were made to return with more time to spend delineating this find.

And return we did.

As shown on Figure 2, a roughly kidney-shaped area of about 18 to 20 square miles was found to contain rounded quartzite clasts that seemed to fit the description

Photo 2. *Makeshift road-signs Farm near southerly extent of Soledad Rojo formation red beds outcrop (Figure 2). This intersection of BLM Roads 704 and 901 can be easily reached by driving 10¼ miles west from SR-78 along Milpitas Wash Road. Photo by W. J. Elliott 2-13-2014.*

Photo 3. *Soledad Rojo formation red beds lag gravel in central portion of outcrop area shown on Figure 2. Photo by W. J. Elliott 2-19-2012.*

Photo 4. *Broken Soledad Rojo formation red beds gravel clasts showing exterior patina as well as fresh quartzite interior. Photo by W. J. Elliott 2-12-2012.*

Photo 5. *Desolate Colorado (Sonoran) Desert. Foreground is underlain by Soledad Rojo formation red beds. An unnamed dry wash and Little Chuckwalla Mountains form the backdrop. View to northwest. Photo by W. J. Elliott 2-14-2014.*

of Colorado River "B" gravel, and to boot, its location, especially along Wiley's Well Road–Milpitas Wash Road, appeared to suggest that the ancestral Colorado River might have turned west at Blythe, California, headed westerly, and then southerly toward Coon Hollow Camp — and beyond.

But the plot thickens

During our most recent foray into this wilderness to finalize our findings, it occurred to us that there seemed to be a somewhat loose correlation between the abundance of rounded quartzite lag gravel on the old abandoned fan surfaces, and the presence of an underlying sedimentary deposit of continental* red beds that also contained rounded quartzite gravel.

Not much site-specific literature was found describing or dealing with these red beds. We did find the following general references: Jennings, 1967; Morton, 1977; and Sherrod and Tosdal, 1991. (Author's clarification: Sherrod and Tosdal (1991, p. 12,409, and Fig. 4 on p. 12,413) show an outcrop of red beds (Tcs) on the west side of the western Palo Verde Mountains, but for some unknown reason, the Tcs symbol is not included in their explanation.)

[Author's Comment: Rivers make great political boundaries but not necessarily geological boundaries. For example, the California and Arizona geological surveys generally confine their work to their own jurisdictions. Universities and regional entities, such as the United States Geological Survey are, however, not usually restricted to arbitrary political boundaries.]

Eberly and Stanley (1978, p. 929) concluded that older red beds in Arizona are probably of "Eocene to Late Miocene (?)" age. In addition, these post-Laramide Orogeny (after approximately 53 million years old) to pre-mid-Tertiary Orogeny (before approximately 17 to 20 million years old) sediments were deposited in a subaerial continental oxidizing environment of broad interior drainage.

A reconnaissance trek was made into Eberly and Stanley's study area (1978, p. 922, 923, 929) to see if the red beds on the Arizona side of the Colorado River looked anything like the red beds we'd found in the western Palo Verde Mountains. For this visit, we utilized previous mapping by Wilson, 1960; Bishop, 1963; Jennings, 1967; Olmsted, 1972; Carr and others, 1980; and Lombard, 1993, to scope out places to look for red beds. Fourteen new individual outcrops of red beds (thirteen in Arizona and one in California) were found, and are briefly described in Appendix A.

With careful observation and thoughtful reflection, we found these scattered red bed deposits to be remarkably similar in color, composition, induration, bedding, and deformation to our red beds in the western Palo Verde Mountains. Although there are individual differences, these deposits appear to be principally a function of sediment provenance (source) and depositional environment (relative distance from sediment source, viz. proximal, medial or distal).

Therefore, there seems to be no compelling reason not to extend to the work of Eberly and Stanley (1978, p. 929) west across the Colorado River to include red beds outcrops in eastern Imperial County (southeastern California).

Soledad Rojo formation

Now back to the red beds in the western Palo Verde Mountains. Based on our visit to Arizona, we believe that, had Eberly and Stanley (1978) extended their work across the Colorado River to include the western Palo Verde Mountains, our eastern Imperial County red beds deposits would fit perfectly into their correlation chart on page 922.

For the sake of having a name to use when discussing our moderately easterly dipping, well-indurated gravelly sandstone red beds in the western Palo Verde Mountains, we propose the *informal* name, *Soledad Rojo* (red solitude) *formation*. One cannot miss the stark bareness and solitude when prowling this chunk of deserted desert real estate (Photo 5). Current critter count in this seemingly lifeless landscape is: several lizards, numerous flighty birds, two quail, one roadrunner, one kit fox, ever-noisy coyotes, and nighttime serenades by melodic hooty owls.

A closer look at our hypotheses

Close scrutiny of several *Soledad Rojo formation* outcrops indicated that they contain scattered gravels that could not be readily distinguished from typical Colorado River "B" gravel (compare Photos 1 and 3). With this observation, we paused to reflect on what we had been doing. It occurred to us to plot the *Soledad Rojo formation* outcrop on the same map as our postulated "B" gravel outcrop.

As it turns out, these old red beds crop out completely within our approximate area of what we were describing as "B" gravel (Figure 2). Additionally, a further look at modern-day drainage patterns showed that clasts eroded from the *Soledad Rojo formation* could have been easily

*The term continental is used herein to describe sedimentary strata that were deposited in an ephemeral inland lake or wind-blown environment, as opposed to an open-ocean or marine environment. Non-marine settings are frequently oxidizing environments where iron bearing minerals weather to red-orange colored secondary minerals. Even in very small concentrations, this "rusty coloration" usually overwhelms and masks the colors of other minerals or rocks.

transported down-slope to all the locations where we found what we thought were Colorado River "B" gravel.

Now for the take-home message from this frustrating, yet fascinating adventure into an area of isolated and desolate desert solitude: KEEP AN OPEN MIND! We expect T. C. Chamberlin (1890) would be proud that we have kept our minds open and used his "Method of Multiple Working Hypotheses."

A lesson in using the method of multiple working hypotheses

In the beginning, our first thought (hypothesis) was to jump on the presumed occurrence of errant Colorado River "B" gravel as being suggestive of the presence of an ancient river channel. This seemed logical since that was what we were looking for to begin with. So why not leap to, and stick with, that conclusion?

Discovery of, and realization that, the *Soledad Rojo formation* lies completely within the aerial outline of the suspected surficial Colorado River "B" lag gravel (Figure 2), and after a short period of denial and bargaining (grief over the death of our first hypothesis), we arrived at our second hypothesis. Maybe there is a mixture of both Colorado River "B" gravel and gravel from the old red beds.

With a good night's sleep and a clear head in the morning, it finally became abundantly clear that a third hypothesis was not only possible, but was a better fit with our findings. With the data at hand, and without prejudice, it became abundantly evident that all the surficial lag gravel we had been mapping was nothing more than erosional debris derived from the *Soledad Rojo formation*.

Further investigation may show that, yet another set of multiple working hypotheses could evolve. "Never, never, never, give up" (Winston Churchill) when looking for the seemingly elusive truth. It is so easy to give birth to an idea and then spend way too much time and expense trying to prove that it is true, when a sober and clear headed assessment could have saved all the trouble and made space for more fruitful endeavors.

CONCLUSIONS

Test, test, and test again

This is what we did in this gravelly exercise. The initial conclusions did change, and for the better. At least the data now seem to make more sense. But, alas, we will keep looking for the elusive real Colorado River "B" gravel — this time with a cautious eye for *faux* gravels that may have been placed there by the devil or by a mischievous troop of Boy Scouts to lead us astray.

For now, we are content to believe that the gravel found in the area shown in Figure 2 is, as far as we can tell at this time, lag gravel eroded out of the *Soledad Rojo formation*, and not Colorado River "B" lag gravel.

ACKNOWLEDGEMENTS

Our collective thanks go to Carole Ziegler, Todd Wirths, and Lowell Lindsay for their patience in allowing us time to prepare this teaching essay. Wendy Elliott's support has been so much appreciated as we doggedly saw this exercise to completion.

POSTSCRIPT

TO BE COMPLETE, WE SHOULD NOTE THAT after this manuscript was submitted for the first edition of *Picacho and the Cargo Muchachos*, we discovered three additional references that, together, could cloud our informal naming of the *Soledad Rojo formation* red beds. These are: Hadley (1942); Berg and others (1982); and Jorgensen and others (1982).

Jorgenson and others (1982, p. 314) extended the use of the Tolbard Fanglomerate name from a red beds outcrop in the Midway Mountains (Eastern Paymaster mining district) to the outcrop in the western Palo Verde Mountains that we had informally named *Soledad Rojo formation* (Figure 2). A field reconnaissance found that, although the Tolbard Fanglomerate may have had a similar genesis as the *Soledad Rojo formation*, there are clearly substantial differences in clast composition, rounding vs. angularity, provenance, induration, and direction of transport. Therefore, it is suggested that separate names be used for these disconnected outcrops, and that additional work be done to establish what, if any, relationship these sediments may have in common (North American Commission on Stratigraphic Nomenclature, 2005, p. 1560, 1566–1570).

Table 1. *Correlation chart. Tertiary stratigraphy in eastern California and western Arizona.*

Approximate Ages	Region				
	Southwestern AZ (3, 8, 11, 12, 13)	**Picacho State Recreation Area, CA (2, 6, 9)**	**Vinagre Wash and Ogilby Road, CA (2, 6, 9)**	**Midway Mtns., CA (1, 2, 4, 6, 7, 9)**	**Western Palo Verde Mtns., CA (6, 7, 9, 11)**
Quaternary	surficial deposits	surficial deposits	surficial deposits	surficial deposits	surficial deposits
~9.54 Ma to ~17 Ma (10)	Unit II (undifferentiated) (3)	Bear Canyon Conglomerate (5, 10)	Bear Canyon Conglomerate	Bear Canyon Conglomerate ("Gravel" 4)	Bear Canyon Conglomerate (6, 9)
~26 Ma to ~54 Ma (3)	Unit I Red Beds (3)	no outcrops	no outcrops	Tolbard Fanglomerate (1, 4, 7)	Soledad Rojo formation
Early and pre-Tertiary	Basement	Basement	Basement	Basement	Basement

Reference Key:
1. Berg and Geis (1982)
2. Dillon (1975)
3. Eberly and Stanley (1978)
4. Hadley (1942)
5. Hughes (1993)
6. Jennings (1967)
7. Jorgensen et al. (1982)
8. Lombard (1993)
9. Morton (1977)
10. Ricketts et al. (2011)
11. Sherrod and Tosdal (1991)
12. Wilson (1933)
13. Wilson (1960)

Ma = million years ago

Outcrops and their locations

A discussion of *Soledad Rojo formation*, Tolbard Fanglomerate, and Bear Canyon conglomerate is below. Location information for each of these geologic units is also noted. Table 1 is an informal correlation chart intended to provide a rough guide to approximate relationships between various Tertiary stratigraphic sequences discussed herein.

The *Soledad Rojo formation* can be found in and around the White Swan barite prospects at the western edge of the Palo Verde mining district (at and in the vicinity of Section 24, T9S, R19E, and Section 19, T9S, R20E, SBBM [United States Geological Survey, 1983, 1988a]). See Figure 2.

The Tolbard Fanglomerate can be found in the Midway Mountains, in and around the Tolbard manganese mines located in the eastern Paymaster mining district (at and in the vicinity of Sections 18 and 19, T11S, R21E, SBBM [United States Geological Survey, 1988b]). This general outcrop area is located approximately 13 miles southeasterly from the *Soledad Rojo formation* outcrop (Figure 2). (*Author's note:* To be complete and to cover our bases, an effort was made to visit every previously identified exposure of early Tertiary red beds at and in the vicinity of our red beds outcrop on the west side of the western Palo Verde Mountains. An approximately two square miles outcrop of "red beds," mapped by Jennings (1967) and by Morton (1977), located about two miles southeasterly from the Tolbard manganese mines and just southeast of Arrowhead Springs, was field checked and determined to be a miss-labeled outcrop of tephra, a pyroclastic volcanic rock.)

The Bear Canyon conglomerate can be found in and around the Picacho State Recreation Area (southern Chocolate Mountains). Specifically, it crops out in Bear Canyon, Carrizo Wash, and Gavilan Wash (at and in the vicinity of Section 20, T13S, R22E, SBBM [United States Geological Survey, 1965b]). This *type area* (Crowe, 1978; and Hughes, 1993) is located approximately 12 miles southeasterly from exposures of the Tolbard Fanglomerate. Although not shown on Figure 2, these exposures of Bear Canyon conglomerate can be found about 6 miles south of the southeast corner of Figure 2.

Tolbard Fanglomerate

Jorgensen and others (1982, p. 313) applied the name Tolbard Fanglomerate to a sequence of red beds in the Eastern Paymaster manganese mining district (Hewett and Fleischer, 1960, p. 23–25; Hewett and others, 1963) in the Midway Mountains (a subset of the Chocolate Mountains). This name appears to be derived from the nearby Tolbard manganese mine which is located within the red beds fanglomerate (Hadley, 1942, Plate 75). Hadley did not name the red beds; he only referred to them as "fanglomerate" for field mapping purposes. It appears that the name Tolbard Fanglomerate was most likely first used by Gardner and others (1960), as referenced by Berg and others (1982, p. 300). Both Berg and others (1982) and Jorgensen and others (1982) apparently picked up on, and used, the name Tolbard Fanglomerate.

Extending the name Tolbard

We have no quarrel with using the name Tolbard Fanglomerate for outcrops in the Midway Mountains—and in fact, it appears more than appropriate. The issue comes with extending the name approximately 13 miles northwesterly (Jorgensen and others, 1982, p. 314) to the completely separate and isolated outcrop of *Soledad Rojo formation.*

A mappable unit

Dunbar and Rodgers (1957, p. 259) spend several paragraphs spelling out and explaining the intended meaning of the use of the term, *formation*. In a nutshell, "A ... succinct definition in common use is that a formation is '*a mappable unit*.'" (See also, North American Commission on Stratigraphic Nomenclature, 2005, p. 1566–1570).

Comparing the Tolbard Fanglomerate and the *Soledad Rojo formation*

With this definition of a mappable unit in mind, it is hard to see how two widely separated fanglomerates qualify for the same name. It is likely that they can both claim a similar high energy genesis and a subsequent period of sub-aerial weathering to produce a similar, though slightly different surficial color cast. But that seems to be about where the similarities end.

For one thing, Jorgensen and others (1982, Figures 2 and 3 on p. 314) show differing clast imbrication directions for the red beds in the Midway Mountains (Figure 2), and in the western Palo Verde Mountains (Figure 3). Apparent average imbrication in the Midway Mountains is north-northwesterly plus or minus about 30 degrees. In contrast, the apparent average imbrication in the western Palo Verde Mountains is east-southeasterly plus or minus about 30 degrees. Although depositional troughs may have had a more or less northwest-southeast orientation, it would appear that energy and provenance directions were vastly different.

In the Midway Mountains, dullish red-brown and tan fanglomerate beds are composed principally of locally-derived intermediate composition volcanic rocks and occasional granitic and metamorphic basement rocks (Hadley, 1942, p. 462; Jorgensen, 1982, p. 313). Similar compositional observations were made by the authors. In the western Palo Verde Mountains, the conglomerate beds are bright red-orange, and are composed principally of intermediate and basic composition volcanic rocks, as well as granitic and metamorphic basement rocks (Jorgensen, 1982, p. 314). Additionally, the authors found a significant percentage of far-traveled quartzite clasts. Distance, imbrication, and compositional differences would, therefore, seem to warrant different names, even though they may be broadly related spatially, likely also in genesis, and in post-depositional weathering and tectonic deformation.

They may also be similar in age. Following the work by Eberly and Stanley (1978) in southwestern Arizona, just across the Colorado River, it is likely that both the Midway Mountains and the western Palo Verde Mountains red beds correlate in age with their "Unit 1 red beds." These Unit 1 red beds are thought to fall into the range of approximately 54 to 26 million years ago (Eocene–Oligocene) (Eberly and Stanley, 1978, Figure 2, p. 922).

A reconnaissance of early Tertiary red beds exposures in western Arizona (Appendix A) was made to obtain a first-hand look at several of the red beds exposures referred to by Eberly and Stanley (1978). Save for some differences, including provenance and environment of deposition, no compelling reasons could be found to suggest that the red beds on the Arizona side of the Colorado River are measurably different in age and geologic history from those on the California side.

Arizona red beds vary from dark-magenta-red to orange-brown, but for the most part, are brick-red in color. Particle and clast sizes vary from silt and clay to boulders up to three feet in diameter; average clast size varies from approximately one inch to six inches in diameter. Clast composition varies widely from outcrop to outcrop, being largely of volcanic and metamorphic origin. Sediments are characteristically well indurated. Bedding is tilted gently to steeply, with average dips falling into the range of 20° to 40°. Outcrop sizes range from small patches of only a few 10s of square feet to very large outcrops that cover up to a square mile.

Other nearby continental clastic sedimentary deposits

Because of the broad regional reconnaissance nature of early geologic mapping (Jennings, 1967; and Morton, 1977), there seems to be a lingering confusion regarding relationships between a number of poorly dated and poorly constrained, clastic continental sedimentary deposits in eastern Imperial County (southeastern California). The following is an attempt to clarify some of that confusion and to pique the interest of some young field-oriented geologists to make thesis or dissertation projects out of this complex and intriguing mid-Tertiary dilemma. (Hint, the rocks are way better exposed in this desert environment than in most humid environments.)

Bear Canyon conglomerate

Jorgensen and others (1982, p. 313) state that, "The Tolbard Fanglomerate is considered to be correlative with the Bear Canyon Fanglomerate described by Crowe and others (1979), and Dillon (1976)." However, there seems to be little or no justification for correlating the red beds in the Midway Mountains, and by association, the western Palo Verde Mountains, with the approximately 17 to 9.45 million year old Bear Canyon conglomerate of Crowe

(1978), Hughes (1993), and Rickets and others (2011). See Table 1, above. For one thing, the Bear Canyon conglomerate is principally buff brown in color, less indurated than the red beds, and is only slightly tilted.

By contrast, the Midway Mountains red beds are extensively faulted and folded, and the western Palo Verde Mountains red beds are tilted into a moderately easterly dipping homocline.

A brief reconnaissance of the Bear Canyon conglomerate was made outside its type locality area (Appendix B). For example, Photo 6 shows a typical exposure of Bear Canyon conglomerate in Vinagre Wash. Compare Photo 6 with Photo 7 of the Tolbard Fanglomerate. In all aspects, there appears to be no clearly defendable correlation between Bear Canyon conglomerate and red beds of eastern California and western Arizona.

Some of the confusion in these Tolbard Fanglomerate–Bear Canyon conglomerate correlations may be derived from an unnamed post-Tertiary "gravel" mapped by Hadley (1942, p. 463, Pl. 75) that *overlies* the Tolbard Fanglomerate.

A review of Jennings (1967) and Morton (1977, Pl. 1) suggests that a *catch-all*, unnamed and undivided non-marine late Tertiary clastic unit mapped by both of these authors may be, at least in part, what was mapped by Hadley (1942, p. 463, Pl. 75) as an unnamed post-Tertiary "gravel." as well as what has now been mapped by Hughes (1993) and Ricketts and others (2011) as the Bear Canyon conglomerate. These two units were identified by Jennings and Morton as follows:

> Jennings (1967): Mc? — undivided Miocene non-marine.
>
> Morton (1977, Pl. 1): Tc — Pliocene non-marine clastic rocks and volcanic conglomerate.

Therefore, there appears to be no compelling reason to make direct correlations or associations between the *Soledad Rojo formation* and Tolbard Fanglomerate red beds and the buff-gray colored Bear Canyon conglomerate.

Photo 6. *Typical exposure of Bear Canyon conglomerate on southeast side of Vinagre Wash. Photo by W. J. Elliott 2-13-2014.*

Photo 7. *Typical exposure of Tolbard Fanglomerate at the Tolbard manganese mine, Midway Mountains, California. Photo by W. J. Elliott 2-13-2014.*

CONCLUSIONS

In conclusion, it is our opinion that the informal name *Soledad Rojo formation* should continue to be used for the red beds outcrop in the western Palo Verde Mountains, and that the name Tolbard Fanglomerate should be limited to use in the Midway Mountains. Distance, differing depositional histories and differing clast compositions are sufficient to suggest the wisdom of using different names for each unit. These two units are likely artifacts of similar genesis and have apparently been subjected to similar post-depositional sub-aerial weathering, erosional histories, and tectonic events.

The age of these two units is most likely Eocene to Oligocene. Therefore, we believe it is inappropriate to correlate them with the Miocene Bear Canyon conglomerate.

Pasted into a correlation chart, the *Soledad Rojo formation* and Tolbard Fanglomerate would likely occupy the same time-stratigraphic slot, but be placed in separate geographic columns. The Bear Canyon conglomerate would also fall into a separate geographic column and occupy a younger time-stratigraphic position (Table 1).

SELECTED REFERENCES

Berg, L., Leveille, G., and Geis, P., 1982, Mid-Tertiary Detachment Faulting and Manganese Mineralization in the Midway Mountains, Imperial County, California, *in* Frost, E. G., and Martin, D. L., eds., Mesozoic-Cenozoic Tectonic Evolution of the Colorado River Region, California, Arizona, and Nevada — Anderson-Hamilton Volume, prepared in conjunction with a symposium and field trip for the Geological Society of America, April, 1982: Cordilleran Publishers, San Diego, California, p. 299–311.

Bishop, C. C., 1963, Geologic Map of California, Needles Sheet: California Division of Mines and Geology, Map Sheet, scale 1:250,000, contour interval 200′.

Crowe, B. M., 1978, Cenozoic Volcanic Geology and Probable Age of Inception of Basin-Range Faulting in the Southeastern-most Chocolate Mountains, California: Geological Society of America Bulletin, v. 89, no. 2, p. 251–264.

Crowe, B. M., Crowell, J. C., and Krummenacher, D., 1979, Regional Stratigraphy, K-Ar ages, and Tectonic Implications of Cenozoic Volcanic Rocks, Southeastern California: American Journal of Science, v. 279, p. 186–216.

Carr, W. J., Dickey, D. D., and Quinlivan, W. D., 1980, Geologic Map of the Vidal NW, Vidal Junction and parts of the Savahia Peak SW and Savahia Peak Quadrangles, San Bernardino County, California: United States Geological Survey, Map I-1126, scale 1:24,000, contour intervals 20 and 40 feet.

Chamberlin, T. C., 1890, The Method of Multiple Working Hypotheses: Science, v. 15, p. 92–96.

Dillon, J. T., 1975, Geology of the Chocolate and Cargo Muchacho Mountains, Southeastern-most California [PhD dissertation]: University of California at Santa Barbara, 575p.

Dunbar, C. O., and Rodgers, J., 1957, Principals of Stratigraphy: John Wiley & Sons, New York, 356p.

Eberly, L. D., and Stanley, T. B., Jr., 1978, Cenozoic Stratigraphy and Geologic History of Southwestern Arizona: Geological Society of America Bulletin, v. 89, p. 821–940.

Gamble, J., Danehy, E. A., and Gardner, M. C., compilers, 1959, Aerial Economic Geologic Map: Southern Pacific Land Company, map scale 1:62,500. (Not available to the authors.)

Gardner, M. C., Gamble, J., Danehy, E. A., and Coonrad, W. L., 1960, Geology and Mineral Resources of Township 11 South, Range 19 and 20 East, San Bernardino Base and Meridian, Imperial County, California: Report submitted to Southern Pacific Land Company. (Not available to the authors.)

Hadley, J. B., 1942, Manganese Deposits in the [Eastern] Paymaster Mining District — Imperial County, California: United States Geological Survey Bulletin 931-S, p. 459–473.

Haxel, G. B., Tosdal, R. M., and Dillon, J. T., 1985, Tectonic Setting and Lithology of the Winterhaven Formation — A new Mesozoic Stratigraphic Unit in Southeastern-most California and Southwestern Arizona: United States Geological Survey Bulletin 1599, 19p.

Hewett, D. F., and Fleischer, M., 1960, Deposits of the Manganese Oxides: Economic Geology, v. 55, n. 1, p. 1–55.

Hewett, D. F., Fleischer, M., and Conklin, N., 1963, Deposits of the Manganese Oxides — Supplement: Economic Geology, v. 58, n. 1, p. 1–51.

Hughes, K. M., 1993, The Conglomerate of Bear Canyon (Miocene), Chocolate Mountains, Southeastern California, *in* Sherrod, D. R., and Nielson, J. E., eds., Tertiary Stratigraphy of Highly Extended Terranes, California, Arizona and Nevada: United States Geological Survey Bulletin 2053, p. 213–216.

Jennings, C. W., compiler, 1967, Geologic Map of California, Salton Sea Sheet: California Division of Mines and Geology, Map Sheet, scale 1:250,000, contour interval 200′.

Jennings, C. W., compiler, 1977, Geologic Map of California: California Division of Mines and Geology, data map series — Map No. 2, scale 1:750,000, contour interval 500′.

Jorgensen, M. R., Natenstedt, C. J, and Trumbly, P. N., 1982, Possible Relationship between Miocene Crustal Extension/Detachment Faulting and the Deposition of the Tolbard Fanglomerate in the Midway and Western Palo Verde Mountains, Imperial County, California, *in* Frost, E. G., and Martin, D. L., eds., Mesozoic-Cenozoic Tectonic Evolution of the Colorado River Region, California, Arizona, and Nevada — Anderson-Hamilton Volume. Prepared in conjunction with a symposium and field trip for the Geological Society of America, April, 1982: Cordilleran Publishers, San Diego, California, p. 313–315.

Lombard, J. P., 1993, Tertiary Stratigraphy of the Laguna Mountains, Yuma County, Arizona, *in* Sherrod, D. R., and Nielson, J. E., eds., Tertiary Stratigraphy of Highly Extended Terranes, California, Arizona and Nevada: United States Geological Survey Bulletin 2053, p. 205–212.

Lucchitta, I., 1979, Late Cenozoic Uplift of the Southwestern Colorado Plateau and Adjacent Lower Colorado River Region: Tectonophysics, v. 61, p. 63–95.

Mayo, D. P., 1993, Tertiary Stratigraphy of the Little Chuckwalla Mountains, Southeastern California, *in* Sherrod, D. R., and Nielson, J. E., eds., Tertiary Stratigraphy of Highly Extended Terranes, California, Arizona and Nevada: United States Geological Survey Bulletin 2053, p. 185–187.

Metzger, D. G., Loeltz, O. J., and Irelna, B., 1973, Geohydrology of the Parker-Blythe-Cibola Area, Arizona and California: United States Geological Survey Professional Paper 486-G, 130 p., Plate 1, map scale 1:250,000, contour interval 200′.

Morton, P. K., 1977, Geology and Mineral Resources of Imperial County, California: California Division of Mines and Geology, County Report 7, 104 p., Pl. 1 — Geologic Map, scale 1:125,000, contour interval 200′.

North American Commission on Stratigraphic Nomenclature, 2005, North American Stratigraphic Code (final acceptance, July 05): American Association of Petroleum Geologists Bulletin, v. 89, n. 11 (November 2005), DOI:10.1306/07050504129, p. 1547–1591. (Accessed on the web 6-23-2013).

Olmsted, F. H., 1972, Geologic Map of the Laguna Dam 7.5-Minute Quadrangle, Arizona and California: United States Geological Survey Map GQ-1014, map scale 1:24,000, contour interval 20′.

Reynolds, S. J., 1988, Geologic Map of Arizona — Map No. 26: Arizona Geological Survey, Tucson, scale 1:1,000,000.

Richard, S. M., and Sherrod, D. R., 1993, Introduction to Tertiary Stratigraphy of the Area South of I-10, Arizona and California, *in* Sherrod, D. R., and Nielson, J. E., eds., Tertiary Stratigraphy of Highly Extended Terranes, California, Arizona and Nevada: United States Geological Survey Bulletin 2053, p. 171–184.

Ricketts, J. W., Girty, G. H., Sainsbury, J. S., Muela, K. K., Sutton, L. A., Biggs, M. A., and Voyles, E. M., 2011, Episodic Growth of the Chocolate Mountains Anticlinorium Recorded by the Neogene Bear Canyon Conglomerate, Southeastern California, U.S.A.: Journal of Sedimentary Research (SEPM – Society for Sedimentary Geology), v. 81, p. 859–873.

Sherrod, D. R., and Nielson, J. E., eds., 1993, Tertiary Stratigraphy of Highly Extended Terranes, California, Arizona, and Nevada: United States Geological Survey Bulletin 2053, 250p.

Sherrod, D. R., and Tosdal, R. M., 1991, Geologic Setting and Tertiary Structural Evolution of Southwestern Arizona and Southeastern California: Journal of Geophysical Research, v. 96, n. B7, p. 12,407–12,423.

Spencer, J. E., Pearthree, P. A., and House, P. K., 2008, An Evaluation of the Evolution of the Latest Miocene to Earliest Pliocene Bouse Lake System in the Lower Colorado River Valley, Southwestern USA, *in* Reheis, M. C., Hershler, R., and Miller, D. M., eds., Late Cenozoic Drainage History of the Southwestern Great Basin and Lower Colorado River Region—Geologic and Biotic Perspectives: Geological Society of America Special Paper 439, p. 375–390.

Strand, R. G., compiler, 1962, Geologic Map of California, San Diego–El Centro Sheet: California Division of Mines and Geology, Map Sheet, map scale, 1:250,000, contour interval 200′.

United States Geological Survey, 1955–79, Laguna Dam, Arizona-California, 7½′ topographic quadrangle map, scale, 1:24,000 contour interval 20 feet.

United States Geological Survey, 1965a, Picacho NW, California, 7½′ topographic quadrangle map, scale, 1:24,000 contour interval 40 feet.

United States Geological Survey, 1965b, Picacho SW, California, 7½′ topographic quadrangle map, scale, 1:24,000 contour interval 40 feet.

United States Geological Survey, 1983, Wiley Well, California, 7½′ topographic quadrangle map, scale, 1:24,000 contour interval 10 meters.

United States Geological Survey, 1988a, West of Palo Verde Peak, California 7½′ topographic quadrangle map, scale, 1:24,000 contour interval 20 feet.

United States Geological Survey, 1988b, Buzzards Peak, California, 7½′ topographic quadrangle map, scale, 1:24,000 contour interval 40 feet.

Wilson, E. D., 1933, Geology and Mineral Deposits of Southern Yuma County, Arizona: University of Arizona Bulletin, v. 4, n. 2; Arizona Bureau of Mines, Geological Series No. 7, Bulletin No. 134, ~223p.

Wilson, E. D., 1960, Geologic Map of Yuma County, Arizona: Arizona Bureau of Mines, University of Arizona, Tucson, Arizona, map scale 1:375,000, contour interval 500 feet.

APPENDIX A: Site Specific Data for Red Beds Exposures in Eastern California and Western Arizona

Field ID #	Field location Name	Latitude/ Longitude	Elevation	Rock Name	Clast Composition	Clast angularity	Clast Size	Distance from Source	Bedding Attitude	Formation	Color	Remarks
1	Sacramento Springs (1 mile N of I-40)	34° 53.61′ 114° 45.79′ W.	1,209′	Gravelly medium to coarse sandstone	60/40 Granitic/ volcanic	Angular to sub-rounded	½″ to 12″	Medial	N60°W, 33°SW	Red Beds	Maroon-red	Ridge former
2	Town of Parker Dam	34° 16.57′ 114° 07.83′ W.	479′	Gravelly med. to coarse sandstone with sandy claystone lenses	Meta-conglomerate, granitic & volcanic	Sub-angular to sub-rounded	½″ to 6″	Medial to Distal	N70° W, 15°SW	Red Beds	Maroon-orange-red	Faulted and fractured
3	Parker Strip Golf Course	34° 13.61′ 114° 10.98′ W.	455′	Fine to medium sandstone with 10–15% gritty sandstone	Volcanic with some granitic	Sub-angular to sub-rounded	¼″ to 2″	Distal	N10°W, 23°SW	Red Beds	Maroon-orange-red	Faulted and folded
4	Lincoln Ranch (east side)	34° 12.14′ 113° 40.84′	911′	Very fine sandstone, silty sandstone and sandy siltstone	n/a	n/a	n/a	Distal, basin center	N50°W, 30°NE	Red Beds	Red-orange-brown	Thinly bedded
5	Mineral Wash (northern exposure)	34° 11.68′ 114° 01.47′	1,083′	Fine to medium sandstone	Scattered metamorphic	Sub-rounded	Up to 3″	Medial, distal	N70°W, 21°NE	Red Beds	Red-orange to chocolate brown	Concretions
6	Mineral Wash (southern exposure)	34° 11.61′ 114° 01.44′	1,085′	Fine to medium sandstone	Scattered volcanic and metamorphic	Sub-rounded	Up to 2″	Medial, distal	N65°W, 50°NE	Red Beds	Pale orange-brown to yellow-ochre	Black sandstone dikes
7	Osborne Wash (north of Osborne Well site)	34° 09.51′ 114° 03.69′	1,117′	Sandstone and breccia	Metamorphic & Volcanic	Angular to sub-round	½″ to 6″	Proximal-medial	N10°W, 35°NE	Red Beds	Red-orange and red	Chaotic structure
8	Shea Road (west of Osborne Well site)	34° 08.62′ 114° 11.37′	596′	Gravelly sandstone	Granitic, Metamorphic, Volcanic	Angular to well Rounded	¼″ to 8″	Medial to proximal	N70°W, 25°S	Red Beds	Red-brown to tan	South side of Osborne Wash
9	Smoketree Valley – north side (north side of BLM Road 564)	33° 22.89′ 114° 56.40′	870′	Gravelly sandstone	Volcanic, quartzite, chert, and gneiss	Sub- to well-rounded	¼″ to 36″	Medial	N50°E, 28°SE	Soledad Rojo formation	Bright red-orange with some red-orange-brown	Ridge former
10	Tolbard Mine	33° 21.54′ 114° 47.77′	1,125′	Gravelly sandstone	Volcanic	Sub- to well-rounded	½″ to 10″	Medial	N25°E, 8°NW N30°E, 8°NW N55°W, 9°SW N10°W, 16°W	Tolbard Fanglomerate	Dull brown and orange to dark gray	Hilltop site
11	Tolbard Mine	33° 12.24′ 114° 47.78′	977′	Gravelly sandstone	Volcanic, some granitic and gneiss	Sub- to well-rounded	1″ to 6″	Medial	N55°W, 22°SW	Tolbard Fanglomerate	Red-orange to red-brown	In road cut
12	Mohawk Mountains (north side)	32° 46.82′ 113° 47.15′	313′	Gravelly sandstone and sandy conglomerate	Granitic and gneiss	Sub-angular to sub-round	½″ to 24″	Proximal	N10°W, 40°W	Red Beds	Red-brown to red-orange	Looks like "grus"
13	Adair Park Road, east side	32° 46.08′ 114° 25.21′	197′	Very fine sandstone, siltstone and claystone	n/a	n/a	n/a	Distal	N55°E, 63°NW	Red Beds	Red-orange and ochre	Live fire range
14	Adair Park Road, west side	32° 46.02′ 114° 25.44′	181′	Very fine sandstone, siltstone and claystone	n/a	n/a	n/a	Distal	N40°E, 30°NW	Red Beds	Maroon-red to gray-green	Live fire range
15	Antelope Hills, north side	32° 42.60′ 114° 00.65′	271′	Gritty sandstone, thin claystone and siltstone	n/a	n/a	n/a	Distal	N50°W, 85°SW	Red Beds	Red-orange-brown	Next to hard-rock quarry
16	Wellton Mesa (west of Antelope Hills, at Radar Hill)	32° 40.85′ 114° 03.15′	326′	Fine to coarse sandstone with thin siltstone and silty claystone	n/a	n/a	n/a	Distal	N50°W, 13°SW	Red Beds	Red-orange and tan-gray	Faulted; located in quarry
17	Baker's Tank (south of Antelope Hills)	32° 37.42′ 114° 00.55′	534′	Gravelly sandstone, sandstone	Granitic and gneiss	Sub-rounded to rounded	2″ to 36″	Medial	N80°W, 65°S	Red Beds	Red-orange and tan	Natural desert watering hole

APPENDIX B: Site Specific Data for Bear Canyon Conglomerate Exposures in Eastern Imperial County, California

Field ID #	Field Location Name	Latitude/ Longitude	Elevation	Rock Name	Clast Composition	Clast angularity	Clast Size	Distance from Source	Bedding Attitude	Formation	Color
1	Vinagre Wash	33° 09.99' 114° 47.55'	794'	Gritty sandy gravel	Granitic, some meta-morphics and volcanics	Sub-round to well rounded	2" to 18"	Proximal	N63°E, 9½°SE	Bear Canyon Conglomerate	Light brown and light gray
2	Tolbard (Eastern Paymaster) Mining District	33° 11.94' 114° 47.73'	895'	Gritstone and sandy gravel	Volcanic	Angular to sub-rounded	½" to 6"	Proximal	N40°E, 10°NW	Bear Canyon Conglomerate	Light brown and light gray
3	Ogilby Road (4¼ miles south of SR-78)	33° 01.58' 114° 14.32'	805'	gritty sandstone, sandy gravel	Gneiss, schist, quartzite, some granitic	Sub-angular to sub-rounded	½" to 12"	Proximal	N35°W, 17°SW	Bear Canyon Conglomerate	Light brown and light gray

CAVE-IN AT HEDGES.

TERRIBLE DISASTER IN THE QUEEN MINE'S SHAFT.

Loose Formation of the Rock Causes the Entombing of Miners at the Bottom.

SEVERAL LIVES REPORTED LOST

A NUMBER OF COTTAGES NEAR BY ARE BURNED.

Constable Burke of That Place in This City—Says a Fall of Earth Wednesday Night Buried Three Men.

(BY THE TIMES' SPECIAL WIRE.)

YUMA (Ariz.,) Dec. 10.—(Special Dispatch.) News has reached Yuma from Hedges, fifteen miles from here, in San Diego county, California, of a terrible disaster in the Queen mine at that place, which is owned by the Gold Cross Mining & Milling Co. Some time last night a cave-in occurred in the main shaft, and as a result several miners have lost their lives and more are injured.

The reports are meager and conflicting, but the generally-accepted one is that four miners are entombed at the bottom of the shaft, with no chance of escape, and four or five have been rescued badly injured. Several parties came in from the mines tonight, but even their reports do not agree, so that it is impossible to gain any definite information.

One man who was interviewed by the Times correspondent said that no one was killed, but three men were dangerously injured, while the account gleaned from yet another was that four dead bodies had been recovered and four miners were still entombed with small chances of escape. This latter report is more generally accepted than the former.

This mine has long been considered dangerous on account of the loose formation of the rock, and the extraordinarly careful timbering required. Several miners have been injured at different times, and one or two lives lost, but if present reports be but partially true, no such disaster has ever occurred there as this one.

The burning of several cottages near the mines in which the accident occurred is also reported, which might be construed to indicate an explosion in the mine, but that is merely theory.

From Los Angeles Times, *December 11, 1896.*

The Gold Mining Town of Hedges in the Colorado Desert, Northwest of Yuma, Arizona. (Unless otherwise indicated, all photographs are circa 1900). Photo courtesy of the Arizona Historical Society, Yuma.

The History of Hedges and the Cargo Muchacho Mining District, Part I

A Case Study of the Lives of Mexican Miners in a Company Town of the Southern California Desert

STEPHEN R. VAN WORMER AND JAMES D. NEWLAND

INTRODUCTION

IN 1900 THE LARGEST TOWN IN PRESENT-DAY Imperial County was the mining camp of Hedges.[1] Consisting of around four hundred inhabitants, the largely Hispanic community occupied a narrow desert canyon of the Cargo Muchacho Mountains in the southeast corner of the Colorado Desert.[2] Prior to the formation of Imperial County in 1907 this area constituted the eastern edge of San Diego County. From the mid-1880s to 1917 the region was the scene of a locally significant, though relatively unknown, mining boom. Originally discovered by Mexican prospectors, serious development of the site began in the early 1890s by corporations with sufficient capital to invest in required machinery and the latest

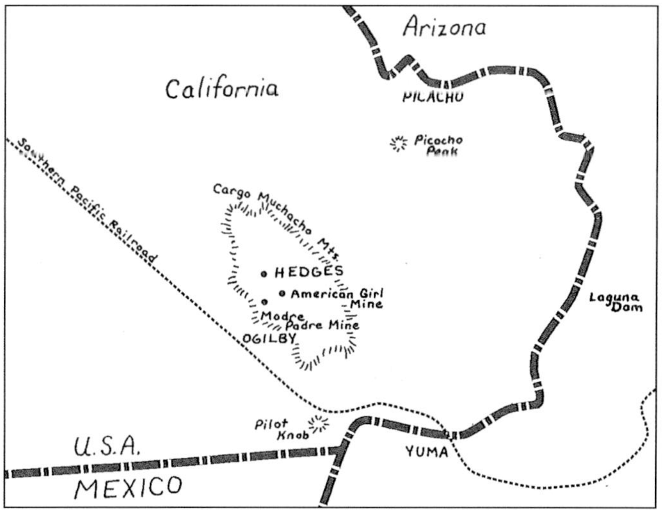

Hedges and Vicinity.

Adapted from the article originally published in *The Journal of San Diego History, San Diego Historical Society Quarterly*, Spring 1996, Volume 42, Number 2, Richard W. Crawford, editor

technology. Eventually, development of the mines brought establishment of the mining camp of Gold Rock, later known as Hedges and then Tumco, a community with a cosmopolitan population containing a large number of Mexican inhabitants.

The importance of Hedges' history is twofold. First, it provides an example of the largely undocumented role of Hispanic pioneers, prospectors, and miners in the development of mining in the California deserts and western Arizona. Second, it provides a case study of the difficult financial and managerial problems faced by mining operators at the turn of the century. Development of Hedges exemplified the merging of corporate American mining interests during the late nineteenth and early twentieth centuries with a Hispanic mining frontier that had existed in the southwest for approximately three hundred years.

Hedges' establishment resulted from the mining boom that swept through the west during the late nineteenth century. Following the Civil War successive waves of pioneers swept over the trans-Mississippi west. Unique among these were the miners. While the frontiersman, trapper, and farmer moved gradually west, the mining frontier eventually traveled from the west coast eastward. Many miners and prospectors originally came to California in 1849 and then followed successive strikes in Nevada, Montana, Arizona, Colorado, and the Dakotas as time passed. Mining technology that developed in California gold fields in the 1850s and Nevada's Comstock Lode in the 1860s also shifted eastward with the mining population.[3]

Unlike the farming frontier, miners occupied some of the western region's most inhospitable locations, establishing major population centers in a matter of weeks or months following a mineral rush. Mining camps, in the form of small towns, constituted urbanized settlements often removed by miles of formidable terrain from the nearest neighboring settled region. These urbanized frontiers formed suddenly, without the preceding stages that occurred in other frontier areas. As a result, miners were instrumental in opening up regions that had not attracted other settlers.[4] Mining settlements have been compared to islands of imported social and cultural environments surrounded by inhospitable wilderness. The settlements were financed, manned, and supplied from urban centers of America and Europe and were often linked on a regional, national, and international scale into vast transportation, communications, demographic, and economic networks.[5]

EARLY PERIOD, 1775–1880

The evolution of the Cargo Muchacho Mining District differed somewhat from the typical mining regions of California, Nevada, and the Rocky Mountains. Mexican miners worked the area for decades before Americans entered the district in the late nineteenth century. In this sense it was more characteristic of mineral regions in Arizona and New Mexico that constituted the northern edge of a Hispanic mining frontier extending north from Mexico, which had been developed in the seventeenth and eighteenth centuries.

Spanish colonists first entered Sonora in the mid-1600s in order to exploit mineral resources. A dozen mines were established by the 1680s.[6] During the late seventeenth century, Jesuit missions were established in northern Sonora, within the boundaries of present day southern Arizona. As missions, presidios, and related settlements expanded, colonists and Indian converts began mining placers and shallow deposits, producing mainly oxidized rich silver bearing galena (lead sulfide), as well as some gold. Mining remained small-scale within present-day Arizona during the Spanish period, and was carried out mostly in the southeastern part of the state. Development was hindered by the high cost of shipping equipment, the limited labor supply, and Apache raids.[7]

In 1775, the Hispanic-Sonoran Frontier reached the junction of the Gila and Colorado rivers, at present-day Yuma Arizona, when Spanish missionaries and soldiers under the leadership of Captain Juan Bautista de Anza established an overland route between Sonora and the Spanish missions on the California coast. To protect the route the Spanish founded a short-lived mission and presidio near Yuma on the site of the present Fort Yuma Indian Reservation from 1780 to 1781.[8] Gold mining began in the Cargo Muchacho region during this period, but lasted only a short time. Originally known as the Sierra de San Pablo, the Cargo Muchachos were first described by Spanish explorer Father Francisco Garcis, who recorded rich looking surface ore in 1776.[9] Spanish colonists worked productive placer deposits near present-day Laguna Dam as well as surface ore in the Sierra de San Pablo in the areas of present day Jackson Gulch and the Madre Valley. In 1781 the Yuma Indians rose in rebellion, destroying the small settlement and closing Anza's route. The region of the lower Colorado River remained closed to European settlement until the conquest of northwestern Mexico by the United States.[10]

War between the United States and Mexico resulted in re-establishment of Anza's former Gila River Route as invading American armies traveled overland to the West

Coast.[11] This trail became extremely important following the discovery of gold in January 1848 at Sutter's Mill in the foothills of the Sierra Nevada Mountains.[12] Within a year 80,000 people had traveled to California from around the world.[13] Gold rush emigrants from the United States and Mexico used the Gila River Trail, which eventually led the Army to establish Fort Yuma where the Gila River joined the Colorado River.[14] Mexicans were among the first to learn of the strike, receiving the news from passing ships calling at the ports of Colima, Guaymas, and Mazatlan. Between six and ten thousand Sonorans followed the old Anza Trail to California, crossing the Colorado at its junction with the Gila River, during 1849 to 1850. Sonorans differed from their Anglo-American contemporaries. They often traveled and worked as family units. In addition, they possessed a knowledge of prospecting and mining techniques so that they found gold where Americans could not. The resulting antagonism against Mexican miners caused discrimination and abuse culminating in the Foreign Miners Tax of April 1850. By 1854 most Hispanics had left the state.[15]

Some Mexican miners who had been run out of the Northern California gold fields began prospecting along the lower Colorado River in the mid-1850s.[16] In 1854 Hispanic prospecting families rediscovered gold placers at Laguna.[17] In 1858 American and Mexican miners encountered placer deposits on the Gila River, twelve miles east of Yuma. A year later prospectors discovered another rich placer field at Chimney Peak, fourteen miles north of Yuma. In June 1860 the *Alta Californian* reported that prospectors on the Gila and Colorado rivers included around 350 Mexicans and only five "Americans." Jose Maria Mendivil and a man named Valenzuela made strikes at Picacho in 1861 where Mexican families were already prospecting. Rich placer deposits were discovered at La Paz, 60 miles north of Yuma, by Jose Maria Remodino in 1862. Mexican prospectors recovered gold from the arid sands using dry washing techniques developed in Sonora.[18] Caution and hostility that resulted from their Northern California experience makes accurate documentation of their activities difficult. They seldom recorded mining claims or discussed their finds with non-Hispanic "*Gringos*."[19]

Although not formerly recorded, evidence existed that Mexican miners also worked in the Cargo Muchachos at this time, digging shallow pits by hand and processing the ore in *arrastras*. In 1891, a visitor to the Madre Valley noted that "... many acres have been burrowed by the [former] Mexican miners, who, digging down until they almost stood on their heads were compelled to be satisfied and quit."[20] It may have been during this period that the mountains received their present name. According to legend, two young Mexican boys came into their family's camp one evening with their shirts filled with gold ore. These *muchachos cargados* (loaded boys) became the namesake for the Cargo Muchacho Range.[21] Another version says the boys found gold while looking for horses that had strayed from an immigrant party on their way to Los Angeles in 1861.[22]

First established in 1862, the Cargo Muchacho Mining District was redefined several times in later years.[23] Continued small scale prospecting and mining characterized the 1860s and 1870s. The inhospitable environment and limited access retarded development of the region. The most important mines at this time were the Padre and Madre claims, located on the eastern edge of the Madre Valley. As previously noted, this area contained rich surface ores mined by Spanish colonists in the 1780s and Mexican prospectors in the 1850s. The Madre and Padre ledges were first formally recorded by William Pulter, James J. Spauer, James Palmer, and Edward Neal in April 1875.[24]

In 1877, the Southern Pacific Railroad completed the Los Angeles to Yuma Link of its transcontinental line. The rails ran within two miles of the Cargo Muchachos, paralleling their western edge. This newly provided access eventually brought large numbers of American prospectors into the region and started a mineral rush during the next decade.[25] Mining activity increased moderately at first, but soon attracted attention from many individuals looking to strike it rich. In February 1877, a northern extension of the Madre Mine was registered to William E. Smith.[26] A second claim to the Padre Mine was filed the following June by James Norton and in July, William Van Arsdale filed a location on the Cargo Muchacho Mine situated 0.75 mile southeast of the Padre and Madre ledges.[27] Van Arsdale also obtained interests in the Padre and Madre mines.[28] The mines proved to be productive. In 1883 it was reported that the Madre was the "principal mine" in the Cargo Muchacho District, followed by the Cargo Muchacho.[29] However, they would soon be eclipsed by other strikes.

GOLD ROCK PERIOD, 1880–1893

By 1880, the railroad had established a siding and section headquarters at Ogilby, approximately 2.5 miles southwest of the Padre Mine.[30] Peter Walters, a railroad employee stationed there, prospected a valley of the Cargo Muchachos three miles north of the Madre Mine that came to be called Gold Rock Canyon. His discoveries soon started a small rush into the area.[31] Of Swedish descent, Walters had a permanent residence in Santa Monica. He filed his first claim, the Grand Mogul, in November 1883. He soon filed a series of additional claims that included the Little Mary in January 1884 and the Black Butte, Three Sisters, Little Pete, Alice Mary, and Carbonate in March of the same year.[32] Assays of ore from the Black Butte were valued at $130 per ton. Walters personally realized $114 after working a "sack" of ore with an *arrastra*. He continued to prospect and filed on the General Lee in December 1884, and the Lady and Horn Silver in January 1885.[33] News of the discovery brought a flood of prospectors to the Cargo Muchachos who filed numerous claims between 1884 and 1886.[34]

Events at Gold Rock Canyon followed a sequence repeated throughout the western United States. Mining communities generally developed in three successive stages. The first was the camp, which formed immediately upon announcement of a major discovery causing a "rush" into the region. The typical mining camp was made up of makeshift temporary structures consisting largely of tents with some brush or log lean-tos and dugouts. If the discovery was substantial and the region continued to develop, the camp phase was usually very short, lasting from a matter of months to one or two years.[35] The early beginnings of Gold Rock Camp represented this type of settlement.

The second phase of mining community development was the boom stage. If the mineral strikes proved valuable more permanent buildings soon appeared to replace the tents and lean-tos. Speed was essential, however, and no time could be spared to build anything beyond a simple, functional, unpainted structure, which resulted in a community of small, undecorated buildings that often gave the settlement a haphazard appearance.[36] If the settlement continued to prosper, it soon passed from the boom period into the third or mature stage. This phase included development of the surrounding region, especially in the areas of transportation and agriculture. Community services such as schools, churches, and fraternal organizations were established. Community architecture became more substantial with larger, more ornate wooden and brick buildings in the business district and residential neighborhoods.[37] The transformation of Gold Rock Camp into the town of Hedges at the turn of the century manifests aspects of the second and third phases of mining camp development.

A first stage mining camp soon developed at the site of Walters' strikes.[38] On December 14, 1884, Gold Rock miners held a meeting that formally established the Ogilby Mining District and christened their settlement Gold Rock Camp. Persons present included Peter Walters, Andrew Rabbiet, J. P. Kennedy, D. D. Field, and Edwin Lundy. In addition, M. C. Era, J. Lambie, Edwin Lambie, and Jerry Long were represented. Officers elected consisted of President Peter Walters, Secretary P. J. Kennedy, and District Recorder Edwin Lundy.[39] In later years the Ogilby and Cargo Muchacho mining districts would be synonymous.

The Yuma newspaper reported creation of the district on January 3, 1885, and noted that two shipments of ore had been sent to San Francisco and milled with favorable returns. Another shipment of "several tons," the third in as many months, was presently in San Francisco. An excavation at Black Butte exposed a body of ore 40 feet thick and 600 feet in length. An open cut 20 feet into the ledge, all in "pay ore," assayed at $100 a ton in San Francisco. Another claim had been opened by two cuts and showed 64 inches of ore that assayed at $200 a ton. The high cost of shipping ore to San Francisco was partially solved when Gillespie and Childs, San Francisco based owners of the Cargo Muchacho Mine, erected a ten-stamp mill on the Colorado River at El Rio (present-day Winterhaven) in 1885.[40]

Establishment of Gold Rock brought accelerated development of the region. Like earlier works at the Madre and Padre ledges, rich ore at Gold Rock was on or near the surface and easily recovered.[41] Samples from Walters' Little Mary Lode assayed at $9,000 and $12,000 a ton in April 1884.[42] Ores worked with *arrastras* yielded $75 a ton.[43] Continued rich assay reports brought substantial outside capital and Walters associated himself with prominent San Francisco and Los Angeles investors.[44] Others also showed interest in Cargo Muchacho mines. In May 1885, J. W. McKay bonded the Racett Gold mines for $90,000. The following November, San Diego newspapers reported that unnamed mining properties in the Cargo Muchacho District had changed hands for $76,000.[45]

Approximately twenty mines operated in the Gold Rock District.[46] By the late 1880s, surface ores were all but exhausted. Faced with increasingly greater expenditures to pursue deep rock mining, Gold Rock operators sold out. In 1889 the *Arizona Sentinel* reported, "Dougherty and Young have purchased the Gold Rock mines and will send the ore to the El Rio Mill. Several carloads of horses, tools and equipment had arrived at the mines last week."[47] Ore bodies in many of the mines continued to become more difficult to work, however, and in September 1890 the paper noted that A. J. Dougherty who owned "some" of the Gold Rock Mines was trying to interest Los Angeles capitalists to invest in erection of a twenty-stamp mill.[48]

The Golden Cross Mine Headframe, Ore Bin, Hoist House. Photo courtesy of the San Diego History Center.

GOLDEN CROSS—FREE GOLD PERIOD, 1893–1905

The Gold Rock operator's inability to fund and run a more extensive hard rock mining enterprise was typical of the situation of many small mining companies in the 1880s. The need for expensive, modern equipment, qualified geologists, and mining engineers, along with unstable national and world metal prices, unionized labor costs, and restrictive land and mining claim laws, all but eliminated the small company and individual prospector from significance in the American mining industry by 1910. Although small companies remained, the shift to technologically-intensive from labor-intensive mining forced profitable hard rock operations into the realm of corporate big business. By 1890 only a large company or corporation could raise the capital to continue profitable mining operations.[49] To gain success in these ventures, mining entrepreneurs needed to handle several factors of production: acquisition of property, affordable labor, large amounts of capital, and technological expertise in the guise of competent mining engineers. This meant that entrepreneurs not only needed to acquire these factors but they needed to manage and organize them to establish a successful operation.[50]

These changes in mining during the decade of the 1890s soon affected the mines of the Cargo Muchachos. As the initial strikes began to wane mining entrepreneurs and promoters brought investments of capital and installation of modern equipment to the older mines, as well as development of new claims throughout the area.

The increased investment in gold mines of the Cargo Muchachos also resulted from several other domestic factors that helped make gold mining throughout the west more profitable during the 1890s. The first and most important was the re-establishment of the gold standard for the United States economy. The gold-silver controversy had been a major economic and political question for the United States since the Depression of 1873. The great success of the Comstock Lode silver mines in the 1870s and early 1880s, along with repercussions from the weak economy led to passage of the Sherman Silver Purchase Act of 1890. The act called for the government to purchase 4.5 million ounces of silver per month. When the severe Depression of 1893 hit, President Cleveland quickly called for repeal of the Silver Act and an increase in government gold reserves to remedy the crisis.

Although Cleveland's initial plan failed to end the depression and caused his political defeat, it created a new demand for gold. Prosperity returned by 1897. This economic upturn, along with a drop in silver production, led to the political defeat of the pro-silver standard Populist Party and re-establishment of gold as the main commodity purchased by the treasury to provide backing for the U.S. economy. For mining entrepreneurs attempting to interest investors into the unstable and technologically experimental frontier industry of hard rock gold mining, the repeal of the Silver Act produced the incentive they needed to obtain new capital.[51] The development resulted in the great interest manifested in the Cargo Muchacho area mines during the 1890s.[52]

In early 1893, outside financiers moved to purchase and develop the mines around Gold Rock Camp from various owners who held title to the claims, including the original locator Peter Walters.[53] The complex of Gold Rock mines consisted of thirty claims with three principle shafts prime with potential; the Golden Cross, Golden Crown, and Golden Queen.[54] In 1892, the Golden Cross alone had produced $24,374 in gold from its small scale operation in which the ore was still hauled by wagon to Ogilby Station and shipped by rail to the El Rio Mill.[55]

Realizing this potential, investors Thomas S. Fuller and William V. Hedges, of the Sioux City, Iowa-based firm of Hedges, Fuller, and Company, along with George G. Mullins and William Rantz, Jr., consolidated the Gold Rock claims and formed the Golden Cross Mining and Milling Company in April 1893.[56] Less than a month later the new company actively began to improve the property. In May, the *Arizona Sentinel* reported that the "... new owners of the Gold Rock mines have just sent 18 men, 28 mules, 4 carloads of machinery and pipe, 2 carloads of lumber, and a carload of supplies..." to the camp.[57] The company had big plans for their new operation including its own stamp mill, and a pump and pipeline to bring water from the Colorado River. Work progressed through the summer and into the fall. In October the company started up the newly built twenty-stamp mill and began processing fifteen dollars a ton ore.[58] Within a year of their establishment, the Golden Cross Mining and Milling Company had also installed a water pumping plant on the Colorado River near Yuma. The plant conveyed between 100 to 125 thousand gallons of water a day through a twelve mile pipeline.[59]

Early returns seemed favorable and by December of that year the company had developed plans to expand the mill with another twenty stamps and had installed a new Huntington Mill which could crush up to twenty-six tons of ore in 24 hours.[60] In March 1894, the Sentinel reported that the additional twenty stamps were ready to start up so that the mill could handle the proposed development of the Golden Cross and Golden Crown shafts. Up to this time, work had proceeded only in the Golden Queen Mine.[61] The newly expanded mill consisted of forty, 850 pound stamps and employed eight men. The machinery crushed one hundred tons of rock that produced one thousand dollars daily.[62]

Early in 1894, reports from the Gold Rock mines painted a bright picture of further expansion and certain success. The Yuma-based *Arizona Sentinel* published several articles on the continued development and good prospects of the new Cargo Muchacho "boom-camp" at Gold Rock. By late March 1894, the camp had already enlarged to a reported thirty-one tents and houses. The settlement's name was officially changed in April 1894 when Golden Cross secretary Thomas Fuller bonded a post office named Hedges in honor of the company's vice president.[63]

The promotion of the area's success led to a "boom" in interest for the Gold Rock mines. Early profits brought added investment and offers to purchase the operation outright. In March, Capt. J. B. Mullen, a 10 percent owner of the Golden Cross Company, said that he turned down an offer of $250,000 for his share.[64] A reported offer of two million dollars was also made for the whole operation.[65] Although those offers were refused, several investors in the operation did sell. Local miner Thomas Johnson, gave up his 4/23rd of a share for seven thousand dollars.[66] In June 1894, the *San Diego Union* reported outside interest in the Cargo Muchachos from as far away as Denver, Chicago, San Francisco, and Kansas City. The Golden Cross Company employed fifty men in their mines and eight more at the mill.[67]

With profits rolling in and success seemingly imminent, Golden Cross announced plans in August 1894 to add an additional sixty stamps to their mill during the upcoming winter. This move would give them a total of one hundred stamps and make the company one of the largest milling operations in the state.[68] However, these rapid expansion plans soon initiated a series of grave managerial mistakes.

By 1894, problems resulting from poor planning began to manifest themselves. The forty-stamp mill had been erected on a low flat in order to have easy access to the numerous mines in the canyon. This was a logical idea for an operation with several main shafts. However, after only a few months, it was obvious that satisfactory disposal areas for the tailings and slag waste was not available at the site. Soon tailing piles had filled the thirty foot deep gully on the south side of the valley. Newly installed elevators raised the tailings to a point where they would flow away from

The 100-Stamp Mill. Ore was conveyed from the mines in horse drawn rail carts on the tracks, in the foreground, to the crusher house, at left center. The crushed ore was carried via the ramped conveyor to the mill for processing. Photo courtesy of the San Diego History Center.

the mill. The pile of sand soon grew to ten feet in height and encroached upon the mill to a point that a bulkhead had to be built so the buildings would not be buried.[69]

In an effort to solve this problem the company decided to build a new sixty-stamp mill on a high hill at the east end of the canyon. In December 1894, the *Arizona Sentinel* reported on the proposed plan, stating that "… in the new year [1895] the company will build a new mill and add sixty stamps. The present forty stamps will continue to operate until the new mill is completed and running, then the other stamps will be moved up, thus losing no time. Eight miles of additional pipeline will also be laid."[70] The planned sixty-stamp mill was to have a capacity of two hundred tons a day. By late summer of 1895, however, plans continued to change and management decided to leave the original forty-stamp mill operating while enlarging the yet to be built sixty-stamp mill to one hundred stamps. Their combined 140 stamps would provide a processing capacity of over three hundred tons a day.[71]

In May 1895, work began on the new mill site, which required the leveling of a large hilltop by drilling and blasting.[72] The estimated construction cost, for what would be one of the largest mills in the state, was nearly $200,000. New investors were sought. H. B. Adsit of Denver took out a $36,000 bond on the mines.[73] Still, prosperity seemed assured. In October the newspapers reported that the forty-stamp mill turned out $33,000 in gold per month and with the opening of the sixty additional stamps in September more profit was expected.[74] By November, Golden Cross ran 140 stamps, employed over two hundred men, and had achieved a daily output of five hundred tons of low grade six to ten dollars a ton ore.[75] Equipped with large 1,000 pound stamps, the new mill had adequate disposal area provided by the slopes and canyon bottom below the high bluff of its site.[76]

By constructing such a large mill, however, Golden Cross over-extended and created financial difficulties that would hinder its success. Ignoring the advice of several of their own engineers, and generally accepted common sense for successful mining operations, company management committed what western mining engineer Almarin Paul defined as the three main errors responsible for the financial failure of hard rock ventures: too much expenditure on speculation; inefficient systems for ore reduction and processing; and, the most costly — especially in the case of Golden Cross — construction of too much expensive machinery on unproven property.[77] The costs to flatten the bluff, ship vast amounts of equipment and supplies, and

Air Compressor and Steam Engine Inside the 100-Stamp Mill. Photo courtesy of the San Diego History Center.

construct the huge mill quickly caught up with the company only to be compounded by problems in processing the low grade ore. Much of the rock contained traces of antimony and arsenic that made it hard to amalgamate on the mercury collection tables in the mill. Therefore, even though milling operations tripled in capacity, profits did not follow. Subsequently, in January 1896, owners of the Golden Cross found the company $125,822.38 in debt.[78] Faced with the complete loss of their operation, the owners agreed to a receivership.

Through mutual agreement, the company's creditors and the Golden Cross Board of Directors decided that a set of trustees should manage the mining operations. A deed of trust was made to operate the mines and apply all the proceeds to the payment of the company's debts. The main trustees were builders and contractors of the one hundred-stamp mill. The Golden Cross Mining and Milling Company transferred title of the Cross, Crown, Queen and all other claims and property to the trustees. The trust deed provided a period of fifteen months to pay off the indebtedness, with a three month trial period to determine if the mines could produce a profit. If no earnings had been realized at the end of the trial period the trustees could sell the mines. Regardless of their profit potential, if at the end of fifteen months the debts had still not been resolved the mines would be sold at auction to pay off any remaining balance.[79] The trustees assigned Mr. Dolan as the new superintendent with a new foreman, Dick Hooper, and two new assistants. They then stated that they believed unwise and unbusiness-like management during the past year had caused the company their great financial embarrassment, and that within a few months they would show that the Gold Rock mines could pay off their debts.[80]

The mismanagement assessment of Golden Cross was echoed a year later by mining engineer Taylor MacCleod. Writing in the *Arizona Sentinel*, he claimed that the problem with the Hedges mines resulted from mismanagement, over expansion, and over extension. The need to keep all 140 stamps running forced the milling of waste along with the pay ore and created a large tailings pile and low profits. The mill foreman quit after management cut his pay. Failure to replace him with a trained and competent individual compounded the problem. However, MacCleod still felt that profit could be made from the large amount of low grade ore left at the mines.[81]

The trustees ran the Golden Cross operation from January 6 to April 15, 1896. At that point they had received $98,754.86 from bullion and incurred $124,567.60 in expenses, leaving a loss of $25,812.74. They, therefore, concluded that it would be impossible to make a profit with the mines and on April 21 advertised that they would be sold at auction on June 1.[82] The Golden Cross Mining and Milling Company immediately filed suit in San Diego County Superior Court against the six trustees, complaining that they had been unprofessional and inefficient in operating the mines. The company requested that the trustees be enjoined from selling the mines and restrained from managing the property. They further requested that the mines be managed by a receiver of the court and that Golden Cross Mining and Milling Company be paid $150,000 in damages. On June 5, 1896 the court granted an injunction restraining sale of the property and appointed W. W. Stewart receiver. The order stated that he "... shall not engage in construction or development work but take out only the large bodies of ore said to be blocked out already."[83]

Reprieved from financial disaster, Golden Cross continued to work, improve, and develop their property. In the summer of 1896 the company installed pipelines into the mines for compressed air powered (Burleigh) hammers.[84] A fire had damaged and shut down the Queen Mine in April. The shaft was re-timbered and reopened by October. At the end of the year Golden Cross had 148 employees and averaged $30,000 a month in profit. The Golden Cross shaft had been sunk to 250 feet, the Golden Crown to 350 feet, and the Golden Queen to 550 feet. Enough ore had been stockpiled to run the one hundred-stamp mill for six months straight.[85] The water pipeline's capacity had been increased to 250,000 gallons a day and the company was utilizing better mining techniques including square set timbering.[86] Even the indebtedness problems seemed under control when receiver W. W. Stewart reported that he expected to be able to pay off the debt at $20,000 a month.[87]

However, the debt owed to the trustees did not decrease. Approach of the fifteen month deadline in the spring of 1897 forced the Golden Cross owners to look for other options. In April, the *Arizona Sentinel* reported that Col. E. W. Stubbins of Denver had engineered a deal in which the Golden Cross mines would be sold to an English syndicate for $1,500,000. According to the report the parties had sixty days to finish the deal. The company was still running full out with 165 men employed producing $40,000 a month in bullion.[88] This report was in error, however. In May, a representative of the English firm spent a week at the camp, inspecting the property with the secretary of the company, Thomas S. Fuller. The offer was actually $750,000 and while Fuller and the receiver W. W. Stewart recommended acceptance, the Golden Cross Company demanded a million dollars.[89]

By June 1897, no part of the indebtedness had yet been paid. The trustees brought suit against Golden Cross in Los Angeles County Superior Court, requesting the property be sold for debts since fifteen months had passed as allowed by the original trust deed. The court decided in favor of the trustees and receiver Stewart surrendered the property.[90] On June 14 the trustees declared the deed of trust to be at fault. On June 21 they advertised that the property would be sold at auction on July 20.[91]

On June 26, 1897 the Golden Cross Mining and Milling Company sold its interests in the Gold Rock mines to the Free Gold Mining and Milling Company of Nevada. This deal turned out to be a last ditch effort by Hedges, Fuller, and the original Golden Cross owners to save the property. Free Gold Mining and Milling Company had been formed under articles of incorporation filed in Douglas County Nevada on June 18 1897. One of the members of the board was the attorney for Golden Cross.[92] The transaction that transferred title had ten major provisions that allowed the original owners of Golden Cross to stay involved in the company, meet their indebtedness, and still receive one million dollars for the property. In exchange for all the claims, deeds, and property of the Golden Cross Company, Free Gold agreed to: 1) challenge the trustees' indebtedness claim of $215,000 in court; 2) take on any pending or future court claims of which they would deduct any costs from the one million dollar sale price; 3) pay for and put into operation "suitable appliances for treating tailings and concentrates" on or near the property; 4) defend the property against all claims or claimants of possession other than the trustees; 5) perform all assessment and patent work; 6) keep the property value up and agree that in case the mines were sold Golden Cross Company would receive any balance due on the one million dollar purchase price; 7) pay off all indebtedness through payment or litigation to lift a mortgage of $215,000; 8) reinvest insurance money that may be received as a result of fire damage to the property; 9) pay T. S. Fuller as the on- site representative of the Golden Cross Mining and Milling Company; and 10) issue stock to W.V. Hedges to make him a director of the Free Gold Company.[93]

The sale was heralded by the *Arizona Sentinel* as not only good for Golden Cross but as "good for Yuma." The Free Gold Company was supposedly a successful mining syndicate with investments in California, Nevada, and Utah. The operation would be headed by Col. Isaac Trumbo of Utah. The *Sentinel* commented that the Golden Cross' mismanagement and bad luck were now in the past. The new company would make the operation solvent and had plans to build a cyanide plant, develop the mines, and provide more jobs for Yuma area men.[94]

On July 7, 1897, Free Gold filed a suit in San Diego County Superior Court, complaining that the trustees were in illegal possession of the property and that they had greatly mismanaged the mines causing fires, flooding, cave-ins, and loss of valuable ore. They requested an injunction against the sale, possession of the property, and management of the mines under court receivership to pay off the indebtedness. The court issued an injunction restraining the sale and appointed Charles W. Pauly as receiver.[95]

Under the court's receivership Free Gold continued to operate the mines at full capacity in 1898. In June Pauly had all 140 stamps at the mills running.[96] When the mines reached one of their most productive periods in the fall, clearing $43,000 in September alone, he reported that $23,000 would be paid toward the still pending indebtedness.[97]

Perhaps the most notable part of the Free Gold's agreement to take over the Golden Cross mines was the plan to build a cyanide plant. Discovery of the cyanide process for milling gold ore constituted one of the most important technological advances in gold mining. Developed in Scotland in 1887 and first used in New Zealand in 1889, it revolutionized the industry. Simply explained, cyanidation consisted of mixing crushed ore or tailings with a low level cyanide solution that dissolved and separated gold from the ore. The solution then passed through zinc shavings or boxes where the gold reformed into a solid, allowing for collection of virtually every atom of gold from tailings, waste, and low grade ore. Suddenly, every other form of gold milling was obsolete. Cyanidation was first used in the United States in 1891, and by that time nearly all major gold mining operations in the world were converting their mills and building the recognizable large vats used in the process.[98]

Cyanidation constituted not only a technological revolution but a financial one also. A company needed only a few men for its milling operations. Importantly, for remote or desert operations, the process needed little water due to the ability to repeatedly reuse the processing solution. Now, many low grade ore veins could be mined with some profit. It is estimated that cyanidation doubled the world's production of gold literally overnight.[99]

Free Gold constructed a large cyanide plant in 1900. One change at the mines was the appointment of the company's director, Isaac Trumbo, as receiver on September 11, 1899. The *Yuma Sun* reported in late 1899 that Trumbo would oversee the erection of a large cyanide plant so that treatment of the nearly 900,000 tons of tailings could begin.[100] On November 2, Trumbo petitioned the court stating that there were 750,000 tons of tailings on the site that were worth a million dollars if treated by the cyanide process. Waste from the stamp mills were augmenting the tailings by four hundred tons a day. He requested permission from the court to install a plant with a capacity of three hundred tons a day. Fuller and Hedges protested the petition as an unnecessary expense. The court ordered installation of the cyanide plant on November 10, 1899.[101]

By December 1900, the three hundred ton cyanide plant was operating. There were still 170 men on the payroll which included whites and Mexicans with the Mexican miners predominating. They worked low grade ore, and the one hundred-stamp mill ran day and night. The forty-stamp mill had been shut down and its buildings incorporated into the cyanide plant operation. A 175 horse power Corliss Engine continued to pump water to the 300,000 gallon reservoir one hundred feet above the town, and wood was delivered to the camp for $3.75 a ton.[102] Over the next few years the cyanide plant would become the focus of the Free Gold operation. In 1901 its capacity was increased to four hundred tons a day by raising the height of the steel vats two feet.[103] In 1902 the *San Diego Union* reported that $400,000 had been made by reworking the accumulated tailings of the forty- and one hundred-stamp mills.[104]

The operation of the cyanide plant was so successful that by 1902 Free Gold had made considerable progress against its debt. Under direction of C. W. Pauly, who had been reappointed receiver in 1901, the number of operating stamps was reduced to twenty. In the spring of 1902 Pauly stated to the court that he felt further development of the mines was unnecessary since returns from the cyanide plant would pay off the indebtedness.[105] The one hundred-stamp mill shut down in April. Former employee Bill Keiser remembered that after the mines and mills ceased operation just a few Mexican workers and their families remained to run the cyanide plant.[106] By September the number of employees had been reduced to fifty.[107]

In May 1903, the receiver reported that the original debt to the trustees had been paid off and their claims closed. The court ordered the receivership to continue until all additional debts had been paid.[108] In February 1904, receiver C. W. Pauly made his twenty-third report on the progress of the operations at Hedges. It showed disbursements of $8,090.93, unreported bullion worth

The Free Gold Cyanide Plant. On the upper left, the solution vats; center, the leaching vats; lower left, the extraction room building and in front of it, sump tanks marked "Poison-No Admittance" in Spanish and English. Photo courtesy of the San Diego History Center.

$4,793, cyanide plant returns of $8,395.65, and other liabilities equaling $5,418.11. He predicted the company would be out of debt in a few months and the receiver dismissed.[109] In June, a new receiver, H. A. Barker was appointed after Pauly resigned. The *Sentinel* noted that the Free Gold operations processed "400 tons of tailings a day." However, the stamp mill would not be restarted until conditions changed and no improvements would be made until the company caught up with the balance sheets.[110] In October, Barker's third receivership report showed that Free Gold continued to make money.[111] On October 29 the court ordered receiver Barker to prepare a final report in preparation for terminating his office. The report was filed and Barker discharged on January 20, 1905.[112]

Following termination of the receivership, operations were shut down and Hedges became a ghost town. In 1907, a report stated "The Golden Cross Mine with its milling plant of 140 stamps and its extensive development yielded nothing and has done nothing since its tailings were cyanided in 1905."[113] A former employee, Dennis Ahern, and his wife remembered visiting the empty camp and saw the houses and buildings closed up with furniture, stoves, and cooking utensils still inside. Even large plate glass mirrors still hung in the saloons.[114]

CONCLUSION

The history of early mining in the Cargo Muchacho Mountains and the chronicle of the Golden Cross and Free Gold mining companies exemplify the merging of corporate American mining with the Hispanic mining frontier, and illustrate the development and pitfalls that faced western mining entrepreneurs at the turn of the nineteenth century. By 1890, mining corporations in Arizona and eastern California were investing in mineralized areas that had first been discovered by Mexican prospectors during the mid-nineteenth century, and later developed by small scale American miners who eventually sold out because they lacked the capitol required to work the more difficult ore bodies. However, corporate enthusiasm and the infusion of large amounts of money did not assure success. Mining had become a complex business that required strict management of production, property, labor, capitol, and technological expertise.[115] Failure to do so brought financial difficulty to many operations of which Golden Cross was typical.

Golden Cross' most notable short-comings were poor planning — placing the forty-stamp mill site on a flat where tailings could not be dispersed — and over speculation — construction of the one hundred-stamp mill on credit without verification that sufficient ore existed to repay the expense. Golden Cross thereby made the three main mistakes that caused most mine failures: excessive expenditure on speculation (credit), use of inefficient ore reduction and processing systems, and investment in too much expensive equipment on unproven property.[116] Efficient management, on the other hand, was exemplified by Free Gold under the receivership of Trumbo and Pauly. Construction of the cyanide plant and closure of the stamp mills represent investment in proven technologies and effective direction, which eventually brought Golden Cross' financial difficulties under control.

NOTES

1. This study was first prepared as part of a Cultural Resource Management Report entitled "History of Hedges Tumco" by Stephen R. Van Wormer and James D. Newland, in Hedges/Tumco: Historic Mining Traditions of Southeastern California, by Michael S. Burney, Stephen R. Van Wormer, Claudia B. Hemphill, James D. Newland, William R. Manley, F. Paul Rushmore, Susan D. Walter, Neal H. Hupel, Jerry Schafer, and Lynne E. Christenson, prepared by Burney and Associates, Boulder, Colorado for USDI-Bureau of Land Management, El Centro, California (1993). Copies are on file at the San Diego Historical Society Research Archives, San Diego, and Arizona Historical Society Archives, Yuma, Arizona. The authors would like to thank American Girl Joint Mining Venture of Winterhaven, California and Burney and Associates of Boulder, Colorado for supporting the research involved in this study. They would also like to express their appreciation to Mr. Frank Love of Yuma, Arizona for sharing his extensive research on mining in the Cargo Muchacho Mountains.
2. The Cargo Muchacho Mining District is located in the Cargo Muchacho Mountains in the southeast corner of Imperial County in Townships 14 and 15 South, Ranges 20 and 21 East, approximately 14 miles northwest of Yuma, Arizona. Paul K. Morton, Geology and Mineral Resources of Imperial County (Sacramento: California Division of Mines and Geology, 1977).
3. Ray Allen Billington, Westward Expansion: A History of the American Frontier (New York: Macmillan Publishing Co., 1974); Donald L. Hardesty, "The Archaeology of Mining and Miners: A View From the Silver State," Special Publication 6, Society for Historical Archeology, (1988).
4. Duane A Smith, Rocky Mountain Mining Camps: The Urban Frontier (Bloomington: Indiana University Press, 1967), 13–14.
5. Hardesty, "The Archaeology of Mining," 1; Smith, Rocky Mountain Mining Camps.
6. The earliest recorded mines in Sonora were at Santiago and San Francisco Javier de Nacatobori. See James E. Officer, Hispanic Arizona, 1436–1856 (Tucson: University of Arizona Press 1987), 4–5. During the late 1640s mining colonies (Real de Minas) were also established at San Pedro de Los Reyes, and Sinoquipe. Continued discoveries resulted in mine development at San Antonio de la Natividad, Nuestra Senora del Rosario de Nacazori, San Juan Bautista and Baccanuche. See Ana Maria Rodriguez Antondo y Martha Ortega Soto, "Capitolo III: Entrada de Colonos Espanoles En Sonora Durante El Siglo XVII," in Historia General de Sonora Vol. II, De La Conquista al Estado Libre de Sonora, Segio Ortega Noriega y Ignacio Del Rio, editors, (Hermosillo, Mexico: Gobenador del Estado de Sonora, 1985), 80. San Juan Bautista became so important that it served as provincial capital for almost a century. See Officer, Hispanic Arizona, 4–5. Continued mineral exploration resulted in discoveries at Cananea, Santa Barbara, Banacare, San Francisco del Yaqi and San Miguel Arcangel in the 1660s, and Los Alamos in the 1680s. See Rodriguez Antondo y Ortega Soto, Historia General de Sonora, 8; Officer, Hispanic Arizona, 4–5.
7. Michael N. Greeley, "The Early Influence of Mining in Arizona," in History of Mining in Arizona, edited by J. Michael Canty and Michael N. Greeley, (Tucson: Mining Club of the Southwest Foundation, 1987), 14. Officer, Hispanic Arizona, 4–5, 7–8.
8. Clifford E. Trafzer, Yuma: Frontier Crossing of the Far Southwest (Wichita: Heritage Books, 1980).
9. San Diego Union, 30 January 1938.
10. J. Ross Browne and James W. Taylor, Reports Upon the Mineral Resources of the United States (Washington D.C.: Government Printing Office, 1867), 13; Frank Love, Mining Camps and Ghost Towns: A History of Mining in Arizona and California Along the Lower Colorado (Los Angeles: Westmoreland Press, 1974), 17; Thomas E. Sheridan, Los Tucsonenses: The Mexican Community in Tucson, 1854–1941 (Tucson: University of Arizona Press, 1986), 11; Paul C. Henshaw, "Geology and Mineral Deposits of the Cargo Muchacho Mountains, Imperial County, California," California Journal of Mines and Geology, Quarterly Chapter of State Mineralogists Report No. 38, (Sacramento: State Printing Office, 1942), 52.
11. Trafzer, Yuma: Frontier Crossing.
12. Sheridan, Los Tucsonenses, 25.
13. Greeley, "Early Influences," 14.
14. Trafzer, Yuma: Frontier Crossing.
15. William Robert Kenny, "Mexican-American Conflict on the Mining Frontier, 1848–1852," Journal of the West 6 (Winter 1967): 582–592; David P. Robrock, "Argonauts and Indians: Yuma Crossing 1849," The Journal of Arizona History 32 (Winter 1991): 21; Officer, Hispanic Arizona, 245, 249; Ralph J. Roske, "The World Impact of the California Gold Rush, 1849–1857," Arizona and the West 5 (Summer 1963): 198–99; Joseph H. Park, "The History of Mexican Labor in Arizona During the Territorial Period," (master's thesis, University of Arizona, 1964), 101.
16. Park, Mexican Labor, 101.
17. Pamela Renner, "La Paz: Gateway to Territorial Arizona," Journal of Arizona History 24 (Fall 1983): 120.
18. Ibid; Love, Mining Camps and Ghost Towns, 29–33, 57; Greeley, Mining in Arizona, 7; Park, Mexican Labor, pp. 105–106.
19. Renner, "La Paz," 121.
20. San Diego Union, 30 April 1891.
21. Henshaw, Geology and Mineral Deposits, 52.
22. Love, Ghost Towns and Mining Camps.

23. San Diego Union, 23 November 1893, 3:1; Cargo Muchacho Mining Records, 1 (1862), San Diego County, Office of Recorder.
24. Miscellaneous Records, 4 (1875), San Diego County, Office of Recorder, 123.
25. Morton, Geology and Mineral Resources, 28–29.
26. Miscellaneous Records, 4:293.
27. Ibid., 293, 336.
28. San Diego Union, 27 November 1887, 1:6.
29. Ibid., 23 November 1883, 3:1.
30. Ibid., 19 September 1971, G2, 6–8.
31. Barbara Rigby Connell, "Cultural Resources Management Plan for Hedges/Tumco Ghost Town, Imperial County, California, A Field Report Submitted in Candidacy for the Degree of Masters of Art Program in Historic Resource Management in Cooperation with The Bureau of Land Management," (Riverside: University of California, 1979), 20.
32. San Diego Union, 23 November 1883, 3:1; 25 January 1884, 3:1; 26 March 1884, 3:1; Cargo Muchacho Mining Records, 4:77, 80, 81, 83.
33. Connell, "Cultural Resource Management Plan," 21; Cargo Muchacho Mining Records, 4:89, 99, 100.
34. Other prospectors included Andrew Rabbiet, Charles Lambie, R. S. Letterwood, George Lumis, Edwin Laundy, and Patrick J. Kennedy. See Cargo Muchacho Mining Records, 4:78–79, 90–95, 110, 143, 150, 152; San Diego Union, 7 December 1884, 3.
35. Smith, Rocky Mountain Mining Camps, 13–14, 42.
36. Ibid., 43–44; Turrentine Jackson, Treasure Hill: a Portrait of a Silver Mining Camp (Tucson: University of Arizona Press, 1963), 23–25.
37. Smith, Rocky Mountain Mining Camps; Jackson, Treasure Hill, 23–25; Glen K. M. Chang, "Leading the Way: Prospectors in the Trans-Mississippi West," Journal of the West 20 (Spring 1981): 31–35.
38. Connell, "Cultural Resource Management Plan," 23.
39. Boundaries of the new district were from Cactus Station, located 2 miles northwest of Ogilby on the Southern Pacific track, southerly along the tracks to Ogilby, thence along a southeast line 9 miles to the Conductor Mine, then 12 miles northwest, and then in a direct line to Cactus Station. Cargo Muchacho Mining Records, 4:97; Arizona Sentinel 3 January 1885.
40. Arizona Sentinel, 3 January 1885; 20 December 1884, 3; 5 December 1885, 3.
41. Connell, "Cultural Resource Management Plan," 23.
42. San Diego Union, 4 April 1884, 3:4.
43. George Spaulding, Gold Mines and Mining in California (San Francisco: George Spaulding & Co., 1885), 286.
44. Walters associates included Dr. A. Sternburn, owners of the Cargo Muchacho Mine and El Rio Mill, J. H. Gillespie and Walter C. Childs; W. T. and W. K. Lambie; and H. McCrabb. Frank Love, Cargo Muchacho Files, Personal Archival Collection, Yuma Arizona; San Diego Union, 4 April 1884, 3, 4; Cargo Muchacho Mining Records, 4:99, 183.
45. San Diego Union, 15 November 1885, 3:2.
46. Connell, "Cultural Resource Management Plan," 24.
47. Arizona Sentinel, 13 April 1889, 3; Love, Cargo Muchacho Files.
48. Arizona Sentinel, 13 September 1890, 3.
49. Mark Wyman, Hard Rock Epic: Western Miners and the Industrial Revolution, 1860–1910 (Berkeley, University of California Press, 1979), 256–257.
50. Richard H. Peterson, The Bonanza Kings: The Social Origins and Business Behavior of Western Mining Entrepreneurs, 1870–1900 (Norman: University of Oklahoma Press, 1977), 23–25, 42.
51. Peterson, Bonanza Kings, 34.
52. The region's expansion was clearly evident in the southern Cargo Muchacho area with the old Padre and Madre and Cargo Muchacho claims and new strikes such as the Pasadena, Blossom, and American Girl mines. In the spring of 1892 Lloyd Tevis, one of the primary investors in Nevada's Comstock Lode, acquired the Padre and Madre claims, combining them under the name of the Con Padre Quartz Mine. Mineral Survey map no. 3416 (1892), San Diego County, Office of Recorder. The neighboring Cargo Muchacho mine also prospered. It was taken over in 1890 by H. H. Blaisdale. He installed a 20-stamp mill and a double pipeline 14 miles in length to the Colorado River. San Diego Union, 30 April 1891. The largest of the southern mines was the American Girl. Work began on a small scale in the early 1890s, but soon accelerated. In 1898 the mine was sold to former California Governor Markham for $300,000. San Diego Union, 1 January 1898, 5:1. At the turn of the century the American Girl mine was the largest in the Cargo Muchachos except for the Golden Cross Mines in Gold Rock Canyon. By 1902 the shaft was 625 feet deep. Ore yielded six to eight dollars a ton. A twelve-mile pipeline brought water from the Colorado River near Yuma. The mill consisted of three sets of Gates (Cornish) rolls that could process 100 tons in twenty-four hours. The accompanying cyanide plant had the same capacity. The mine employed fifty men, and a camp of around twenty structures had developed. Register of Mines (1902); San Diego Union, 6 January 1902, 6:3; United States Geological Survey, (Map of) Yuma Quadrangle (1904). The investment in these mines illustrates the interest western mining entrepreneurs had in the Cargo Muchacho District. Most of the new investors in these mines were from San Francisco and Los Angeles. As previously noted, Lloyd Tevis had been one of the primary investors in Nevada's Comstock Lode and a successful mining entrepreneur throughout the west. Peterson, Bonanza Kings.
53. Deed Book 15 (1893): 315; 21:143–146, San Diego County, Office of Recorder.
54. J. J. Crawford, Twelfth Report (Second Biennial) of the State Mineralogist, Two Years Ending September 15, 1894 (San Francisco: California State Mining Bureau, 1894).
55. Edward B. Preston, California Gold Mill Practices, California State Mining Bureau Bulletin, No. 6 (San Francisco: California State Mining Bureau, 1894), 83.
56. Articles of Incorporation, Golden Cross Mining and Milling Company, no. 797 (15 April 1893), San Diego History Center Research Archives; Deed Book 15 (1893): 313–314.
57. Arizona Sentinel, 13 May 1893, 3.
58. Engineering and Mining Journal (14 October 1893): 403; Love, Cargo Muchacho Files.
59. Crawford, Twelfth Report, 240–41; San Diego Union, 10 June

1894, 5:1; 20 September 1894, 5:3.

60. Arizona Sentinel, 2 December 1893, 3.
61. Ibid., 17 March 1894, 3.
62. Crawford, Twelfth Report, 240–41; San Diego Union, 10 June 1894, 5:1; 20 August 1894, 5:3.
63. Arizona Sentinel, 17 March 1894, 3; 24 March 1894, 1; 7 April 1894, 3.
64. Ibid., 17 March 1894, 3.
65. San Diego Union, 10 June 1894, 5:1.
66. Deed Book 15 (1894): 315–316. Johnson undoubtedly used the money to develop his new American Girl claim.
67. San Diego Union, 10 June 10 1894, 5:1.
68. Ibid., 20 August 1894, 5:3.
69. Crawford, Twelfth Report.
70. Arizona Sentinel, 8 December 1894, 3.
71. Ibid., 7 September 1895, 3.
72. Engineering and Mining Journal, 4 May 1895, 420; Arizona Sentinel, 1 June 1895, 2.
73. Engineering and Mining Journal, 31 August 1895, 204–205; Arizona Sentinel, 7 September 1895, 3; Harold Weight and Lucile Weight, "Tumco: Two-Time Ghost Town of the Cargo Muchachos." Calico Print (1951), 4, on file at Arizona Historical Society Archives, Yuma Arizona.
74. Arizona Sentinel, 5 October 1895, 3.
75. Ibid., 2 November 1895, 2.
76. J. J. Crawford, Thirteenth Report (Third Biennial) of the State Mineralogist For The Two Years Ending September 15, 1896 (San Francisco: California State Mining Bureau, 1896), 337–338.
77. Peterson, Bonanza Kings, 100.
78. Arizona Sentinel, 11 January 1896, 3.
79. The trustees included James Spiers, of Fulton Engineering and Shipbuilding; and Joseph Sloss of Mill, Sloss and Scott, both of San Francisco; L. W. Blinn of Blinn and San Pedro Lumber Company; J. M. Johnston of Union Hardware and Metal Company; A. Haas of Haas, Baruch and Company; and C. Seligman of M. A. Newark and Company; all of Los Angeles, Deed Book 21 (1896): 129–138; Arizona Sentinel, 11 January 1896, 3; L. W. Blinn, Affidavit of L. W. Blinn (1897), San Diego County, Superior Court, case no. 10215, San Diego Historical Society Research Archives, San Diego.
80. Arizona Sentinel, 11 January 1896, 3.
81. Ibid., 15 May 1897, 1.
82. Blinn, Affidavit.
83. Superior Court, case no. 9683, Complaint and Judgment (1896), San Diego Historical Society Research Archives, San Diego; Blinn, Affidavit.
84. Arizona Sentinel, 15 August 1896, 3.
85. Ibid., 7 November 1896, 2.
86. Crawford, Thirteenth Report, 337–338.
87. Arizona Sentinel, 7 November 1896, 2.
88. Ibid., 3 April 1897, 3.
89. San Diego Union, 26 May 1897, 5:1.
90. Blinn, Affidavit.
91. Superior Court, case no. 1,0215, Complaint; Blinn, Affidavit.
92. Blinn, Affidavit, 1897.
93. Cargo Muchacho Mining Claim Book 6 (1897): 183–190.
94. Arizona Sentinel, 3 June 1897, 3.
95. Superior Court, case no. 10215, Complaint.
96. San Diego Union, 8 June 1898, 5:4.
97. Ibid., 4 October 1898, 5:1.
98. Otis E. Young, Western Mining: An Informal Account of Precious-Metals Prospecting, Lode Mining, and Milling on the American Frontier from Spanish Times to 1893 (Norman: University of Oklahoma Press, 1970), 283–286; Dr. A. Scheidel, "The Cyanide Process: Its Practical Application and Economic Results," in California State Mining Bureau Bulletin, No. 5, (San Francisco: State Mining Bureau, 1894), 67, 96.
99. Ibid.
100. Yuma Sun, 17 November 1899, 1; Love, Cargo Muchacho Files.
101. Superior Court, case no. 10,215, Petition of Isaac Trumbo (1899).
102. Yuma Sun, 10 August 1900, 1; San Diego Union, 15 December 1900, 8:3; Love, Cargo Muchacho Files.
103. Superior Court, case no. 10,215, Petition of C. W. Pauly (1901).
104. San Diego Union, 6 September 1902, 8:1.
105. Superior Court, case no. 10,215, Petition of C. W. Pauly (1902).
106. Weight and Weight, "Tumco Two-Time Ghost," 4.
107. Weight and Weight, "Tumco Two-Time Ghost," 4. In April 1903, the Arizona Sentinel reported that Free Gold mines were sold to Gold Fields of California, Ltd., an English company. A debt of one million dollars owed to Frank A. Gibson due on July 1, 1905 was assumed by the buyers. The sale price was $1,987,600 with a down payment of $40,000 and the balance given in paid up shares in the company. They promised to expand the cyanide plant capacity from 200 to 1000 tons daily. See Arizona Sentinel, 29 April 1903, 2. This turned out to be a false report.
108. Arizona Sentinel, 17 February 1904, 3.
109. Ibid.
110. Ibid., 15 June 1904, 3.
111. Ibid., 5 October 1904, 3.
112. Superior Court, case no. 10,215 (1907).
113. George Otis Smith, Mineral Resources of the United States: Calendar Year 1907 (Washington D.C.: United States Geological Survey, 1908), 234.
114. Weight and Weight, "Tumco: Two-Time Ghost", 4.
115. Peterson, Bonanza Kings.
116. Ibid., 100.

In the late 1980s, the mining camp of Hedges filled a small narrow canyon in the Cargo Muchacho Mountains. The schoolhouse is located at center. The 40-stamp mill is at upper left. The superintendent's elegant house is at the right rear on the hillside above the company compound. Photo courtesy of the San Diego History Center.

The History of Hedges and the Cargo Muchacho Mining District, Part II

A Case Study of the Lives of Mexican Miners in a Company Town of the Southern California Desert

BY STEPHEN R. VAN WORMER AND JAMES D. NEWLAND

INTRODUCTION

IN 1900, SAN DIEGO COUNTY STRETCHED FROM the Pacific Ocean eastward to the banks of the Colorado River. The most active gold mines in the county were not the more commonly known mines of the Cuyamaca Mountains near Julian, but at Hedges, located in the Cargo Muchacho Mountains of present-day eastern Imperial County.[1] As a company town, the settlement owed its existence to the Golden Cross and Free Gold Mining and Milling Companies, whose histories have been presented in the first part of this series [*Journal of San Diego History*, Spring 1996]. The locality became largely a society of Mexican mining families. In spite of lives that emphasized family and religious values, they lived in a harsh, unsafe, and often violent environment. The social history of the Hedges community illuminates the lifestyle led by many Mexican miners who emigrated to this country to work and live in company towns throughout Arizona, New Mexico, and southeastern California during the late nineteenth and early twentieth centuries.

Adapted from the article originally published in *The Journal of San Diego History, San Diego Historical Society Quarterly,* Summer 1996, Volume 42, Number 3, Richard W. Crawford, editor.

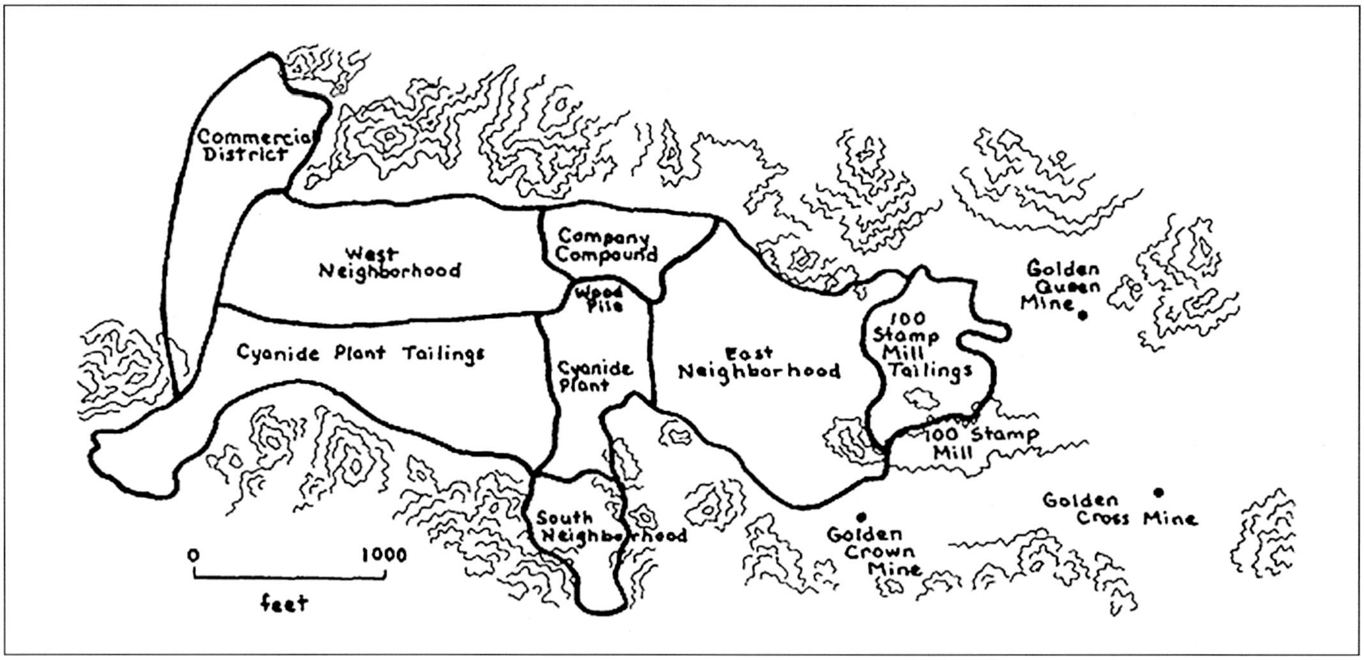

Hedges Townsite Neighborhoods, Districts, and Major Mine Locations.

Situated in a narrow desert canyon, in 1894 the town of Hedges consisted of approximately two dozen unpainted structures scattered randomly across the valley floor. By the end of the decade, Hedges had grown into a large community of numerous buildings that extended from the mouth of the valley eastward into the canyon for approximately a mile. The town had taken on many aspects of second and third phase mining camps, with more permanent structures: a church, school, distinct neighborhoods, business districts, and social organizations.

Hedges was a company town, run from a company compound on a small terrace in the center of town. This type of community was common among western mining camps where development of remote locales in undeveloped regions required operators to organize and supply transportation, housing, food and supplies, as well as other requirements to support a large work force.[2] At Hedges, the company supplied the community with a running water system, electrical lights, boarding house, miner's club and reading room, housing for the management, and a hospital. The store was in a frame building. Adjoining it were the office of the receiver and a sleeping apartment for commercial travelers. Residences for the management consisted of four small wooden dwellings and three brick houses. The court appointed receiver occupied a seven-room structure, originally known as "Fuller's House." The miner's club and reading room, actually a company saloon located in an adobe building, carried a general stock of liquors and tobacco and provided billiard and pool tables. The hospital was located in a stone building next to the miners club and had a resident doctor by 1897.[3]

In 1900, the *San Diego Union* noted that "... the buildings of the Free Gold Mining Company are well and neatly constructed, and the front entrance of each bears a sign designating the department."[4] A stage ran daily to the railroad station at Ogilby, six miles to the south where connections were available with Yuma to the east and to points further west. A forty-stamp mill and cyanide plant were located directly south of the company compound. Together the compound, mill, and plant formed a line through the middle of the camp that divided it into east and west halves.

Residential neighborhoods occupied by Mexican families were located to the east and west of the company compound and up a small canyon behind the cyanide plant vats. Here, families built their own houses based largely on vernacular Sonoran architectural traditions. Structures consisted of a variety of materials including adobe, ocotillo wattle and daub, stone, and board and bat. Exterior ramadas, patios, and kitchens and livestock corrals were common. Crowding the houses closely together in the canyon bottom along narrow dusty streets created a lack of space not very conductive to sanitary living conditions. Privies, refuse heaps, and livestock corrals stood adjacent to residential structures.

Overlooking the town from a high bluff just beyond the eastern edge of the East Neighborhood stood a one-hundred stamp mill and its large tailings pile. The mines were at the extreme east end of the canyon beyond the mill.

On the western edge of town, at the entrance to the narrow valley a third segment of Hedges society existed. Here a small district of private businesses formed that provided services which competed with and augmented those provided by the company. The district included two small barber shops run by "Spanish Americans," Malquiador Lopez and Juan Cano; a meat market and stockyard run by two brothers, Ed and Frank Hodges; and numerous saloons.[5] At various times saloons were owned and operated by H. C. Adams and Billy Horan, Samuel Wilson, Frank Norton, Mabel Lee, and a man named Brown. Adams and Horan employed two Mexican "sporting women" in their establishment and had a roulette table.[6] Wilson's place consisted of two adjacent rooms, one used as a bar, the other as a restaurant. A kitchen was attached to the rear and four rooms "out back" housed prostitutes.[7] In 1900, he employed two barkeepers and three "sporting girls." Two of the ladies claimed to be from France and one from Pennsylvania. Next to Wilson's place lived a Chinese cook, probably employed in his restaurant.[8]

DEMOGRAPHICS

The town of Hedges resembled mining communities in Arizona and Sonora Mexico due to the fact that, by far, the majority of the inhabitants were Mexican. The transition of the mining industry between 1870 and 1890 from small individually owned operations to corporate enterprises provided work for large numbers of laborers in the border regions. Mine owners, who realized laborers from northern Mexico possessed traditional mining knowledge, encouraged recruitment of Mexican *mineros*. These miners became the overwhelming majority of the mining labor force in Arizona and played an indispensable role in development of the industry.[9]

Mineros made up the majority of the work force at Hedges, some coming directly from Mexico while others had migrated from previous jobs in Arizona. Individuals of Mexican descent made up about 80 percent of the town's population. This is reflected in the school census of 1896 and 1897. In 1896, there were 117 children at Hedges under

The Majority of the Work Force at Hedges Consisted of Mexican Laborers. A group of miners pose before the company store, May 1896. Photo courtesy of the San Diego History Center.

eighteen years of age. Only twenty-two (19 percent) were non-Hispanic while ninety-five (81 percent) were from Mexican families. In 1897, there were 125 children. Twenty-seven (21 percent) were non-Hispanic and ninety-eight (79 percent) were from Mexican families. San Diego county directories also indicate a large Mexican population for the years 1905–1906. They listed 102 households: twenty-one non-Hispanic households (21 percent) and eighty-one (79 percent) with Hispanic surnames.[10]

Some writers have claimed that, at its height, the population of Hedges reached two to four thousand.[11] This seems unlikely. In 1898, the population of children peaked at 209.[12] By 1901, the number had fallen to 184.[13] If each household listed in the 1905–06 directory averaged two adults, this number (204) combined with approximately 200 children would indicate a population of approximately 400 people. This seems a realistic estimate. In 1900, the population for all of Yuma Township, of which Hedges was by far the largest community, was less than 700.[14]

The 1900 census gives what can probably be considered the most accurate profile of the community at that time. The census returns list 382 people residing in Hedges in June of that year. Hispanics made up 83 percent of the population (318). Individuals of Anglo-American descent constituted 15 percent (57) of the community, while six Chinese males and a single black man employed as a porter represented 1.6 and 0.3 percent, respectively, of the inhabitants.

The makeup of the Hispanic and Anglo-American populations differed significantly. Among the Hispanics, 1 percent (4) were over seventy years of age; 53 percent (168) were between the ages of eighteen and sixty-nine; while 46 percent (146) were less than eighteen years old. Overall, females made up 41 percent (131) of the Hispanic population and males 59 percent (187). Among adults between the ages of eighteen and sixty-nine, females constituted 37 percent (62) and males 63 percent (106) of individuals of Mexican descent. Among adult females, none were divorced, 11 percent (7) were widows, 15 percent (9) were unmarried, and 74 percent (46) were married. Male marital status differed slightly: 1 percent (1) were divorced, 8 percent widowed (8), 44 percent unmarried (47), and 47

These Workers in the Extraction Room of the Free Gold Cyanide Plant were Probably Among the Better Paid skilled Laborers at Hedges. The man at left holds zinc shavings used to recover gold from the cyanide solution. Photo courtesy of the San Diego History Center.

percent (50) married.[15] Such a distribution meant that the Mexican population was relatively young, with a high proportion of families with children, and consequently a large number of dependents.[16]

The Anglo-American community, in contrast, exhibited a high ratio of males to females with few married males and, therefore, very few families or dependents. Males made up 88 percent of this group overall (50), and 90 percent (44) of the fifty individuals between eighteen and seventy years of age. This age group constituted 87 percent of the Euro-American population, while a single individual seventy years old made up 2 percent, and six individuals less than eighteen years of age represented 11 percent of Hedges' non-Hispanic inhabitants.[17] This group was consequently more mobile, economically active, and less tied to the demands of family and dependents than the Mexican population.[18]

A social hierarchy existed at Hedges in which Anglo-Europeans held the skilled and professional jobs while Mexican *mineros* occupied the majority of unskilled lower paying positions. This treatment of Mexican labor typified mining camps throughout the southwest. Managers and investors recruited from the eastern states held the skilled jobs opened only to union members. Mexicans, who could not join the union, filled the majority of unskilled positions.[19] At Hedges, of sixty professional and skilled individuals, only sixteen (27 percent) were of Mexican descent. The single black porter constituted 2 percent of the skilled work force, three Chinese cooks made up 5 percent, and the Anglo-American population, which constituted only 14 percent of the camp's inhabitants, held 66 percent of the professional and skilled occupations.[20]

Significant wage differences existed between skilled and unskilled labor. Hoist operators, managers, and other top positions received four dollars a day. Other skilled positions such as timbermen, machine operators, and chuck tenders received from three to three and a half dollars a day.[21]

Social stratification was also reflected in educational opportunities. In 1900, of the 122 school-aged children listed in the school records for the entire district, only seventy-five actually attended school. More significantly, of those who attended school, fourteen (20 percent) attended private schools.[22] Needless to say, these students were not from Mexican families. More children are listed in the 1900 school census than the federal manuscript census returns for the same year because the district also included children at American Girl and other Cargo Muchacho mining camps as well as Ogilby.

TOWN LIFE

Life in these communities could be hard. Workers in a company town often found themselves at the mercy of mine owners who supplied services that usually benefitted the company while neglecting the needs of employees.[23] Under the operation of the Golden Cross Mining and Milling Company several abuses were reported. One miner complained that the company should call itself the "White Man's Slavery Company." According to newspaper accounts, employees were not provided with their own housing but were charged a dollar a day for the privilege of sleeping on company property. All were required to trade at the company store on credit since wages were not paid until an employee quit or was fired. Goods in the store were of poor quality and sold at high prices. Boots cost four dollars a pair and shirts two dollars and fifty cents. An additional dollar a month was deducted from their wages for the services of a doctor who, in 1895, did not even have a shanty for his practice.[24]

The company store often enhanced the profitability of the mining enterprise. Wages paid out soon found their way back to the company through the store. By paying miners their wages in goods, operators kept labor costs to a minimum.[25] At Hedges, laborers often spent 50 to 60 percent of their weekly wages at the store. Selected examples from Golden Cross Company's pay vouchers for 1897 are provided on Table 1.[26] Since most of the money probably went for groceries, and these laborers were *mineros* supporting their families, the expenditures may not be unreasonable considering the times. Few working class people were able to save money while supporting a family in the United States at the turn of the century. In 1900, the average wage in the country was seven dollars a week.[27] Discrimination at Hedges appeared to be most evident in the inability of the majority of Mexican laborers to obtain higher paying skilled or managerial positions rather than low wages for the menial jobs they did do.

The company's lack of concern for its employees and their families was also manifested in other areas. By 1896, the mines were expanding and the town was growing rapidly. On January 22 of that year citizens of Hedges petitioned the San Diego County Board of Supervisors to establish a public school in the town, claiming there were over fifty school-aged children.[28] T. S. Fuller, secretary of the Golden Cross Mining and Milling Company, wrote to the Board of Supervisors on January 30, protesting the petition. He stated: "I am informed that certain persons of this place have busied themselves in preparing and

Table 1. *Examples of Laborers' Accounts Taken from Golden Cross Mining and Milling Company Pay Vouchers.*

NAME	Date	Wage	Days	Total	Board	Hospital	Store	Balance
M. Jarmitto	9/10/97	$2.00	12	$24.00	$1.00	$1.00	$20.00	$2.00
V. Alverez	9/10/97	$2.00	3	$6.00	$1.00	$1.00	$15.00	($9.00)
P. Acosta	9/10/97	$2.00	12	$24.00	–	$1.00	$12.50	$10.50
Whung Yu	9/3/97	$14.00	wages	$14.00	–	–	–	$14.00
M. Baxter	9/2/97	$3.00	1	$3.00	–	$1.00	–	$2.00
Ah Jim	9/3/97	$43.05	wages	$43.05	–	–	–	$43.05
P. B. Warden	9/3/97	$3.50	33	$115.00	$33.00	$1.00	–	$81.50
J. McKain	9/3/97	$3.50	31	$108.50	$31.00	$1.00	–	$76.50
R. Ocha	8/3/97	$2.50	31	$77.50	–	$1.00	$30.00	$46.50
J. Shay	8/30/97	$2.50	31	$77.50	$31.00	$1.00	$5.00	$40.50
M. Villacosa	7/30/97	$2.00	25.5	$51.00	–	$1.00	$22.50	$27.50
J.M. Martinez	7/30/97	$2.00	17.25	$34.50	–	$1.00	$30.00	$3.50
A Romero	7/23/97	$2.50	21	$52.50	–	$1.00	$40.00	$11.50
J. M. Lead	7/23/97	$3.50	14	$49.00	$16.00	$1.00	$5.00	$27.00
H. Rodriquez	7/22/97	$2.00	18.25	$36.50	–	$1.00	$27.50	$8.00
J. M. Ilen	7/20/97	$3.50	14.5	$50.75	$16.00	$1.00	$1.00	$32.75
J. Romero	7/20/97	$2.00	16.5	$33.00	–	$1.00	$25.00	$7.00
HJ. Kenne	7/20/97	$3.00	15	$45.00	$19.35	$1.00	$7.50	$17.15
G. Villa	7/19/97	$2.00	14	$28.00	–	$1.00	$17.50	$9.50
C. Lopez	8/31/97	$2.00	29	$58.00	–	$1.00	$37.50	$19.50

SOURCE: Superior Court Case # 1,025

"Muckers" Loading Ore Carts in the Open Pit of the Golden Cross Mine. Photo courtesy of the San Diego History Center.

forwarding to you a petition for the organization of a school district. . . ." Fuller argued that none of the petitioners were taxpayers and the company was not "... disposed to assume the burden of additional taxation for school purposes." He further noted, "This company has no children to educate and is not disposed to pay for a school for the benefit of a floating and changing population such as is peculiar to a mining camp."[29] When the San Diego Board of Supervisors took up the matter in March, the Golden Cross Company's attorney, H. L. Titus, again protested the petition. However, the board overruled his arguments and established the Hedges School District.[30]

The varied ethnicity of Hedges created a diverse cultural community. Traditional American culture of the company management, professionals, and skilled laborers was reflected in their celebration of certain holidays. For example, in February 1896, the Knights of the Drill held a party to celebrate the construction of their union hall. "An elegant supper was served by Mrs. Roberts, between 11 p.m. and 1 a.m." At daybreak some of the guests witnessed "... two of Yuma's fair daughters climbing the hills on burros. And a most charming sight they were."[31] The newspaper noted that in November 1900, "Thanksgiving was observed in the usual manner. The boarding house gave an excellent dinner of turkey and fixings."[32]

The Mexican inhabitants held to their traditional culture. Spanish was the primary language of the town. Data from the 1900 census indicate the family was the foundation for Mexican society at Hedges. As in many Mexican communities, it served as the basic social unit that provided food, clothing, and shelter and also as the primary institution "... for the transmission of values, knowledge, and beliefs from one generation to another. Within a web of close kin, children were raised, money and goods shared, and old people cared for."[33]

Religion formed the other basic social institution of Hedges' Mexican population. Mexican families worshipped at a small chapel described as "the smallest Catholic Church in the nation."[34] It was a white painted rectangular frame structure on the eastern end of town with a small cross on the roof. The building had undoubtedly been constructed by the *minero* families. Catholicism constituted an essential force in the Mexican community, and permeated almost every aspect of culture, from marriage to health care. It was a force Mexican people shaped themselves, "... taking what they needed in the way of ritual and belief from the church hierarchy and embellishing it with their own traditions."[35] The pious at Hedges worshipped on their own without the regular services of a priest. Elderly people could be heard praying in their homes as miners walked to work in the early morning hours.[36] The small church constituted the physical presence of religious faith in the community and, like almost all Hispanic churches, was undoubtedly always open for the use and comfort of the faithful even without a resident priest and regular services.

Another social institution relied upon by Mexicans at Hedges was the Alianza Hispano Americana. In order to preserve their culture and lifestyle in the face of a rapidly growing Anglo-American population in the southwest, Mexican Americans formed fraternal mutual aid associations with the goal of preserving their culture. Known as *mutualistas*, the first to be formed was the Alianza Hispano Americana, founded in Tucson in 1894. Its goal was to preserve the dignity of the Hispanic race and to teach the working class temperance and moderation. The society nourished cultural pride among its members and offered low cost life insurance benefits and social activities.[37]

The Alianza soon grew from a mutual aid society created in southern Arizona to a fraternal insurance society that served Hispanic communities throughout the United States and Mexico.[38] In existence by 1898, Hedges' Alianza Lodge Number 8 was among the first ten chapters to be established.[39] Meetings were held at the school house. Lodges were extremely popular in mining towns like Hedges where Mexican miners earned low wages and most needed to provide for a family, making savings difficult. The Alianza's insurance provided for funeral services and paid benefits to a family if the breadwinner died.[40]

Minero families at Hedges celebrated traditional Mexican holidays as well as other festivities. As part of the larger Hispanic community living throughout the Cargo Muchacho mining camps and at nearby Ogilby, Hedges' Mexican residents participated in a variety of social events. On September 16, 1891, Mexican Independence Day, a large crowd from the entire region celebrated at the Cargo Muchacho Mine with dancing, races, free dinner, and sports.[41] September 16 was also regularly celebrated at Hedges.[42] The town had a reputation for celebrations and dances still remembered by descendants of its Mexican population.[43]

Unlike the large copper mining towns of Arizona, where management forbade socializing between Hispanics and other "whites,"[44] social gatherings at Hedges apparently resulted in a relaxation of class hierarchy. On April 20, 1895 the Yuma newspaper noted:

Hedges' Alianza Hispano Americana Lodge No. 8, Meeting Inside the Schoolhouse, circa 1902. Photo courtesy of the Arizona Historical Society, Yuma.

Those that attend the ball this evening at the Golden Cross will no doubt have a pleasant time. The people of that place have won a reputation for hospitality second to none here abouts. We hope a good crowd from here will be present, as the members of the Golden Cross Dance Club will be pleased to see and entertain all who go on a grand scale.[45]

Given the scarcity of available Anglo women in the camp it seems unlikely that the Golden Cross Dance Club would have been closed to Mexicans. In November 1896, the Miner's Union sponsored a dance.[46] In spite of union members' opposition to *mineros* joining their ranks, they could not have excluded them from this type of social gathering since their daughters and sisters were the only available women in camp. Attempts to import ladies of their own heritage from nearby Yuma would have been largely unsuccessful since Mexican families also dominated that community.[47]

ACCIDENTS AND CRIME

Untimely death constituted another aspect of life at Hedges. Family life, religious worship, and social festivities of the community functioned against a backdrop of accidental and violent death that must have added an almost fatalistic aspect to the lives of the inhabitants.

Accidents were a constant reminder of the precarious nature of life in western mining camps, and Hedges was no exception. Mining has always been a dangerous occupation where accidents are common.[48] In April 1895, a large cave-in occurred at the Golden Queen Mine. Luckily, no one was in the shafts at the time. Results of later incidents would not be so fortunate. Allen McLean, the underground foreman at the Golden Cross, was smashed by loose rock in September 1896 while he inspected a recent blast to determine the timbering needed. His back was broken and his body crushed. A doctor and priest arrived from Yuma to attend while he died in "agony."[49] Three months later a cave-in at the 400 foot level of the Queen trapped four miners who were later rescued.[50] In September 1900, papers reported that an accidental explosion at the Gold Rock mines killed William Ames and Daniel Bean.[51] Francisco Gutierrez was killed in June 1901 when he was buried in a tailings pile while loading ore carts for conveyance to the cyanide plant. Gutierrez, a risk taker, had been told repeatedly not to work next to the steep sandy tailings banks after being buried up to his waist twice before.[52]

Fire was another concern for the community. The large number of flimsy structures crammed together in the valley bottom combined with open fires and the close proximity of industrial activity compounded the hazards. In December 1896, two dwellings caught fire. Luckily no one was injured and the *Arizona Sentinel* noted they represented a "small loss" for the company.[53] The previous year, the situation had been less fortunate when a Yuma newspaper reported in October 1895 that "... the pretty little daughter of Manuel Oleata was burned to death when her clothing caught fire while she played around a small outdoor fire." The doctor was sent for while the parents started with the child for Ogilby. The girl died at Ogilby before the doctor could reach her. She was buried in Yuma.[54] In February 1903, Juliana Cilla, an elderly woman about seventy, burned to death when her small canvas covered cabin incinerated on the early morning of February 8th.[55]

Harsh environmental conditions combined with health and social problems to cause tragedy. A relatively minor incident occurred in April 1895 when a doctor from Yuma responded to a summons to treat a man suffering great pain from the bite of a small red spider.[56] One can only speculate the individual suffered from an allergic reaction. On May 8, 1897, J. A. Gassir, an asthma sufferer, died of morphine poisoning. Gassier often took the drug because of his condition and had experienced several close calls in recent months.[57]

On the morning of October 19, 1899, a twenty-three-year-old prostitute known as Georgia or Etta Winters was found dead in her room. Georgia "boarded" at Mabel Lee's place in the "stingaree" or commercial district. Mabel evidently had a series of small rooms associated with her saloon where the "girls" resided and entertained, similar to Wilson's establishment. Georgia had risen about seven on the morning of October 19. Around eleven she went to her room. A short while later one of the other "girls" passed by and saw her reclining on the floor with her face toward the wall. As Georgia often rested by stretching out on the floor, her fellow inmate thought nothing of it. Later that day a male patron became concerned when he found Georgia's door locked and could not raise her. He broke down the door to find a Mexican miner, well known in the stingaree for "treating" everyone on payday, inside with his pants down and Georgia's dress pulled up over her head. Georgia's patron, recognizing she was dead, threw the inebriated miner into the street where a gathering crowd began to kick and beat him. The company doctor was summoned and determined Georgia had been dead for some time. She had no signs of violence on her body and the cause of death was undetermined. The miner, who was quite intoxicated, claimed he thought she was drunk and passed out on the floor.

Georgia had arrived in Hedges on the second of the previous month with "Miss Lee." The madam had met the girl in Ventura, apparently destitute, without money or belongings. Mabel had offered to pay her train fare to Hedges and the opportunity to occupy one of her cribs in order to "make something." The young girl drank regularly and had discussed killing herself with laudanum several times after her arrival. At the coroner's inquest Mabel Lee denied knowing Georgia Winters, claiming her only contact with the girl had been a business transaction. Her fellow inmates at Miss Lee's also denied any knowledge of her, her habits, or companions. The jury determined that Georgia or Etta Winters, "true name unknown" had died of unknown causes.[58]

Lawlessness added another dimension to life in Hedges. Although a sheriff and constable resided in town, outrageous violence, often encouraged by the catalyst of alcohol, gambling and prostitution, occurred regularly. Violence was perpetuated in the context of other crimes as well as resorted to as a means of resolving personal problems between individuals. In August 1894, the body of a well dressed man was found murdered four miles from town. The same month a reporter noted that "young" Ignacio Huegria had been forced to leave town to escape the vengeance of "... the father of the girl he was reported to have ruined."[59]

Saloons of the Commercial District became the locations of many violent acts. The first murder of the camp occurred at Wilson's Saloon on the evening of December 16, 1895. Among the patrons in the crowded room, a prospector named Mulachy stood at the bar drinking with two brothers, Juan and Tirco Gonzales. Around seven o'clock Henry Randolph, another prospector, entered and sat at a table. Mulachy walked to Randolph's table and accused him of the murder and the claim jumping of a friend of his on the desert. The antagonist had been drinking heavily all day and "... Randolph told him to drop the subject.... Mulachy then engaged in some vile epithets."[60] He continued his harassment and repeated his slurs until Randolph stood up and knocked him to the floor. Still standing at the bar, Juan Gonzales pulled a "six shooter" out of his hip pocket. Sam Wilson grabbed his arms attempting to pin him. The gun went off firing over the heads of the crowd. Juan broke free of Wilson's grip and fired a shot at Randolph who had started backing toward the front door. Randolph returned fire, hitting him in the stomach. As Juan slumped to the floor, Tirco Gonzales rushed to his collapsing brother's side. Twisting the pistol from Juan's hand he pointed it at Randolph while backing out the door and pulled the trigger twice. The weapon, however, misfired. Randolph fired through the door, hitting Tirco, who fell over some beer kegs stacked on the front porch and into the street, a bullet in his heart. Randolph walked out of the saloon past his second victim and started up the street. In spite of his wounds, Juan Gonzales ran out after him, pulled the pistol from his dying brother's hand and fired after Randolph as he escaped. Juan died twenty-four hours later, survived by a destitute mother. A coroner's inquest determined Randolph had killed in self-defense.[61] Randolph continued to prospect in the area through the turn of the century and filed the Inglewood, Mayate, Telephone Man, and Telephone Girl claims.[62]

Four months later, on March 8, 1896, another slaying occurred at Wilson's. Benjamin Serrano, who had been in the camp for two weeks visiting his brother, arrived at the saloon on a Sunday evening about seven. Inside, patrons were drinking and dancing. Serrano took up with a local miner, Abril Rivero, and Jesues Amenta, one of four prostitutes who resided and plied her trade in a room behind the saloon. At one point in the evening another patron, Manuel Boboun, had words with Benjamin, and accused him of spreading rumors that Boboun had stolen thirty-five dollars from Jesues. Around midnight Benjamin, Abril, and Jesues decided to eat and sat down at a table in the adjoining room while they waited to be served. Boboun came in from the bar, walked up to Rivera and asked accusingly, "*Estas senando con este desgraciado* (Are you taking supper with this disgraceful lowlife)?" In the same instant he picked up a pitcher of milk that had been set on the table and threw it at Serrano, then moved toward him as he drew a dagger. Serrano stood up, drew his pistol and shot Boboun twice. He died the following morning. A coroner's inquest determined Serrano had killed in self-defense.[63]

By December of the same year, the town had gained a reputation for homicide when the *Arizona Sentinel* reported that "Gold Rock has added another murder to her already long list."[64] Bernardo Contreras and Joaquin Fuentes were in town participating in a drunken orgy. When Contreras tried to force Fuentes to eat a watermelon, the latter slashed him across the stomach with a knife. Contreras died a few hours later. Both men were considered "hard characters" and had been feuding for several days. Fuentes was convicted of manslaughter.[65]

The following December, a shooting took place at the saloon of Horan and Adams. Deputy Sheriff Joe Smarr came into the bar and began to search a miner named Juan Flores. The deputy shot Flores, claiming that the miner had made a quick motion with his arm as if attempting to pull a knife. Even though Flores's companions testified that Juan had not made any sudden moves the coroner's inquest found Smarr had killed in self-defense.[66] Another incident occurred in the same saloon when Billy Horan tried to break up an argument over a card game. One of the Mexican antagonists pinned his arm to the wall with a knife. Horan's partner, Adams, shot and killed the man, saving Billy's life.[67]

In April 1899, still another shooting occurred when Constable August Jarick entered a saloon and ordered a card game closed. A heated argument broke out between the constable and the bartender Crawford. Crawford suddenly broke off the exchange and started to go behind the bar. Jarick, apparently afraid he was going for a gun, shot and wounded him.[68]

On December 18, 1899 violence combined with a language barrier led to the death of an innocent bystander. The incident began when a man named Charley Salamon was knifed in Frank Norton's saloon. Norton and a companion, Epimenio Melendez, armed themselves and attempted to follow Salamon's fleeing assailant. After a while they split up. Melendez, hiking alone, came over the crest of a hill and saw someone walking away from him. He thought this was the fugitive when in reality it was a black man named John Lee, employed as a barber in Juan Cano's shop. Melendez, who could not speak English, shouted four times at Lee to stop while pointing a rifle at him. Lee, who did not understand Spanish, evidently felt threatened. He drew a gun and started to run. Melendez fired three shots, mortally wounding the barber. Melendez was sentenced to prison for manslaughter.[69]

After 1900, the shootings ended, perhaps as a result of a reduced work force. In December of that year a newspaper report described Hedges as an "... orderly and prosperous camp.... Everything is well regulated and is run on exact methods so that there is far less noise and lawlessness here than is usually found in mining camps."[70]

The lack of gunplay, however, did not signify an end to all violence. On January 1, 1900, James Kane was arrested for assault and battery upon Thomas Gallan. In May, Juan and Refugio Varlen were charged with assault and battery.[71] In July 1901, Dolores Pistrano was also charged with battery, and in 1902 Jose Pistrano was charged for larceny.[72] In general, the violence at Hedges mirrored the experience of many other western mining camps where saloons, liquor, and prostitution provided an explosive environment in which beatings, knifings, and gunfights occurred on a regular basis.[73]

CONCLUSION

After the shutdown of Free Gold's operation in 1905 Hedges lay abandoned. In 1909 the mines were reopened by the United Mines Company whose directors included T. S. Fuller, W. V. Hedges and Gail Borden (of canned milk fame) along with several others.[74] Laborers reoccupied the town, now called TUMCO after the initials of The United Mining Company. A school opened in the fall of 1910.[75] The venture, however, was short-lived. United Mines closed and abandoned the mines and town by the summer of 1911.[76]

By the 1920s, Hedges had become a place of legends. According to popular accounts and local residents it had been a town of two to four thousand people, with few families or children, numerous Chinese, a large red light district, and at least seven saloons. A hundred and fifty Chinese were said to have been buried in a cave-in, and the brutal violence that actually had occurred was romanticized into swashbuckling adventure.[77]

The true story of Hedges needs no embellishment. It is a saga of real people struggling for survival and success in a harsh desert frontier environment. The town was a diverse place, where residents led lives that emphasized family, religious, and community values, and where children were important enough that families joined together to provide them with a school in spite of the protests of company managers. Yet, community and family life occurred against a backdrop of crime, prostitution, accidents, and violence.

Finally, in exemplifying the merging of corporate American mining with the Hispanic mining frontier, which was discussed in Part I of this series, and providing a profile of the lives of Mexican miners in a company town at the end of the nineteenth century, the history of Hedges is a case study of a still largely undocumented aspect of western mining. During the last half of the nineteenth century, American mining companies replaced traditional Spanish-Mexican mining methods with modern machinery and technology in the southwest, but continued to rely heavily on the Hispanic mining population for labor. With only a few exceptions, the important role of Hispanic prospectors and *mineros* has remained undocumented by western mining historians.[78]

Yet Mexican miners did not disappear from the west after passage of the California Foreign Miners Tax in 1850. As independent prospectors and labors, Hispanics were often the overwhelming majority of the mining work force in Arizona, New Mexico, and portions of the deserts of eastern California and southern Nevada. In many communities, including Hedges, *minero* families constituted a majority of the population and played an indispensable role in western mining development.

NOTES

1. The area of San Diego County between the Colorado River and the eastern edge of the Peninsular Range became Imperial County in 1907. See Owen C. Coy, California County Boundaries: A Study of the Division of the State into Counties and the Subsequent Changes in their Boundaries (Berkeley: California Historical Survey Commission, 1923; revised edition, Fresno: Valley Publishers, 1973), 113.
2. James B. Allen, The Company Town in the American West (Norman: University of Oklahoma Press, 1966), 6; Hector Galan and Paul Espinosa, "Los mineros," transcript of the film produced by Hector Galan, Galan Productions, for the American Experience, WGBH, Boston (1991).
3. C. W. Pauly, Petition of C. W. Pauly, San Diego County, Superior Court, case no. 10,215 (1897), San Diego History Center Research Archives, San Diego; Isaac Trumbo, Final Report of Isaac Trumbo, Receiver, Superior Court, case no. 10,215 (1901).
4. San Diego Union, 15 December 1900, 12:3.
5. San Diego County Directory 1900, San Diego History Center Research Archives; U. S. Census Bureau, Population Manuscript Returns of the Twelfth Census of the United States (1900) for San Diego County, National Archives and Record's Center, Laguna Niguel, CA.
6. U.S. Census Bureau, Twelfth Census, 1900, Population Manuscript Returns.
7. Jesues Amenta, "Testimony of Jesues Amenta in Inquisition by the Coroner's Jury in the Matter of the Inquest Upon the Body of Manuel Boboun" (1896), Coroner's Inquests, San Diego History Center Research Archives, San Diego.
8. U.S. Census Bureau, Twelfth Census, 1900, Population Manuscript Returns.
9. Michael E. Casillas, "Mexican Labor and Strife in Arizona, 1896–1917," (master's thesis, University of New Mexico, 1979), 223; Galan and Espinosa, "Los mineros"; Mario T. Garcia, Desert Immigrants: The Mexicans of El Paso, 1880–1920 (New Haven: Yale University Press 1981); James E. Officer, Hispanic Arizona, 1436–1856 (Tucson: University of Arizona Press 1987).
10. Hedges school records 1896, 1897, School District Records, San Diego Historical Society Research Archives; San Diego County Directories 1905–1906, San Diego History Center Research Archives.
11. Donald C. Miller, Ghost Towns of California (Boulder, Colorado: Pruett Publishing Company, 1978), 168–170; Peter Odens, "Last of Tumco's Residents 'Love It'" Imperial Valley News, 17 January 1980.
12. Hedges school records, 1897.
13. Ibid., 1901.
14. U.S. Census Bureau, Twelfth Census, 1900, Population Manuscript Returns.
15. Ibid.
16. For an analysis of demographic patterns similar to those at Hedges see Thomas E. Sheridan, Los Tucsonenses: The Mexican Community in Tucson, 1854–1941 (Tucson: University of Arizona Press, 1986), 134.
17. U.S. Census Bureau, Twelfth Census, 1900, Population Manuscript Returns.

18. For an analysis of demographic patterns similar to those at Hedges see Sheridan, Los Tucsonenses, 134.
19. Galan and Espinosa, "Los mineros"; Diane North, "'A Real Class of People' in Arizona: A Biographical Analysis of the Sonora Exploring and Mining Company 1856–1863," Arizona and the West 26 (Fall 1984): 261–174; Casillas, Mexican Labor, 6–10. Billy Horan, "When Tumco was Hedges," Calico Print (1951), on file Arizona Historical Society Archives, Yuma.
20. San Diego County Directory, 1900; U.S. Census Bureau, Twelfth Census, 1900, Population Manuscript Returns.
21. Alex Sortillon, interview by William Ballinger, 9 April 1941, Tumco File, Colvocoresses Files, Arizona Geological Survey, Tucson; Pay vouchers for the Golden Cross Mining and Milling Company (1897), San Diego Superior Court, case no. 10,215.
22. Hedges school records, 1900.
23. Galan and Espinosa, "Los mineros."
24. Arizona Sentinel, 7 December 1895, 3; Frank Love, Mining Camps and Ghost Towns: A History of Mining in Arizona and California Along the Lower Colorado (Los Angeles: Westmoreland Press, 1974), 80.
25. Casillas, Mexican Labor, 21; Galan and Espinosa, "Los mineros."
26. Pay vouchers (1897).
27. Thomas J. Schlereth, Victorian America: Transformations in Every Day Life, 1876–1915 (New York: Harper Collins, 1991), 78; Time Life Editors, This Fabulous Century, Sixty Years of American Life, Vol. I: 1900–1910, (New York: Time Life Books, 1969).
28. Hedges school records, 1896.
29. Thomas Fuller, Affidavit of Thomas Fuller, Superior Court, case no. 10,215 (1896).
30. Hedges school records, 1896.
31. Arizona Sentinel, 29 February 1896, 3; Barbara Rigby Connell, "Cultural Resources Management Plan for Hedges/Tumco Ghost Town, Imperial County, California, A Field Report Submitted in Candidacy for the Degree of Masters of Art Program in Historic Resource Management in Cooperation with The Bureau of Land Management," (Riverside: University of California, 1979), 42–43.
32. San Diego Union, 15 December 1900.
33. Sheridan, Los Tucsonenses, 152.
34. San Diego Union, 15 December 1900.
35. Sheridan, Los Tucsonenses, 153.
36. "Inquisition by Coroner's Jury in the Matter of the Inquest Upon the Body of Juliana Cilla," 10 February 1903, Coroner's Inquests.
37. James McBride, "The Liga Protectora Latina: A Mexican-American Benevolent Society in Arizona," Journal of the West 14 (Fall 1975): 82–90. Kaye Lynn Brigel, "Alianza Hispano-Americana, 1894–1965: A Mexican American Fraternal Insurance Society," (Ph.D. diss., University of Southern California), (Ann Arbor Michigan: University Microfilms #74-23,572 1974), 1; Sheridan, Los Tucsonenses, 111–113.
38. Sheridan, Los Tucsonenses, 111–113.
39. Arizona Sentinel, 16 December 1898, 1; McBride, "Liga Protectora."
40. McBride, "Liga Protectora."
41. Arizona Sentinel, 21 November 1891, 3; Frank Love, Cargo Muchacho Files, Personal Archival Collection, Yuma Arizona.
42. Connell, "Cultural Resource Management Plan," 45.
43. Galan and Espinosa, "Los mineros."
44. Ibid.
45. Arizona Sentinel, 20 April 1895, 3.
46. Ibid., 22 February 1896, 3.
47. Joseph H. Park, "The History of Mexican Labor in Arizona During the Territorial Period," (master's thesis, University of Arizona, 1961).
48. Galan and Espinosa, "Los mineros"; Richard Lingenfelter, The Hardrock Miners: A History of the Mining Labor Movement in the American West, 1863–1893 (Berkeley: University of California Press, 1974).
49. Arizona Sentinel, 19 September 1896, 3.
50. Love, Cargo Muchacho Files; Arizona Sentinel, 12 December 1896, 3.
51. Love, Cargo Muchacho Files; Yuma Sun, 7 September 1900, 5.
52. "Inquisition by Coroner's Jury in the Matter of the Inquest Upon the Body of Francisco Gutierrez," 20 June 1901, Coroner's Inquests.
53. Love, Cargo Muchacho Files; Arizona Sentinel, 7 September 1896, 5:1.
54. Arizona Sentinel, 5 October 1895, 3.
55. "Inquisition by Coroner's Jury in the Matter of the Inquest Upon the Body of Juliana Cilla," Coroner's Inquests, 10 February 1903.
56. Arizona Sentinel, 20 April 1895, 3.
57. "Inquisition by Coroner's Jury in the Matter of the Inquest Upon the Body of J. A. Gassir" (1897), Coroner's Inquests.
58. "Inquisition by Coroner's Jury in the Matter of the Inquest Upon the Body of Etta Winters, Sometimes Called Georgia Winters, Whose True Name is Unknown," 22 October 1899, Coroner's Inquests.
59. San Diego Union, 20 August 1894, 5:3.
60. Arizona Sentinel, 21 December 1895.
61. "Inquisition by Coroner's Jury in the Matter of the Inquest Upon the Body of Tirco Gonzalez" (1895), Coroner's Inquests; Arizona Sentinel, 21 December 1895; Love, Mining Camps and Ghost Towns, 117.
62. San Diego Union, 2 April 1899, 5:1; 1 December 1895, 5:1; 23 October 1902, 8:2.
63. "Inquisition by Coroner's Jury in the Matter of the Inquest Upon the Body of Manuel Boboun," 10 March 1896, Coroner's Inquests; Arizona Sentinel, 14 March 1895: 3; Love, Mining Camps and Ghost Towns, 117–118.
64. Arizona Sentinel, 26 December 1896, 3.
65. "Inquisition by Coroner's Jury in the Matter of the Inquest Upon the Body of Bernardo Contreras," 15 August 1896, Coroner's Inquests; Arizona Sentinel, 15 August 1897, 3; Love, Cargo Muchacho Files.
66. "Inquisition by Coroner's Jury in the Matter of the Inquest Upon the Body of Juan Flores," 19 December 1896, Coroner's Inquests; Love, Mining Camps and Ghost Towns, 118.
67. Horan, "When Tumco Was Hedges."
68. Love, Ghost Towns and Mining Camps, 118; Arizona Sentinel, 21 April 1899, 1.
69. "The People vs. Epimenio Melendez," Transcript of Preliminary Examination (1899), San Diego County Justice Court, San Diego

History Center Research Archives, San Diego.

70. San Diego Union, 15 December 1900, 8:3.
71. Isaac Trumbo, Report of Justice of the Peace (1900), Hedges Documents File, San Diego County Historical Society Research Archives.
72. Warner, Reports of Justice of the Peace (1901–1902), Hedges Documents File, San Diego County Historical Society Research Archives.
73. For a case study analysis of violence in western mining camps see Roger D. McGrath, Gunfighters, Highwaymen, and Vigilantes: Violence on the Frontier (Berkeley: University of California Press, 1984).
74. Arizona Sentinel, 20 December 1909, 3.
75. Common School Reports for Imperial County (1910), Sacramento: California State Archives.
76. Ibid., 1911. From 1912 to 1917 Seeley Mudd and the Queen Mining Company both attempted to find profitable ore where so many had already failed before them. In the mid-1920s Carl Walker purchased the Gold Rock claims. He lived with his family in the abandoned ghost town and later homesteaded nearby Walker Ranch, which became a popular camping spot and desert antique and rock shop. Several entrepreneurs and mining companies leased the mines from Walker and attempted small scale operations between 1929 and 1948 including Earnest Riggs, the Lacey Manufacturing Company, the Sovereign Development Company, and TUMCO Mines Incorporated. Since 1949 no serious work has been conducted at the mines, however the sharp increase in the price of gold during the 1980s opened a new era of mining in the Cargo Muchachos at the former Madre y Padre and American Girl claims and plans are now (1994) underway to reopen the Golden Cross Mine.

    For a detailed account of mining in the Cargo Muchachos during the Twentieth Century see, Stephen R. Van Wormer and James E. Newland "History of Hedges Tumco" in Hedges/Tumco: Historic Mining Traditions of Southeastern California, by Michael S. Burney, Stephen R. Van Wormer, Claudia B. Hemphill, James D. Newland, William R. Manley, F. Paul Rushmore, Susan D. Walter, Neal H. Hupel, Jerry Schafer, and Lynne E. Christenson, Cultural Resource Management Report prepared by Burney and Associates, Boulder, Colorado for USDI-Bureau of Land Management, El Centro. Copies on file at the San Diego History Center Research Archives, San Diego, and Arizona Historical Society Archives, Yuma, Arizona.
77. Miller, Ghost Towns of California; Odens, "Last of Tumco's Residents 'Love It'."
78. Such important works on western mining as Lingenfelter, The Hardrock Miners; Peterson, Bonanza Kings; Young, Western Mining, and Black Powder and Hand Steel: Miners and Machines on the Old Western Frontier (Norman: University of Oklahoma Press, 1976); and Jane Eppinga, "Ethnic Diversity in Arizona's Early Mining Camps," History of Mining in Arizona, Vol. II, edited by J. Michael Canty and Michael N. Greeley (Tucson: Mining Club of the Southwest Foundation and American Institute of Mining Engineers, 1991) have failed to discuss the role of Hispanic prospectors or hardrock miners in the development of western mining during the late 19th and early 20th centuries. Notable exceptions are Galan and Espinosa, "Los mineros"; and Park, The History of Mexican Labor in Arizona.

Steven Van Wormer is a self-employed historical and archaeological consultant. He received an M.A. in history from San Diego State University in 1986 and has over thirty-three years of experience in conducting historical and archaeological studies in greater Southern California. Mr. Van Wormer has published articles in several scholarly journals including the *Southern California Quarterly*, *Pacific Coast Archaeology Society Quarterly*, and *California and Great Basin Anthropology*.

James D. Newland is the Supervising Historian and Manager of the Resources & Interpretive Services Sections of California State Parks' Southern Service Center in San Diego. He is the author of two books, hundreds of technical studies, numerous scholarly and popular historical articles and is an award-winning columnist with the La Mesa Patch web-paper. Mr. Newland earned his B.A. degree in Social Sciences and his M.A. in Public History at San Diego State University.

Once a large gold mining operation, Tumco Mine today [1969] is silent with only a few buildings standing as reminders of its once active past. Photo by Jack Pepper.

Mysterious Cargo Muchachos

BY JACK PEPPER

SWEATING AND CURSING THE 120-DE-gree heat the miners hacked at the iron ore deposits under the blistering sun which turned the rock formations into an inferno. Beside them worked captive Indians— slaves to the white man and his insatiable quest for gold. Indians who would soon revolt, kill their masters and dump the gold into the Colorado River.

The time was the 1770s and the place the Cargo Muchacho Mountains in California's Imperial County. Today the Cargos and the nearby Chocolate mountains are pocketed with more recent mining operations and there are a few — but not many — more people in the area. There still remain hundreds of forbidding canyons and washes which probably never have been transgressed by white man.

Although there is no documentary evidence, it is believed that the Cargos were the site of the first discovery of gold in California, three quarters of a century before Marshall found nuggets in the Mother Lode country and started the stampede of the '49ers.

J. Ross Browne and William Blake, both early explorers and historians, said gold was known along the Colorado River in the 1770s, and Paul Henshaw, writing in the *California Journal of Mines and Geology* in 1942 stated mining was first carried on in the Cargo Muchachos in 1780 with the founding of the Spanish settlements on the Colorado River.

Although what Spanish records might have existed were lost when the Indians revolted and destroyed the mission pueblos, present day members of the Yuma (or Quechan) Indians still talk of their ancestors who were forced to dig gold for the Spaniards. As Fray Salmeron reported in 1629, the Spanish soldiers "out of greed for

This article originally appeared in *Desert Magazine*, Volume 32, Number 1, January 1969.

silver and gold would enter hell to get them." This particular hell was in the form of *tepustete de color*, found in the region. *Tepustete* is goldbearing iron ore found in Sonora and in which the early Spaniards thought would be found gold in the Colorado River area.

In 1775 Father Francisco Garces, the dedicated Jesuit missionary who spent his life exploring and documenting the Southwest, commented on the *tepustete* and said his Indian interpreter indicated there was much gold in the area. At the time they were camped near the present town of Yuma.

Five years later Father Garces, three other missionaries and some settlers, allegedly "protected" by Spanish soldiers, established two mission-pueblos on the Colorado River. This was in 1780. Father Garces was not interested in gold, but evidence indicates the Spaniards had been mining gold for 10 years in the area with Indian slave labor. One year after the mission-pueblos were founded, on July 17, 1781, the Indians revolted, killed Father Garces and the other missionaries, and captured those soldiers and settlers who were not slain. The Indians made no distinction between the white men who sought their salvation and those who sought gold.

Today the evidence of these Indians can be found in the distinct rock trails across the mountains and desert, petroglyphs and broken pieces of pottery (potsherds), a few arrowheads and the mysterious scratches on the hundreds of rocks at Indian Pass.

These interesting Indian artifacts and landmarks, coupled with old mines which produced millions of dollars in gold during their operation in the late 1800s and the early 1900s, and the availability of semi-precious stones make the Cargo Muchacho Mountains an area rich in adventure.

And you can see and visit the majority of the interesting places in a passenger car. The once rutted roads have now been graded and are kept up by the county. Al Pearce, whose article on the vanishing Bighorn Sheep appeared in last month's *Desert*, and his wife and I explored the area for three days recently. We could have spent a week and still only covered a small section.

Vandals have dug up and desecrated the graves at Tumco Mine and shot holes into a sign warning it is a criminal offense to commit such acts. If further acts of destruction continue the mine may be closed to the public.

The road from Indian Pass through Gavilan Wash to the Colorado River is spectacular and colorful. It should not be attempted except in back country vehicles due to several steep hills and soft sand.

Al Pearce, explorer and writer, uses a metal detector to look for old coins near one of the many adobe ruins found in the Cargos. The area is also a favorite hunting ground for rockhounds.

Although there are many old mines in the Cargo Muchachos which are abandoned and may be explored, some have recently been reactivated and are posted. Even if there are no signs of life at the posted mines, do not trespass.

We made our headquarters at the Gold Rock Trading Post. To reach Gold Rock, turn left off Highway 80 from El Centro to Yuma, just before Winterhaven and just after 80 ceases to be a freeway. A good gravel road takes you to the former railroad town of Ogilby and the railroad tracks. Once a flourishing community, Ogilby today consists of cement slabs and three lonely graves.

Cross the railroad tracks at Ogilby and watch for the sign to Gold Rock. Keep left and five miles down the graded road you will reach the trading post. An asphalt road is being built parallel to the graded road to Gold Rock and goes between the trading post and Tumco Mine. If you take this road turn left at the entrance to the Tumco Mine — seen on the right — and two miles down the road you'll arrive at the post.

Hosts at Gold Rock are Bob and his wife, Margrete, better known as "Scotty," who have an interesting and vast collection of rocks and mining equipment, plus Indian artifacts which they have collected during their years of exploring. They are a wealth of information and will direct you to places of interest and rock hunting areas.

They also have soft drinks and sandwiches and a wide area where you can pitch your tent or park your camper. Bob Walker is the son of Carl and Margaret Walker who established the trading post and were the last owners of the Tumco Mine.

Tumco Mine was more than a mining operation; it was a gold camp and company town. Named from the initials of the corporation, The United Mines Company, it had a peak population of 2500 during its heyday. Started in 1892, it flourished until 1909 when it was shut down for one year. New owners operated it on a more limited scale from 1910 until its demise in 1916. It is located only two miles from the Walkers.

As you enter Tumco Mine keep to the right on the gravel road. Although there are several washes by driving slowly a passenger car can easily maneuver. The main road will take you up to the main glory hole where you can turn around. Don't let children wander by themselves as there are unfenced holes in the area.

In the middle of Tumco and across the tailings to the left you will see the main building shown in our color photograph with this article. Just below this building and across the old mining road is the graveyard. To get there walk across the tailings and keep to your left.

Vandals have dug up several of the graves and shot bullets into a sign which states it is a criminal offense to desecrate graves. Vandals have also torn down the parts of the houses of Tumco and dumped trash into the mine shafts. For this reason the Walkers finally gave up their claim to the mine and any liability. If further vandalism continues, however, it may be closed to the public by Imperial County and the Bureau of Land Management.

Tumco was one of several mines which operated at the same time. Others included the American Girl Mine, Cargo Muchacho, Piacho, Micatalc Mill, Vitrifax Hill (which produced the vital kyanite during World War II) and the Padre y Madre Mine. Only the Tumco can be reached by passenger car, however, and many of the other mines in the area once again have been claimed and are off limits to visitors. If you have 4-wheel drive vehicles or dune buggies and are exploring the other mining areas, respect the "Keep Out" signs; there are many places to explore where you will not trespass on private property.

Another interesting side trip is to the area of the mysterious scratched rocks in Indian Pass, easily accessible in a passenger car. From Gold Rock go north 9.3 miles and turn right on a good gravel road which takes you across the flatlands. Ahead on the right you will see Picacho Peak — a landmark for prospectors. If you stop to explore the washes along the way you will discover what look like graves. Actually they are small gun emplacements used by soldiers during maneuvers in World War II.

Approximately 11 miles from the turnoff you start into Indian Pass, reaching the mysterious rocks at about 19 miles from the turnoff. They are at the top of a hill and from here the road descends sharply downward to Gavilan Wash. Unless you have a 4-wheel drive vehicle or a dune buggy DO NOT go down the hill.

If you do have a back-country vehicle, the road through Gavilan Wash takes you to the Colorado River and the now abandoned 4-S Ranch. A Boy Scout camp is located at the end of the road. North along the Colorado are several interesting washes. The road south follows along the banks of the Colorado and through the hills from which you get a panoramic view of the river. Unfortunately there is heavy brush to the river's edge so there are few camping areas.

Approximately 10 miles south from the Boy Scout camp is a small campground and launching ramp. At one time it was a private enterprise, but today is operated by the state. There are no services available here. The campgrounds can also be reached over a passenger car road from Winterhaven (see map). Along the passenger car road from Winterhaven are several old mines, including the Picacho Mine. The section is a favorite hunting ground for rockhounds.

The area approaching Indian Pass and the Pass itself is also rock hunting territory. I am told you can find petrified palm root and fiber, blue dumortierite, silicified fossils, jaspers, agates and chalcedony.

At the top of Indian Pass park your car on a level spot above Gavilan Wash. The scratched rocks are only a few yards off the road to your left. They cover a large area and can easily be found. Also through the area you will find Indian rock rings and Indian trails.

The scratches on the rocks are not very deep but they must be quite old as they are covered with patina (desert varnish). When Malcolm Rogers, an archeologist, visited the area many years ago and dug into the rock circles it is reported he found conventional petroglyphs along with the marked stones. However, as far as I can determine, there are no petroglyphs in the area today.

Who scratched the rocks and when? There are numerous theories. One is that Indians gathered at the pass for religious ceremonies and kept time to their chants by striking the boulders with sharp rocks. Another is they were scores for some kind of game or contest of physical skill. Historians say Apaches once hunted mountain sheep in the area.

A scratched rock from Indian Pass. There are many theories as to why and when the rocks were marked. Indian trails and rock rings are found in the same area.

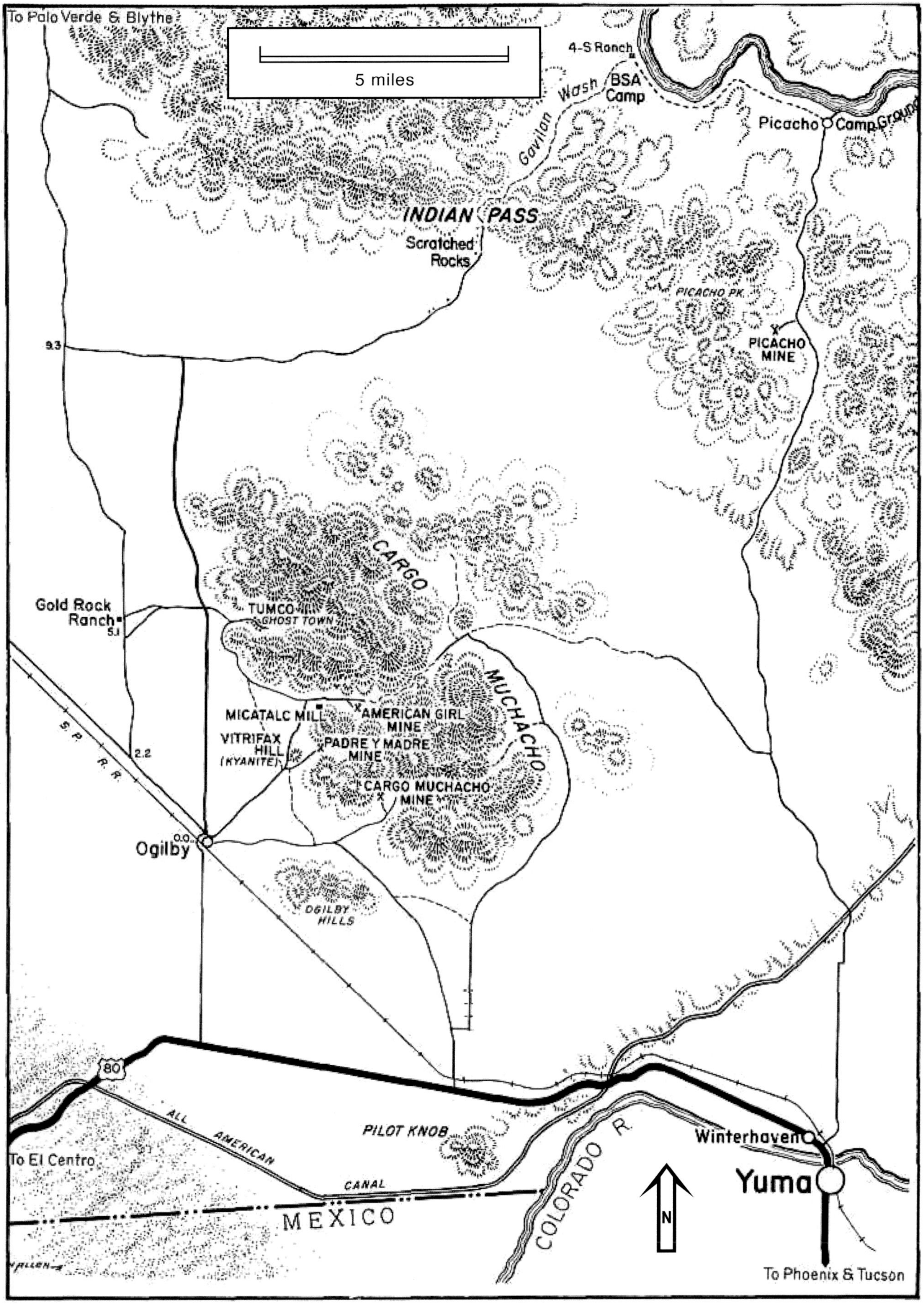
To Palo Verde & Blythe
5 miles
4-S Ranch
BSA Camp
Gavilan Wash
Picacho Camp Ground
INDIAN PASS
Scratched Rocks
PICACHO PK.
PICACHO MINE
9.3
CARGO
MUCHACHO
Gold Rock Ranch
5.1
TUMCO GHOST TOWN
MICATALC MILL
AMERICAN GIRL MINE
VITRIFAX HILL (KYANITE)
PADRE Y MADRE MINE
CARGO MUCHACHO MINE
S. P. R. R.
2.2
0.0
Ogilby
OGILBY HILLS
80
ALL AMERICAN CANAL
PILOT KNOB
To El Centro
COLORADO R.
Winterhaven
Yuma
MEXICO
N
To Phoenix & Tucson

The most logical theory, in my opinion, was advanced by Harold Weight in an article on the rocks in *Desert Magazine* in February, 1949. The author points out the pass lies close to the western boundary of land once claimed by the Yuma Indians. It also marks the natural division point between the valleys and river bottoms of the Colorado and the desert.

It is then possible the camp on the pass was sort of a border station and trading area for various Indian tribes and the marks tin the rocks would be a tally of the persons traveling through the pass or a tally of traded goods. After all, the white man has toll roads, so why not the Indians?

The scratched rocks of Indian Pass are just one of the mysteries of the Cargo Muchacho Mountains and the nearby Colorado River. Who knows what other unsolved riddles are hidden within the rocks and washes of the area? Maybe someday you too will find another page out of the past in this country of fascinating history.

Part of an old Indian trail winds through the rocks covered with patina or "desert varnish." The trails wind for miles and can be seen from the Colorado River to the Chocolate Mountains.

Geothermal Power in Imperial Valley

Only as clean as the Corporations and Fluid Source allow

FRED CAGLE

GEOTHERMAL ENERGY IN THE IMPERIAL Valley is a rising star among government agencies and environmentalists who see it as a source of pollution-free renewable energy. It has several attractive features not shared by wind and solar energy facilities. One of these is that the area actually occupied by a geothermal plant is typically quite small. A 100 megawatt plant may require only 20 acres. It is also a fact that geothermal plants are extremely efficient since the energy is reliably available day and night. Wind and solar energy facilities typically have an annual energy production that is only 20–30% of their rated maximum. It is easy to understand the current interest in developing geothermal energy.

In June 2008, the Department of the Interior announced plans to make more than 190 million acres of federal land available for geothermal energy development in twelve western states. This could potentially provide 5,540 megawatts (MW) of new capacity by 2015 and another 6,600 MW by 2025. In the Imperial Valley, the Bureau of Land Management recently (2009) completed the Truckhaven Environmental Impact Statement and began work on two more to consider geothermal energy in the West Chocolate Mountains. Additionally, the Bureau of Land Management has issued five pending leases and another twelve competitive leases totaling 16,000 acres. The total for geothermal energy production within the Imperial Valley may be about 134,000 acres, although only 4,808 acres are currently developed and producing.

Once thought to be all but negligible, the environmental effects of geothermal energy production can indeed be significant. Among the problem areas are gaseous effluents (hydrogen sulfide, mercury, radon, ammonia) and aqueous solutions that contain heavy metals as well as radioactive materials. Additionally, pulling out vast amounts of heat and brine from deep within the earth may cause land to slump and subside, warping the surface into hills and valleys. Deep reinsertion of the waste geothermal fluids could ameliorate many environmental impacts but are complicated by silica deposition on equipment and in the underground reservoir.

There are three types of geothermal plants operating today:

1. Dry steam plants pipe hot steam directly from the ground to the turbines. The only one operating in the United States is The Geysers facility in northern California.

2. Flash steam plants pull deep (10,000 feet), very hot water (550 degrees Fahrenheit) into low-pressure tanks where it flashes into steam to drive turbines. The steam is then condensed into a liquid brine that requires disposal.

3. Binary cycle plants pass moderately hot geothermal water through a heat exchanger, vaporizing another liquid such as isobutene, which then drives the turbines.

Flash steam plants, the kind found in the Imperial Valley, bring up hot water brines with levels of toxic and radioactive substances that are often sufficiently high to be considered hazardous. When the resulting steam is then condensed into a brine it may contain up to 30% of its weight as dissolved solids. This hot brine is then treated in order to separate out the solids, known as filter cake. At a temperature of about 200 degrees Fahrenheit, the resulting brine is then sent to outdoor pools where still more solids precipitate out. Finally, the remaining fluids are re-injected into the underground reservoir. The solid residues are tested on site by the company and any hazardous filter cake is sent to a hazardous waste facility. The company assumes that all the remaining filter cake is non-hazardous by standards of the Environmental Protection Agency.

A very large proportion of the national geothermal capacity is in California, and it is the southern sites that produce the most waste. While dissolved solids represent less than 5% by weight of production brines in most geothermal areas, this figure is closer to 30% in the Imperial

Adapted from *The Desert Report*, December 2010, California/Nevada Desert Committee of the Sierra Club

Valley. A typical Imperial Valley plant in 2000 pulled up 410 tons of production brine per day which then produced 30 tons of filter cake and another 10 tons of solids from the brine pool. About 360 tons of remaining fluids are then re-injected into the underground reservoir (minus some cooling water). The estimated total annual production of this solid waste in the Imperial Valley, which may not be injected back into the ground, is between 40,000 and 59,525 tons.

The disposal site in Imperial County is a good example of the potential solid waste needs. This facility contains 400,000 cubic meters of geothermal solid waste (about 816,000 tons) which is approximately 20 years worth of filter cake from plants operating in the Imperial Valley. The site occupies 100,000 square meters with a waste depth of four meters. Closed sites have about 2.4 meters of soil added to reduce radon emissions. The proposed geothermal development by the Bureau of Land Management is going to require significant increases in waste facilities. Much of this material is classified as hazardous, requiring transport to Class 1 waste facilities in central California. Extensive development of geothermal energy in California may require the development of extensive Class 1 and 2 facilities depending on the composition of the solid wastes.

In addition to the heavy metals which can be concentrated in the solid wastes remaining from geothermal energy production, the process can also enhance naturally occurring radioactive materials, such as radon, which are present in the original production brine.

There are, of course, numerous laws and regulations applicable to geothermal waste which are intended to preclude any environmental harm. As in many other instances where pollution occurs, it is more a question of the integrity of the producer, the enforcement of the laws and regulations by the relevant authorities, and the siting of the operation which determines the greenness, in addition to public health and safety, of the facility. All geothermal sites are not the same, especially in the production of solid wastes, the enhanced radioactive materials, and how they are handled.

Table 1: *Typical characteristics of geothermal materials in the Imperial Valley:*

	Brine (mg/L) (liquid)	Filter Cake (mg/kg) (solid)	Pipe Scale (mg/kg) (solid)
Arsenic	15*	300	30,000**
Antimony	0.9	10	1,000**
Lead	100*	30	1,000**
Copper	5	250	250
Cadmium	1.2**	0.2	500**
Zinc	370*	130	400

* Exceeds STLC (soluble threshold limit concentration
** Exceeds TTLC (total threshold limit concentration)
For example, hazardous waste limits for arsenic are:
TTLC 500 mg/L and STLC 5 mg/L as set forth in California EPA Title 22 California Code of Regulations Chapter 11

Investigations of geothermal facilities in the Imperial Valley, conducted by the Department of Toxic Substances Control (DTSC) in 2005, have resulted in fines totaling $910,000 against Cal Energy Operating Corporation of Calipatria, California, a subsidiary of MidAmerican Energy Holdings in Omaha, Nebraska. Violations included the illegal storage, treatment, and disposal of hazardous waste; the failure to label and cover hazardous waste containers; and failure to properly train employees. Recent investigations (2010) by DTSC may uncover further violations. In 2007, a geothermal plant was fined over $230,000 by the California Regional Water Quality Control Board for exceeding the levels of lead, arsenic, and copper in wastewater sent to the Salton Sea. There have also been air quality problems found in the steam associated with cooling towers in the Salton Sea region. As shown in the photo, most of the land surrounding the geothermal plants is producing multiple crops of hay and vegetables.

Although many of these violations were corrected, landowners who have geothermal plants sited on their property are concerned about the permanent loss of the land after the life of the facility. Members of the Environmental Justice Task Force, on a recent bus tour of Imperial Valley, had a firsthand view of many of the geothermal power plants and their footprint on the local environment.

Geothermal plants ideally can be clean and less detrimental to the environment than other types of energy production, especially if sited in areas of low effluent solids. From the number of leases proposed by the Bureau of Land Management, it would appear that there may be many areas which meet this criteria and would cause minimal damage to the environment. It is important that every proposal be evaluated individually on its own merits.

Fred Cagle, Ph.D. and PA, was a former board member of the Desert Protective Council, a Sierra Club Life Member, and served on the Governor's Advisory Committee for the Salton Sea. Professional commitments included environmental and occupational medicine. His principal interests included effecting changes in ecosystem health and justice. Fred passed away in April 2011.

GEOLOGIC TIME SCALE

EON	ERA	PERIOD (SYSTEM)		EPOCH (SERIES)	Age estimates of boundaries (in mega-annum [Ma] unless otherwise noted)
PHANEROZOIC	Cenozoic	Quaternary (Q)		Holocene	11,700 ±99 yr
				Pleistocene	2.588
		Tertiary (T)	Neogene	Pliocene	5.332 ±0.005
				Miocene	23.03 ±0.05
			Paleogene	Oligocene	33.9 ±0.1
				Eocene	55.8 ±0.2
				Paleocene	65.5 ±0.3
	Mesozoic	Cretaceous (K)		Late	99.6 ±0.9
				Early	145.5 ±4.0
		Jurassic (J)		Late	161.2 ±4.0
				Middle	175.6 ±2.0
				Early	199.6 ±0.6
		Triassic (Ʀ)		Late	228.7 ±2.0
				Middle	245.0 ±1.5
				Early	251.0 ±0.4
	Paleozoic	Permian		Lopingian / Guadalupian / Cisuralian	299.0 ±0.8
		Carboniferous	Pennsylvanian	Late / Middle / Early	318.1 ±1.3
			Mississippian	Late / Middle / Early	359.2 ±2.5
		Devonian		Late / Middle / Early	416.0 ±2.8
		Silurian		Pridoli / Ludlow / Wenlock / Llandovery	443.7 ±1.5
		Ordovician		Late / Middle / Early	488.3 ±1.7
		Cambrian		Late / Middle / Early	542.0 ±1.0
Precambrian					~4600

Reference: U.S. Geological Survey Geologic Names Committee, 2010, Divisions of geologic time—major chronostratigraphic and geochronologic units: U.S. Geological Survey Fact Sheet 2010–3059, 2 p.